Biofuels Production

Scrivener Publishing
100 Cummings Center, Suite 541J
Beverly, MA 01915-6106

Publishers at Scrivener
Martin Scrivener (martin@scrivenerpublishing.com)
Phillip Carmical (pcarmical@scrivenerpublishing.com)

Biofuels Production

Edited by

Vikash Babu, Ph.D, Ashish Thapliyal, Ph.D & Girijesh Kumar Patel, Ph.D

Scrivener
Publishing

WILEY

Co-published by John Wiley & Sons, Inc. Hoboken, New Jersey, and Scrivener Publishing LLC, Salem, Massachusetts.
Published simultaneously in Canada.

For general information on our other products and services or for technical support, please contact our Customer Care Department within the United States at (800) 762-2974, outside the United States at (317) 572-3993 or fax (317) 572-4002.

Wiley also publishes its books in a variety of electronic formats. Some content that appears in print may not be available in electronic formats. For more information about Wiley products, visit our web site at www.wiley.com.

For more information about Scrivener products please visit www.scrivenerpublishing.com.

Cover design by Kris Hackerott

Library of Congress Cataloging-in-Publication Data:

ISBN 978-1-118-63450-9

Printed in the United States of America

10 9 8 7 6 5 4 3 2 1

Contents

Preface

"Anybody who has been seriously engaged in scientific work of any kind realizes that over the entrance to the gates of the temple of science are written the words: 'Ye must have faith."

Max Planck

The true sign of intelligence is not knowledge but imagination.

Albert Einstein

Biofuels is an emerging area for research now a day because existing fossil fuels are likely to diminish within few years and many governments would like to reduce their dependence on fossil fuels. Hence, developing biofuels and alternative energy sources is among the priority of many nations. In the present scenario, the need of hour is to utilize latest scientific approaches and amalgamate them with proper utilization of natural resources and then the technology can impact the life of common man. New technological intervention requires effective co-ordination between different organizations like Universities, Academic Colleges, Research institutions, Government Agencies, Non-governmental Organization and people participation, the end users of technology. This book is an effort to provide latest information on recent scientific methodologies involving biofuel.

Lots of research papers and review articles are available on the internet covering different type of biofuels. It is difficult to comprehend those developments in a single review article. Therefore, there is a need to collect scattered information in a single book with recent advancements.

Each chapter in this book is contributed by experts of their field. In this book, methods of biofuels production such as biodiesel, biomethane, Bioethanol, Biobutanol and Biohydrogen production have been summarized. Apart from production methods, their global scenario, recent advancements, processing, microbial metabolic engineering for biofuel production and role of micro-organisms in biofuels production, have been well summarized. By all the efforts of contributors, this book will be very helpful for all the graduate, post graduate students and researchers who are working in this area.

The editors are thankful to all the contributors for their co-operation. Finally the authors solicit suggestions for improvement and enlargement of this book from the researchers, students and readers.

Vikash Babu
Ashish Thapliyal
Girijesh Kumar Patel

List of Contributors

Vikash Babu
Department of Biotechnology,
Graphic Era University,
Dehradun-248002, India

Girijesh Kumar Patel
Department of Biotechnology,
Graphic Era University,
Dehradun-248002, India

Ashish Thapliyal
Department of Biotechnology,
Graphic Era University,
Dehradun-248002, India

M.D. Berni
Interdisciplinary Centre of Energy Planning,
State University of Campinas – UNICAMP, Brazil.

I.L. Dorileo
Interdisciplinary Centre of Energy Planning,
Federal University of Mato Grosso - UFMT, Brazil.

J.M. Prado
LASEFI/DEA/FEA (School of Food Engineering) / UNICAMP
(University of Campinas),
R. Monteiro Lobato, 80, Campinas, 13083-862, SP, Brazil

T. Forster-Carneiro
LASEFI/DEA/FEA (School of Food Engineering) / UNICAMP
(University of Campinas),
R. Monteiro Lobato, 80, Campinas, 13083-862, SP, Brazil

M.A.A. Meireles
LASEFI/DEA/FEA (School of Food Engineering) / UNICAMP (University of Campinas),
R. Monteiro Lobato, 80, Campinas, 13083-862, SP, Brazil

Pramod Kumar
Department of Biotechnology,
Indian Institute of Technology Roorkee,
Roorkee-247667 (India)

Divya Gupta
Department of Biotechnology,
G. B. Pant Engineering College, Pauri Garhwal, Uttrakhand-246001, India.

Ajeet Singh
Department of Biotechnology,
G. B. Pant Engineering College,
Pauri Garhwal, Uttrakhand-246001,India.

Ashwani Sharma
Computer-Chemie-Centrum,
Universität Erlangen-Nürnberg Nägelsbachstr. 25, 91052 Erlangen, Germany.

Anshul Nigam
IPLS (renamed as BUILDER),
 Pondicherry University, Puducherry-605014, India.

Virendra Kumar
Institute of Genomics and Integrative Biology,
Mall Road, Delhi-110007, India

Purnima Dhall
Institute of Genomics and Integrative Biology,
Mall Road, Delhi-110007, India

Rita Kumar
Institute of Genomics and Integrative Biology,
Mall Road, Delhi-110007, India

Shubhangini Sharma
CDL, Intas Biopharmaceutical,
Ahmedabad, Gujrat, India.

Reena
Institute of Genomics and Integrative Biology,
Mall Road, New Delhi, India.

Anil Kumar
National Institute of Immunology,
New Delhi, India.

Pallavi Mittal
ITS Paramedical College,
Ghaziabad, UP, India

A.S Panwar
Department of Biotechnology,
HNB Garhwal University,
Srinagar (Garhwal), Uttarakhand, India

J Jugran
Department of Biotechnology,
HNB Garhwal University, Srinagar (Garhwal), Uttarakhand, India

G.K Joshi
Department of Biotechnology,
HNB Garhwal University, Srinagar (Garhwal), Uttarakhand, India

Navneet Singh Chaudhary
Department of Biotechnology, Sir Padampat Singhania University
(SPSU), Udaipur-313001, Rajasthan, India

Mohammad J. Taherzadeh
School of Engineering,
University of Borås, Borås, Sweden

Patrik R. Lennartsson
School of Engineering,
University of Borås, Borås, Sweden

Oliver Teichert
Lantmännen Agroetanol AB,
Norrköping, Sweden

Håkan Nordholm
Lantmännen Agroetanol AB,
Norrköping, Sweden

Sapna Jain
Department of Biotechnology,
Institute of Biomedical Education and Research,
Mangalayatan University, Aligarh-202145, India

Mukesh Kumar Yadav
Department of Biotechnology,
Institute of Biomedical Education and Research,
Mangalayatan University, Aligarh-202145, India

Ajay Kumar
Department of Biotechnology,
Institute of Biomedical Education and Research,
Mangalayatan University, Aligarh-202145, India

Komal Saxena
Department of Biotechnology,
Institute of Biomedical Education and Research,
Mangalayatan University, Aligarh-202145, Uttar Pradesh, India.

Avinash Kumar Sharma
Department of Biotechnology,
Institute of Biomedical Education and Research,
Mangalayatan University, Aligarh-202145, Uttar Pradesh, India.

Ashish Deep Gupta
Department of Biotechnology,
Institute of Biomedical Education and Research,
Mangalayatan University, Aligarh-202145, Uttar Pradesh, India.

Lalit Agrawal
CSIR-National Botanical Research Institute,
Rana Pratap Marg, Lucknow, 226 001, Uttar Pradesh, India.

Sumita Srivastav
Department of Physics,
Government Post Graduate College, Uttarkashi, Uttarakhand

Prashant Anthwal
Department of Biotechnology,
Graphic Era University, Dehradun, Uttarakhand

Tribhuwan Chandra
Department of Biotechnology,
Graphic Era University, Dehradun, Uttarakhand

Ruchika Goyal
Department of Biotechnology,
Graphic Era University, Dehradun-248002, India

Introduction to Biofuels

Pramod Kumar[1] and Vikash Babu[2,*]

*[1]Department of Biotechnology, Indian Institute of Technology Roorkee,
Roorkee-247667 (India)*
[2]Department of Biotechnology, Graphic Era University, Dehradun-248002

Biofuels mark their presence since the discovery of fire and have been very profoundly used for ages. The ancient raw material for biofuel is wood, exploited in solid form and having several usages with major applications in cooking and heating. Later on, the evolved form of biofuel came into existence as a form of liquid oil that was used from the time immemorial to light up homes and paths for everyday life. Olive and whale oils are some of the ancient types of biofuels employed for this purpose, mostly derived from plants and animals, they were in use for a very long period of time until the application of kerosene replaced them [1, 2]. Moreover, other forms of biofuel started prevailing from the late eighteen century; ethanol is one of the most exploited biofuels for its remarkable application especially to the transportation sector [3]. Corn derived ethanol was first employed for early transportation, mainly in cars. Subsequently, several other feed stocks were employed as sources for biofuel extraction such as plants like peanuts, hump, grains and potatoes [4]. Biodiesel, a later discovered form of biofuel, came into existence only in the twentieth century [5]. Presently, these two classes are the largest exploited biofuel types.

The twentieth century was an era of exploration and the use of resources with concern to the availability of reserves was not a big question. However, with rising populations and urbanization, finding the energy solution has become an area of prime importance. Major energy thrusts are required from the transportation, industrialization

Corresponding author: vikash.iitroorkee@gmail.com

Dr. Vikash Babu, Dr. Ashish Thapliyal & Dr. Girijesh Kumar Patel (eds.) Biofuels Production,
(1–10) 2014 © Scrivener Publishing LLC

and agricultural sectors. Fossil fuels are key sources to bear the burden of entire need but ever increasing demand and limited stock of such fuels force us to employ alternative approaches for the renewable and sustainable production of energy [6]. Hence, biofuels are considered one of the remarkable solutions for this problem. Fuels that are obtained from biological material and have been recently taken out from their natural growing places or are by-products of living organisms are placed in a class of biofuels contrary to the fossil fuel derived products that are extracted from fossilized organisms buried for millions of years under the earth's crust and converted to a form of fuel due to high pressure and temperature. Because of renewable nature and the immense possibility of improvement and engineering, biofuels are becoming a promising source of energy contrary to the limited and localized availability of fossil fuels. Moreover, biofuels are a possible solution to the dependence on foreign energy sources and are also suitable to circumvent environmental concerns. There is big hidden potential in biofuel based energy sources, especially when it is combined with efficient agriculture and scientific application that enables it to provide mankind with various raw materials required for food fiber and energy [7].

There are several forms of fuel that can be produced from biomass, referred to in general as biofuel, that cover liquid forms of fuel such as ethanol, methanol or biodiesel and gaseous forms like methane and hydrogen. On the basis of application and feedstock utilization, the biofuel can be summarized in two stages, first generation biofuel and second-generation biofuel [8]. The most predominant types of first generation biofuels are ethanol, fatty acid methyl ester (FAME or biodiesel) and pure plant oil (PPO). The most common form of biofuel exploited worldwide is bio-ethanol with global production increases from 17 thousand million liters in the year 2000 to 68 thousand million liters in 2008 [9, 10]. The key feedstocks for production of ethanol are sugarcane, wheat, sugar beet, rapeseed, soybean and palm oil [11]. Most of ethanol's worldwide production is contributed by the United States and Brazil by using corn or sugarcane as main feedstock, while Europe produces from potato, wheat or sugar beet. For biodiesel, a major producer is Europe, where Germany is the leader whose production meets 3% of the entire German fuel requirement [12]. Rapeseed is exploited as the most widely used feedstock for an approximate contribution of 70% of European biodiesel production followed by soy that contributes 17% of the production. A smaller portion of production is obtained from sunflower and palm oil [13].

Pure plant oil is a relatively new biofuel resource and it has been gaining importance recently due to early limited local productions. The key features associated with this class are economic value and the feasibility to produce high yield ratios for per hectare production. These properties make it suitable for the markets of developing countries. Some good examples are observed in countries like Malaysia and Indonesia because of the low cost of labor and production in comparison to the countries of Europe and North America. Recently, imports from these countries have gained importance [14]. The main advantages around first generation biofuels are curbing the release of CO_2 and domestic energy security. However, the availability of raw material, adverse effects over the biodiversity and competition for farmlands are major setbacks. Furthermore, the major concerns associated with first generation biofuels are the sustainability of resources from which they are produced as well as their direct competition for food crops and environmental threats related to ecosystems.

There is now well established analysis for biofuels that they should be very efficient in terms of reduction of emissions and net life cycle of green house gas (GHG) emissions that should certainly meet the criteria of social and environmental sustainability. Except for bioethanol from sugarcane, none of the first generation biofuels appear to be fruitful for a future transport fuel mix. All these concerns gave rise to the next stage of biofuel production, so-called second generation biofuels [15]. Due to the choice of feedstock and cultivation technology, second generation biofuels have immense advantages like the consumption of waste and the use of abandoned land, so the second generation of biofuels paves the way for immense application in the biofuel generation that can also satisfy the economical, social and environmental criteria. But unjustified use of second generation biofuel can also compete with regular food crops and may lead to unsustainable resources. Hence, it is of the utmost importance to set some benchmarks for their exploitation like minimum life cycle GHG reductions, land use changes and strict limits for social as well as economic standards.

The criteria to exploit non-food biomass is well addressed by second generation biofuels with the application of several strategies like the application of feedstocks having lignocellulose material that can come from bi-products of agriculture, such as rice husk, corn rub, and sawdust and residues that comes from the forest, such as sugarcane bagasse etc. [16]. According to a report from the US EPA in 2009, cellulosic ethanol is far more promising than any of the first generation biofuels, except bioethanol from the sugarcane of Brazil. Moreover, high organic content of the waste and sludge also possess good potential for application as a feedstock due to the presence of a reasonably high proportion of carbohydrates and proteins. The application of anaerobic digestion to sludge makes it very useful for bioenergy production [17]. The slight modification in the processing of waste can yield other compounds like acetic acid and related organic acids, having additional economic advantages [18]. The acetic acid and organic acid are very important industrial intermediaries that act as carbon sources for the growth of several kinds of microbes that can produce a range of biofuels and chemicals. Apart from these sources, with the recent advancement of microbial engineering, the algal derived feedstock is also showing remarkable potential for the production of biofuel [19]. Because of the versatile nature of algal growth condition, it possesses huge potential to grow over almost any kind of stringent environmental condition like saline water, wastewater, coastal sea water and non-arable lands [20]. Moreover, algal biomass is especially suitable for the high yield of lipids required for production of biodiesel [21]. Because algal feedstock production has very limited competition from regular food crops, it makes it further suitable for biofuel generation.

In sum, a cumulative approach involving specific applications regarding based on the sources of feedstock, availability of land, labor, socioeconomic conditions and selecting a suitable kind of biofuel possesses an immense possibility to meet the needs of fuel in an efficient and sustainable way.

1.1   Global Scenario of Biofuel Production and Economy

The possibility lying in biofuel based energy solutions has gained worldwide attention now that it is visible in the form of policies made by several governments to cut their dependency on fossil fuels by using an environmentally friendly approach. Some of the leading nations in line for promotion of biofuel are the United States, Brazil, E.U. member countries, Canada, China and India. The US government has made one of the most ambitious projections of biofuel by advocating the three-fold increase of bioenergy in the duration of the next ten years [22]. Under the name 'biofuel' two major commodities lie, these are bioethanol and biodiesel. The feedstocks for bioethanol production are mainly sugar, corn, soybean, wheat and sunflower whereas jatropha, vegetable oil, palm, rapeseed and soybean are raw materials for biodiesel. Bioethanol is the most forward standing type of biofuel that is ready to replace gasoline and is now part of the many government's policies for biofuel application, as observed from some of the big countries like Brazil with a mandatory use of 22% bioethanol, 10% in several state of USA and China. Moreover, the hydrous bioethanol (96 percent bioethanol with 4% of water) is also promoted in these countries for extensive use [23]. According to the US Energy Independence and Security Act of 2007, it is envisioned that the renewable energy contribution by bioethanol and other biofuels will increase to 36 billion gallons annually by 2022. Moreover, the US EPA (the United States Environmental Protection Agency) is now permitting the mixing of 10% ethanol due to an amendment of the Clean Air Act. The related effects of these policies are visible in the form of increased corn ethanol production in the US market [24]. The impact of policies made for ethanol production and uses is giving positive outcomes observed for last two decades with respect to the social cost and benefits produced by monitoring biofuel related taxes, tariffs and credit effects of the agricultural sector [25]. This is visible in the form of an average price reduction of 14 ¢ per gallon analyzed from data obtained over the period of 1995–2008 [26]. Such policies demonstrate the linkage of the agricultural and energy markets as is visible in the form of exceeding the mandatory level of ethanol due to high petroleum prices and corn yield [24].

The Agricultural Trade Office of São Paulo projected that total ethanol production for the year 2012 will be 25.5 billion liters. That is followed by total production of 21.1 billion liters in 2011 that is approximately 24.9 percent of the total worldwide biofuel despite the crisis phase that appeared due to many reasons [27, 28]. Brazil is an example of setting a competitive market where successful application of bioethanol is commercialized without subsidy using sugarcane as feedstock [29], moreover Brazil is the largest exporter of bioethanol and the second largest producer after the US

The European Union is the third largest biofuel producer implementing the projection of biofuel by a contribution of five percent share by 2015 and a further rise to ten percent by 2020 [30]. The EU2009 directive has an explicit link to correlate the consumption and production of biofuel to make it a sustainable industry [31]. Moreover, the EU has taken a safe way by prioritizing the import of 30 percent of feedstock or biofuel to reduce the price pressure in

EU feedstock. For the anticipated biofuel production in 2012, the E.U. requires 10.3 MMT of sugar beet and 9.7 MMT of vegetable oil and animal fat. For bioethanol, in comparison to the US and Brazil, the E.U. is only a minor producer and for volume contribution bioethanol shares 28 percent of the total biofuel market in the road transport sector [32]. While for biodiesel, the E.U. is the largest producer with 60 percent of the market share [33]. In the European Union, Germany is the largest biofuel producer followed by France as the second largest producer [34]. France also has an ambitious goal to reach a biofuel share of 10 percent by 2015 [35].

For Canada, the mandatory renewable fuel content has been shouted to 5 percent by 2010 as per the recommendations of federal legislation in 2008 [36]. Moreover, the federal mandate also implemented a law requiring two percent renewable content in diesel by 2011. For bioethanol production, Canada has reached almost 2 billion liters of production per year [37], where main feedstock for the bioethanol production in Canada is corn and wheat, the biodiesel is preferably made from canola [38]. Canada has also been subsidized to import biodiesel by a tariff of 6.5 percent as most favored nation and three percent by general preferential tariff [36]. The policies made earlier to make bioethnol an energy alternative are now visible in form of reasonable rise of ethanol production and its sustainability. But due to the limited biofuel production for the short as well as the medium term, it appears that Canada may not become a major player for ethanol production in the near future.

China is also enacting mandates for the blending of ethanol up to 10 percent by 2020. With the policy of using non grain based feedstock for second generation biofuel production, there are five Chinese provinces Heilongjian, Jilin, Liaoning, Anhui and Henan working on this mandate [39]. There are a total of five ethanol producing plants in China; out of these, four use grain based feedstock (corn and wheat) and the other one uses tuber of cassava. Grain based ethanol production was 2,103 million liters while the cassava based production was 152 million liters. On the other hand, Chinese biodiesel production was estimated to be 3,408 million liters where the main input materials employed are used/waste kitchen oil and vegetable oil crusher residues [40].

India is producing the anticipated amount of bioethanol sufficient to meet the two percent blending target for the year 2012. The production of biodiesel from the plant jatropha is presently insignificant. As per the national biofuel policy approved by the government of India in 2009, it was promoted to blend 20 percent of biofuel to the fossil fuels by the end of 12th five year plan (2017). Present targets of mixing five percent of bioethanol is in direct relation to the surplus production of sugar for last three years, whereas due to the unavailability of suitable feedstock for biodiesel production, it is improbable to achieve the 20 percent replacement of biodiesel. To meet the need of the 5 percent blending of biodiesel rule for the year 2011–12, 3.21 million tons of biodiesel from 3.42 million hectares of land is required [41]. With an average production of 2.5 tons per hectare the jatropha is considered the most promising feedstock with a 30 percent biodiesel recovery rate. At this rate there is a requirement of 18.6 million hectares of cultivation of jatropha by 2017 to support the 20 percent blending target. Moreover, the biomass based energy production, especially the grid quality power, is working reasonably well at an estimated 31,000 MW with the bagasse surplus contribution of 10,000 MW to fuel factories of sugar, petrochemical, distillery, mills of rice, sugar and textile. With the projection of a 29 percent rise in ethanol production for the year 2012 by giving 2.1 billion liters as a cascading result of higher sugar and related higher molasses production [41], India is achieving reasonable progress towards the sustainable biofuel market.

References

1. A. Arsuf and V. Sussman, "The Samaritan Oil Lamps from Apolonia-Arsuf" T.A. 10, pp.71–96, 1983.
2. L.S. Russell, *A Heritage of Light: Lamps and Lighting in the Early Canadian Home*, University of Toronto Press, 2003.
3. N.A. Otto, Apparatus for heating the ignition tube of gas motor engines, US patent 388301, Assigned to N.A. Otto, August 21, 1888.
4. Ford Predicts Fuel from Vegetation, *New York Times*, Sept. 20, p. 24, 1925.
5. M. Fangrui and M.A. Hanna, *Bioresource Technology*, 70, 1–15, 1999.
6. B. Kamm, P.R. Gruber and M. Kamm, *Biorefinery industrial processes and products. Status and future direction*, vols. 1. Weinheim, Wiley-Vch Gmbh and Co KGaA; 2006.
7. C.V. Stevens and R. Verhe, *Renewable bib resources scope and modification for non-food application*, Chichester, John Wiley and Sons Ltd, 2004.
8. S.N. Naik, V.V. Goud, P.K. Rout and A.K. Dalai, *Renewable and Sustainable Energy Reviews*, Vol. 14, p. 578–597, 2010.
9. M. Balat, *Energy Exploration and Exploitation*, Vol. 25, p. 195–218, 2007.
10. Biofuels Platform, Production of Biofuel in the EU. Biofuels Platform, Lausanne. http://www.biofuels-platform.ch/en/infos/eu-bioethanol.php. 2010.

11. M. Balat, and H. Balat, *Applied Energy* Vol. 86, p. 2273–2282, 2009.
12. Status Report Biodiesel, Biodiesel Production and Marketing in Germany 2005, Union for the Promotion of Oil and Protein Plant, Berlin, Germany 2005.
13. Biofuels Annual, EU Annual Biofuels Report, Global Agriculture Information Network, GAIN Report Number:NL0019, 2010.
14. S.H. Gay, M. Mueller and F. Santuccio, Analysing the implication of the EU 20–10-20 targets for world vegetable oil production, paper presented in '07th EAAE Seminar" Modelling of Agricultural and Rural Development Policies". Sevilla, Spain, January 29th -February 1st, 2008.
15. E.H.S. Ralph, W. Mabee, J.N. Saddler and M. Taylor, *Bioresource Technology*, Vol. 101, p. 1570–1580, 2010.
16. G. Chaudhary, L.K. Singh, and S. Ghosh, *Bioresource Technology*. Vol. 124, p. 111–118, 2012.
17. Y. Chen, J.J. Cheng and K.S. Creamer, *Bioresource Technology*. Vol. 99, p. 4044–4064, 2008.
18. H. Rughoonundun, C. Granda, R. Mohee and M.T. Holtzapple, *Waste Management*. Vol. 30, p. 1614–1621, 2010.
19. P. M. Schenk, S.R. Thomas-Hall, E. Stephens, U.C. Marx, J.H. Mussgnug, C. Posten, O. Kruse and B. Hankamer, *Bioenergy Research*, Vol. 1, Issue 1, 20–43, 2008.
20. A. Singh, P.S. Nigam and J.D. Murphya, *Bioresource Technology*, Vol. 102, Issue 1, p. 10–16, 2011.
21. L. Yanqn, M. Horsman, N. Wu, C. Lan and N. D. Calero, *Biotechnology Progress* Vol. 24, p. 815–820, 2008.
22. M.F. Demirbas and M. Balat, *Energy Conversion and Management*, Vol. 47, p. 2371–2381, 2006.
23. W.D. Walls, F. Rusco and M. Kendix, *Energy Policy*, Vol. 39, p. 3999–4006, 2011.
24. W. Thompson, S. Meyer and P. Westhoff, *Energy Policy*, Vol. 37 Issue 2, p. 745–749, 2009.
25. M.S. Luchansky and J. Monks, *Energy Economics*, Vol. 31, p 403–410, 2009.
26. X. Du and D.J. Hayes, *Energy Policy*, Vol. 37, Issue 8, p. 3227–3234. 2009.
27. Renewable Fuels Association, Accelerating industry innovation: 2012, ethanol industry outlook. In: Renewable Fuels Association, http://ethanolrfa.3cdn. net/ d4ad995ffb7ae8 fbfe_1vm62ypzd.pdf. 2012.
28. R. Colitt and S. Nielsen (2012-03-13). "Brazil Ethanol Drive Falters on Domestic Supply Shortage". *Bloomberg Businessweek*. Retrieved 2012-03-22.
29. S. Macrelli, J Mogensen and G. Zacchi, *Biotechnology for Biofuels*, 5:22, 2012.
30. Directive 2009/28/EC of the European Parliament and of the Council of 23 April 2009 on the promotion of the use of energy from renewable sources and amending and subsequently repealing Directives 2001/77/EC and 2003/30/EC.
31. P. Genovesi, *Current Opinion in Environmental Sustainability*, Vol. 3, Issues 1–2, p. 66–70, 2011.
32. EU-27 Biofuel Annual. EU biofuel annual 2012.
33. R. Steenblik, BIOFUELS – AT WHAT COST? Government support for ethanol and biodiesel in selected OECD countries, The Global Subsidies Initiative (GSI) of the International Institute for Sustainable Development (IISD), Geneva, Switzerland, 2007.

34. R. Brand, Networks in renewable energy policies in Germany and France, Berlin conference on the human dimension of global environmental change: greening of policies—policy integration and interlinkages, Berlin, 3–4 December, 2004.
35. European Commission (EC). Green Paper, "A European strategy for sustainable, competitive and secure energy". Brussels, March 8, 2006.
36. Global Subsidies Initiative (GSI-2009), Biofuels-At What Cost? Government support for ethanol and biodiesel in Canada. Available online at: http://www.globalsubsidies.org/files/assets/oecdbiofuels.pdf, verified May 10th 2011.
37. Growing beyond oil delivering our energy future, a report card on the Canadian renewable fuels industry, Canadian renewable fuel association, November 2010.
38. GAIN (Global Agriculture Information Network) report number CA0023, Canada Biofuel Annual 7/13/2010.
39. C.A. Tisdell, Working Paper – "The Production of Biofuels: Welfare and Environmental Consequences for Asia 2009".
40. GAIN (Global Agriculture Information Network) report number 12044, China-Peoples republic of Biofuel Annual 07/09/2012.
41. GAIN (Global Agriculture Information Network) report number IN2081, India, Biofuel Annual 6/20/2012.

2

Advances in Biofuel Production

M.D. Berni[1], I.L. Dorileo[2], J.M. Prado[3], T. Forster-Carneiro[3],* and M.A.A. Meireles[3]

[1]*Interdisciplinary Centre of Energy Planning, State University of Campinas – UNICAMP, Brazil. mberni@unicamp.br*
[2]*Interdisciplinary Centre of Energy Planning, Federal University of Mato Grosso - UFMT, Brazil.*
[3]*LASEFI/DEA/FEA (School of Food Engineering) / UNICAMP (University of Campinas), R. Monteiro Lobato, 80, Campinas, 13083–862, SP, Brazil*

Abstract

The main driving forces for the development of biofuels are the instability of world oil prices, security of energy supply, global warming, and the creation of new opportunities for agriculture. Interest in the commercial production of biofuels for transport was renewed in the mid-1970s, when ethanol began to be produced from sugarcane (Brazil) and corn (United States) for blending mandates, which define the proportion of biofuel that must be used in road transport fuel. Nowadays over 50 countries have adopted blending targets. However, biofuels are still little represented in comparison to fossil fuels. This chapter will discuss the advances in the production of first, second and third generation biofuels, mainly sugar- and starch-based ethanol, conventional biodiesel, biogas and biomethane, cellulosic ethanol, syngas, bio-oil from pyrolysis and hydrothermal process, hydrogen, and the biorefinery concept. The production of biomass for biofuels and the future trends of biofuels development also are discussed in this chapter.

Keywords: Biodiesel, biofuels, biogas, biomass, bio-oil, ethanol, hydrogen

Corresponding author: taniaforsterc@gmail.com

Dr.Vikash Babu, Dr. Ashish Thapliyal & Dr. Girijesh Kumar Patel (eds.) Biofuels Production, (11–58) 2014 © Scrivener Publishing LLC

2.1   Introduction

Concerns with the instability of world oil prices, security of energy supply, global warming, and the creation of new opportunities for agriculture, are stimulating the search for sources of energy that are clean, sustainable and competitive with fossil energy. These are the main driving forces for the development of biofuels, which have become some of the most promising forms of energy to ensure a sustainable energy matrix [1, 2].

Biofuels started to be produced in the late 19th century, when bioethanol was derived from corn and Rudolf Diesel's first engine ran on peanut oil. Until the 1940s, biofuels were seen as viable transport fuels, but falling fossil fuel prices stopped their further development. Interest in the commercial production of biofuels for transport rose again in the mid-1970s, when ethanol began to be produced from sugarcane in Brazil and then from corn in the United States. In most parts of the world, the fastest growth in biofuel production has taken place over the last 10 years, supported by ambitious government policies. Besides energy security and sustainable agriculture concerns, the reduction of CO_2 emissions in the transport sector has become an especially important driver for biofuel development. One of the most common support measures is a blending mandate, which defines the proportion of biofuel that must be used in road transport fuel and is often combined with other measures such as tax incentives [3]. Over 50 countries have adopted blending targets or mandates and several more have announced biofuel quotas for future years [4, 5]. Therefore, biofuels are gaining importance among the alternatives to fossil fuels.

However, biofuels are still little represented compared to fossil fuels (Figure 2.1). Their large scale production ultimately depends on advances in productivity, in order to mitigate any negative effects associated with them, such as decreases of indigenous forests or the increase in price of agricultural commodities due to land use. In light of this, a worldwide technological race is taking place to develop biofuels from second and third generations, whose main support programs are carried out by the United States (US) and the European Union (EU).

Biofuels are derived from renewable biomass sources. The conversion of sunlight into chemical energy is one of the most important processes to sustain life on the planet. The process of converting solar energy into chemical energy responsible for the reproduction of plants involves the consumption of O_2 and the production of CO_2 and plant resources. The term "biomass" is used to name

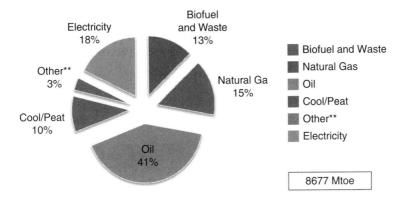

Figure 2.1 Percentage of sources of energy in year 2010. Adapted from IEA [6].

the vegetable resources used for the production of bioenergy. The main sources of biomass are forests, agricultural crops and residues resulting from agroforestry and livestock industries.

Biomass is classified as modern or traditional, according to its origin and type of processing. Traditional biomass is associated with the production of energy using resources from unsustainable management and techniques that are characterized by low efficiency and high emission of pollutants. Modern biomass is obtained by proper management, using technologies that guarantee high efficiency of production and conversion processes, ensuring biofuels of high quality, such as ethanol, biogas and bio-oil from vegetable oils, reforested wood, industrial and municipal waste, etc. [7].

Biomass is one of the oldest energy resources used by humanity. Although there are no precise figures, it is estimated that one third of the world population depends on traditional biomass as their main source of energy (wood, agricultural, livestock and forestry residues, among other sources), so that about 90% of the world consumption of biomass is of traditional biomass. In some regions of Africa, Asia and Latin America families use traditional biomass to meet their energy needs, mainly for cooking. In these cases, the use of biomass is habitually inefficient, resulting in the depreciation of natural resources and damage to the health of the operators of the cooking systems. The quality of the energy services provided by this type of application is usually poor, and it requires intensive labor to perform the activities of collecting and transporting the fuel. These activities are usually undertaken by women and children. Furthermore, the production of fuels from traditional biomass sources can aggravate the problem of deforestation, increasing the pressure on the local ecosystem and the net emissions of greenhouse gases (GHG). Despite these drawbacks, billions of people still use traditional sources of biomass to supply their energy needs because they are more accessible and less expensive. Dry biomass is easily obtained and stored, and its use has cultural roots in many societies. Moreover, in the absence of this feature, many countries would have to increase their energy imports and many needy families would have to spend more money to acquire other forms of energy [7].

Biofuels are energy vectors produced from modern biomass. Their conversion occurs by means of physical, chemical and/or biological processes. The production of liquid biofuels to replace petroleum (especially diesel and gasoline) has attracted particular interest, and is considered a promising alternative to the energy market. The American National Biofuels Action Plan points to five strategic areas in the biofuels production chain that should have federal investment: production of raw materials; logistics of the distribution of raw materials; the best conversion process; fuel distribution systems; and technologies for the efficient use of fuel. The US Congress passed in 2007 the "Renewable Fuel Standard" (RFS), as part of the Energy Independence and Security Act (EISA). This act aims to reduce gasoline consumption by 20% by 2017. This would be achieved by increasing the domestic production of biofuels. To reach this proposal it would be necessary to have an annual increase of the production of biofuels of 35 billion gallons.

Europe has also stimulated the production of biofuels. The Common Agricultural Policy (CAP), revised in 2003, provides incentives for farmers that select European species for energy purposes, such as rapeseed and sugar beet. The new policy provides a special remuneration of € 45 per hectare to produce bioenergy, and in these areas food crops cannot be grown [8].

Global biofuel production reached 105 billion liters in 2010, which represents an increase of 17% compared to around 90 billion liters in 2009 (Figure 2.2). However, this figure stands far behind the 35 billion gallons increase proposed by the EISA in order to reduce gasoline consumption. Therefore, there is still much to be done to achieve the objective of turning biofuels into viable replacements for oil-based fuels.

The United States leads the production of ethanol, and also is the major producer of biofuels (ethanol + biodiesel) in the world. On the other hand, the European Union is the major biodiesel producer, accounting for about 60% of its world production. Brazil is the second

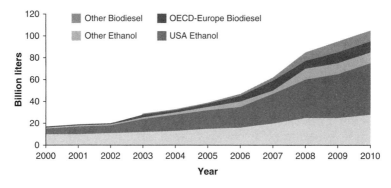

Figure 2.2 Global biofuel production in the 2000–2010 period. Adapted from IEA [9].

largest ethanol producer in the world, and is the second largest bio-fuels producer in the world. The Brazilian Alcohol (ethanol) Program (ProÁlcool) was launched in 1975 as a policy to reduce the country's dependence on oil import. It encouraged the production of ethanol from sugarcane, which today is an important crop for economic development of Brazil, for the substitution of gasoline in transport. Today around 50 % of Brazilian fleets run with flex fuel engines, which run on gasoline, ethanol, or a mixture of them both. Furthermore, sugarcane is a source of residue biomass (bagasse), which is used for electric energy production when burned, but which could also be used in second generation processes for the production of bioethanol [10].

It is estimated that 86 billion liters of ethanol and 19 billion liters of biodiesel were produced worldwide in 2010. Ten years ago, the production of biofuels did not surpass 20 billion liters (Figure 2.2.) [9]. However, despite all the environmental advantages presented by biofuels, the expansion of their production is limited by their high production cost when compared to their competitor: fossil fuels (diesel and gasoline). One of the main factors that makes the production of first generation biofuels more expensive is the cost of the raw material. The raw materials used in the production of biofuels are typically of high added value, such as corn and sugarcane that are used for bioethanol production, and soybeans that are used for biodiesel production. Thus, the cost of the products is usually high. Because of that, with the exception of sugarcane ethanol produced in Brazil, biofuels still require subsidies to enable their production [11].

In 2011, bioethanol was projected to overtake the animal feeding industry as the largest corn consumer in the US, helping the production margins and decreasing the need for subsidies. The Renewable

Fuel Standard (RFS), by the US Environmental Protection Agency (EPA) provides a guaranteed market of 50 billion liters in 2012, but the industry nearly matched this goal in 2010, suggesting that the mandate alone would not support the existing market. The corn ethanol mandate increases to 57 billion litters in 2015.

Brazilian sugarcane ethanol would likely become more prevalent in the US if American ethanol subsidies and tariffs were removed. Sugarcane bioethanol is cheaper and more efficient to produce, although there are worries that its production may indirectly lead to deforestation. Brazil has plans to build 100 new sugarcane mills by 2019, increasing capacity by 66 %, to increase Brazilian ethanol production and exports in the coming years.

Considering this entire scenario, there is much to be done in the biofuels field to improve economic feasibility. Technological advances in the production and conversion of biofuels can lead to more competitive fuels. The development of chemical and biological sciences, along with new crops for energy production, new enzymes and artificial simulation of biological processes (anaerobic digestion, fermentation, etc.) can reduce their production [12]. Therefore, heavy investments are being made for the development of these technologies. The objective of this work is to review some of the most recent advances in the production of biofuels.

2.2   Advances in the Production of First, Second and Third Generation Biofuels

The term biofuel refers to liquid and gaseous fuels produced from biomass. There is considerable debate on how to classify biofuels [3]. They are commonly divided into first-, second- and third-generation biofuels, but the same fuel can be classified differently depending on whether technology maturity, GHG emission balance or feedstock is used to guide the distinction. This work uses a definition based on the maturity of a technology, and the terms "conventional" and "advanced" for classification [9].

2.2.1   Conventional and Advanced Biofuels

There are a number of technologies for energy conversion from biomass, suitable for applications in small and large scales. Considering the convention adopted by the International Energy Agency (IEA)

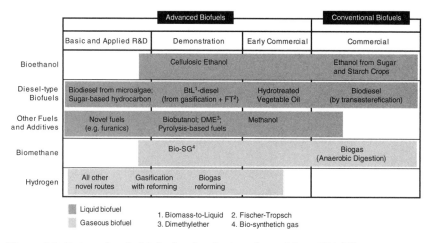

	Advanced Biofuels			Conventional Biofuels
	Basic and Applied R&D	Demonstration	Early Commercial	Commercial
Bioethanol		Cellulosic Ethanol		Ethanol from Sugar and Starch Crops
Diesel-type Biofuels	Biodiesel from microalgae; Sugar-based hydrocarbon	BtL[1]-diesel (from gasification + FT[2])	Hydrotreated Vegetable Oil	Biodiesel (by transesterefication)
Other Fuels and Additives	Novel fuels (e.g. furanics)	Biobutanol; DME[3]; Pyrolysis-based fuels	Methanol	
Biomethane		Bio-SG[4]		Biogas (Anaerobic Digestion)
Hydrogen	All other novel routes	Gasification with reforming	Biogas reforming	

Liquid biofuel
Gaseous biofuel

1. Biomass-to-Liquid 2. Fischer-Tropsch
3. Dimethylether 4. Bio-synthetich gas

Figure 2.3 Status of main biofuel technologies adapted from IEA [9].

for the classification of biofuels, Figure 2.3 presents a summary of the conventional and advanced biofuels, illustrating technologies and processes for obtaining them. It can be noted, by the number of applications, that special attention has been given to liquid fuels [9].

Conventional biofuel technologies include well-established processes that are already producing biofuels at a commercial scale. These biofuels, commonly referred to as first-generation, include sugar- and starch-based ethanol, oil-crop based biodiesel and straight vegetable oil, as well as biogas derived from anaerobic digestion. Typical feedstocks used in these processes include sugarcane and sugar beet, starch-bearing grains like corn and wheat, oil crops like rapeseed, soybean and oil palm, and in some cases animal fats and used cooking oils.

Advanced biofuel technologies are conversion processes that are still in research and development, pilot or demonstration phases, commonly referred to as second or third generation. This category includes hydrotreated vegetable oil (HVO), which is based on animal fat and plant oil, as well as biofuels based on lignocellulosic biomass, such as cellulosic ethanol, biomass-to-liquids (BtL) diesel and bio-synthetic gas (bio-SG). It also includes algae-based biofuels and the conversion of sugar into diesel-type biofuels using biological or chemical catalysts [9, 13].

The second-generation biofuels can be produced from waste materials resulting from industrial production processes, agriculture or agro-forestry. They constitute alternatives to reduce the cost of production of bioenergy and to decrease the competitiveness with food. The production of cellulosic ethanol, which is one of the most promising sources of "clean and cheap energy", can, in principle, use as input any raw material containing cellulose and hemicellulose (such as bagasses, straws, hulls, etc.). However, significant technological advances are needed in this field, since these technologies are not economically feasible yet. The processes are complex and involve the use of technologies that are still embryonic.

From the third-generation biofuels, the production of biodiesel from microalgae cultures is a promising alternative form of bioenergy at an affordable cost, using soils that are not of high value for food production. However, this technology is still in the laboratory stage. In the near future, it will contribute to the large scale production of biofuel.

The GHG emission balance depends on the feedstock and processes used, and it is important to realize that advanced biofuels performance is not always superior to that of conventional biofuels. Nonetheless, environmental interest propels their development so they can become competitive with both fossil fuels and conventional biofuels.

Next, the main processes used to produce first-, second- and third-generation biofuels are presented in more depth.

2.2.2   First Generation Biofuels

First generation biofuels are those that have currently reached a stage of commercial production. In general, they come from food crops. The first generation biofuels use agricultural feedstocks as inputs to their production, which is the case of ethanol from sugarcane and biodiesel from vegetable oils, for instance. Table 2.1 shows an overview of production technologies of first generation biofuels. Next, the advances in their production are presented.

2.2.2.1   Sugar and Starch Based Ethanol

In the sugar-to-ethanol process, sucrose is obtained from sugar crops such as sugarcane, sugar beet and sweet sorghum, and it is subsequently fermented by yeast into ethanol, also generating other metabolic by-products such as carbon dioxide. The ethanol is then recovered and concentrated by a variety of processes. The

Table 2.1 Technologies for producing first generation biofuels.

Biofuel type	Specific name	Feedstock	Conversion technologies
Biodiesel	Biodiesel from energy crops: methyl and ethyl esters of fatty acids	Oil crops (soybean, rapeseed, palm, etc.)	Cold and warm pressing extraction, purification, transesterification
	Biodiesel from waste	Waste, cooking/ frying oil	Hydrogenation
Bioethanol	Conventional ethanol	Sugar beet, sugarcane	Direct fermentation of juice
	Starchy ethanol	Corn, wheat and other grains	Enzymatic hydrolysis, fermentation

process of ethanol production from sugarcane includes the following steps (Figure 2.4):

- Milling: the biomass is cleaned and milled;
- Must preparation: water is mixed to the sugarcane juice and molasse to adjust the concentration of sugar for subsequent fermentation;
- Fermentation: yeast is added to the mixture, converting sugars to ethanol and carbon dioxide;
- Centrifugation: the liquid and solid fractions are separated;
- Distillation: the ethanol contained in the liquid fraction is separated from the water, with a purity of approximately 95.6 % (hydrated ethanol);
- Dehydration: the hydrated ethanol goes through a process to remove the remaining water (azeotropic distillation, extractive distillation or molecular sieving), yielding the anhydrous ethanol;
- Denaturation: the ethanol to be used for fuel is then denatured with a small percentage of additives, such as methanol, isopropanol, acetone, methyl ethyl ketone, methyl isobutyl ketone, etc., to make it inappropriate for human consumption.

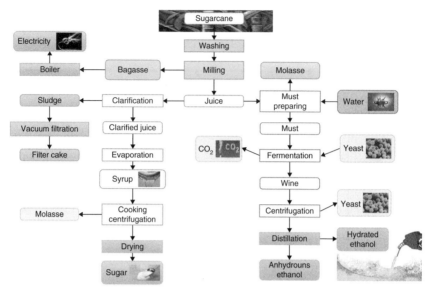

Figure 2.4 Stages of production in sugar mills and distilleries. Adapted from Prado and Meireles [14].

For the production of first generation bioethanol from corn, the process used for sugarcane needs to be adapted because the starch must undergo a pretreatment so that it is hydrolyzed into sugar prior to fermentation. In the milling step the raw material is comminuted into finer particles. These particles are then blended with water and enzymes (α-amylase), and the mixture is cooked at high temperatures (140–180°C) to liquefy the starch. The mixture goes through saccharification, where a gluco-amylase enzyme is added to convert the starch molecules into fermentable sugars. The fermentation and distillation steps are similar to the ones used in sugarcane processing. The high cooking temperatures imply high operating costs, which makes starch-based ethanol economically less advantageous when compared to sugarcane-based ethanol.

The first generation ethanol can be used as a pure fuel or can be blended with gasoline and other fuels.

The most recent advances in first generation ethanol have come from improving ethanol production by *S. cerevisiae*, particularly by genome shuffling and global transcription machinery engineering. Among the most significant developments in the fermentation field are the implementation of very high-gravity technology, the use of lignocellulosic hydrolysates as feedstock and the application of high-cell-density continuous processes. Such technologies benefit from the selection and engineering of more robust yeast strains, with tailored properties for each of the processes. [15].

2.2.2.2 Conventional Biodiesel

Several processes are under development aiming to produce fuels with properties very similar to diesel and kerosene, one of which is biodiesel. Biodiesel is defined by the American Society for Testing and Materials (ASTM) as a "synthetic liquid fuel, originating from renewable raw material and consisting of a mixture of alkyl esters of long chain fatty acids derived from vegetable oils or animal fats". It may also be defined as "derivative from renewable biomass that can replace, partially or completely, fossil fuels in internal combustion engines or for generating another type of energy".

Conventional biodiesel can be produced from raw vegetable oils derived from soybean, rapeseed, palm oil or sunflower, among others, as well as animal fats and used cooking oil. Vegetable oils transesterified with alcohols are the most common form of biodiesel. Transesterification aims at modifying the molecular structure of the vegetable oil, so that its physicochemical characteristics are similar to mineral oil. The transesterification reaction is the conversion of vegetable oil into methyl or ethyl esters of fatty acids, which constitute biodiesel. The reaction occurs in the presence of an acid, basic or enzymatic catalyst, and a short chain alcohol. The most used alcohols in this type of reaction are methanol, which is toxic and originates from fossil fuels, and ethanol, a GRAS solvent. The fatty acid reacts with the short chain alcohol forming the ester (biodiesel) and water. Furthermore, glycerin is formed as a byproduct; it has application especially in the cosmetics industry. The transesterification reaction may be slow when conducted at room temperature, but it can be accelerated by using heat and/or catalyzers, especially basic, which can be recovered at the end of the reaction and reused. In a typical mass balance, for each 100 kg of crude vegetable oil 10–15 kg of alcohol is required to produce 100–105 kg of biodiesel and 10 kg of glycerol [16].

The great advantage of using the transesterified oil (biodiesel) is the possibility to replace the mineral diesel without the need to modify the engines. Biodiesel is one of the biofuels that has some of the most compatible characteristics with fossil fuels (petroleum diesel). For example, the high heating value of biodiesel (39–41 MJ/kg) is comparable with petrodiesel (43 MJ/kg); and other important parameters like flash point, cetane number and kinematic viscosity are similar to its fossil alternative [17]. Therefore, these fuels are blendable with fossil fuels at any proportion, can use the same infrastructure and are fully compatible with engines in heavy duty vehicles.

Current global biodiesel production, from different fatty raw materials, reaches about 6 billion liters per year and represents 10% of total biofuel production. In Brazil, government mandates for blending biodiesel with gasoline has promoted a production increase from around 100 million m^3 in 2006 to more than 1.4 billion m^3 in 2009, although the sustainability of biodiesel production is still to be demonstrated [1].

The most recent advances in first generation biodiesel production includes moving from methanol use to ethanol use due to the toxicity of the former; testing new catalysts; and using different raw materials, especially non-edibles. Typically, a more saturated fat allows better biodiesel properties, especially regarding the cetane number and stability, although they present higher melting or dripping points, which can be a problem in colder climates. Therefore, soybean and castor have limited feasibility, whereas tallow and palm oil represent more suitable alternatives [1]. Therefore, there are still a number of approaches to be explored that can lead to the optimization of biodiesel production and thus decrease its cost.

2.2.2.3   Biogas and Biomethane

Biogas can be produced by anaerobic digestion of feedstocks such as organic waste, animal manure and sewage sludge, or from dedicated green energy crops such as maize, grass and wheat. The anaerobic digestion process, or biomethanization, represents an attractive treatment strategy for reducing the contamination of the different biosolid residues and may benefit society by providing a renewable biofuel source from different organic substrates. Basically, the microorganisms are retained in the bioreactor, with separation between the acidogenic and methanogenic bacteria. The biometanization process of biomass is accomplished by a series of biochemical transformations, which can be separated into a first step where hydrolysis, acidification and liquefaction take place and a second step where acetate, hydrogen and carbon dioxide are transformed into methane. The biogas product from biomethanization contains between 60–80% of methane. Biogas is often used to generate heat and electricity, but it can also be upgraded to bio-methane by removing CO_2 and hydrogen sulfide (H_2S), and inject-ing it into the natural gasgrid. Biomethane can also be used as fuel in natural gas vehicles.

There are a large number of factors that affect biogas production efficiency such as pH, temperature, and inhibitory parameters (like high organic loading) [18]. The recent trends in biogas production focus on optimizing these operational conditions to improve yields and decrease costs.

2.2.3 Second Generation Biofuels

Instead of only using readily extractable sugars, starches or oils as in the previous generation, second generation biofuels do not use edible sources as raw materials. They focus on different feedstocks and their parts in order to explore a broader range of raw materials. One example is bioethanol produced from lignocellulosic biomass from various non-edible sources using all the parts of the biomass. The raw materials can be agricultural residues such as straw and stover, residues from forestry, or biomass crops such as grasses (e.g. switchgrass) and wood from short rotation forestry. All these raw materials can be converted into biofuels via biochemical routes using enzymes and/or microorganisms, including genetically modified microorganisms that have been developed specifically for this purpose [19]. As second generation biofuels use different bio-conversion pathways, they apparently avoid the "fuel versus food" dilemma. However, they can compete with the use of agricultural lands which could be used to grow food crops [20].

The main processes for the production of second generation bio-fuels are shown in Figure 2.5. Biomass conversion is conducted via two generic approaches: thermochemical decomposition including gasification, bio-carbonization, liquefaction and thermal decom-position (pyrolysis) processes; and biological digestion, essentially referring to microbial digestion and fermentation. While biological processing is usually very selective and produces a small number of discrete products in high yield using biological catalysts, ther-mal conversion often gives multiple and complex products, in very short reaction times, and inorganic catalysts are often used to improve the product quality or spectrum [21].

The term "second generation biofuel" is defined mainly on the basis of the feedstocks and conversion technologies. However, there

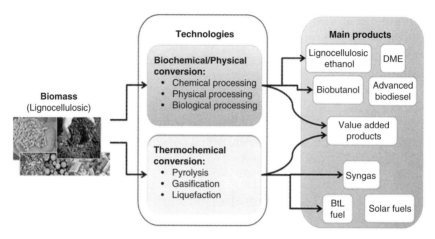

Figure 2.5 Main processes for production of second generation biofuels. Adapted from Bacovsky *et al.* [22].

is no precise definition; therefore, some biofuels cannot be allocated to a particular "generation" (e.g. biomethane), while other products claim to be third generation (fuels from CO_2 fixing bacteria). The main second generation biofuels that have been studied in the last years include:

- Hydrotreated vegetable oils (HVO): they are not strictly second generation because the raw material is (currently) first generation;
- Hydrotreated esters and fatty acids (HEFA) fuels, also referred to as bioJet: also based on HVO, they were first developed for application in aviation, and now they are applied to different biofuels;
- Cellulosic ethanol: chemically, there is no difference between cellulosic ethanol and conventional bioethanol; however, the raw material is made of cellulose in second generation, whereas in first generation simple sugars are directly fermented into bioethanol.

For more than three decades, biofuels produced from lignocellulosic sources have been receiving much attention. The extensive database in the literature confirms that the developments in this field have reached a near commercial stage [23]. Next, the main second generation biofuels are presented.

2.2.3.1   Cellulosic Ethanol

Lignocellulosic biomass is considered an important non-edible alternative for the production of cellulosic ethanol. It is also cheap, abundant, and fast-growing. Lignocellulosic biomass consists of a combination of lignin, cellulose and hemicellulose. The lignocellulosic complex provides the structural framework making up most of the plant matter. A wide variety of fuels can be derived from lignocellulosic material via the biological or chemical synthesis of products from the biological or chemical breakdown of cellulose, hemicellulose and lignin. Second generation bioethanol focuses on the production of ethanol via these routes. There is also interest in producing other chemicals and fuel components.

Fermentation of simple sugars from sugar crops and from hydrolysis of starch crops to produce ethanol is a commercial and widely used first generation process. Routes from lignocellulosic materials to ethanol are more complicated than those from sugar and starches, as lignocellulosic materials contain more complex sugar polymers, such as cellulose and hemicellulose, which are more difficult to break down [24]. Because of that, the second generation bioethanol production from biomass requires additional processing steps (Figure 2.6).

The cellulose (source of hexoses such as glucose) as well as hemicellulose (mainly source of pentoses such as xylose) is not accessible to the traditional ethanol producing microorganisms. Therefore, these fractions must be hydrolyzed prior to fermentation. The main purpose of hydrolysis is splitting the polymeric structure of lignin-free cellulosic material into fermentable sugar monomers [22]. To obtain lignin-free cellulosic material a pretreatment is required. The pretreatment aims to separate the biomass into cellulose, hemicellulose and lignin fractions via physicochemical or biochemical routes; this process can also sometimes hydrolyze the hemicellulose into simple sugars. Then the cellulose undergoes hydrolysis to generate fermentable sugars. The sugars derived from hemicellulose and cellulose are then fermented into ethanol using yeasts [25].

The breaking down of the lignocellulosic complex is usually achieved by high energy-consuming biochemical conversions of the cellulose and hemicellulose components of the biomass into fermentable sugars. The cellulose hydrolysis stage can be carried out by chemical routes, using acidic or basic catalysts, or by biological routes, using enzymatic catalysts. Acidic hydrolysis is the most

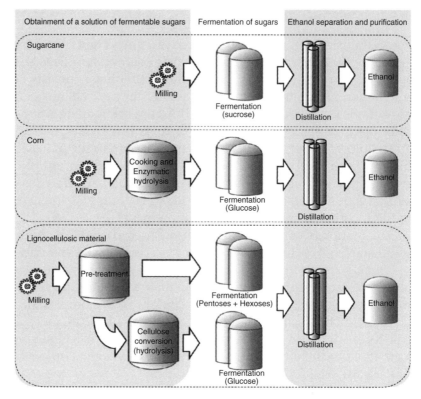

Figure 2.6 Processing steps for second generation bioethanol production from lignocellulosic biomass compared to first generation processes. Reprinted from Biotechnology Advances, 28, S.I. Mussato, G. Dragone, P.M.R. Guimarães, J.P.A. Silva, L.M. Carneiro, I.C. Roberto, A. Vicente, L. Domingues, J.A. Teixeira, Technological trends, global market, and challenges of bio-ethanol production, 817–830, 2010, with permission from Elsevier [15].

well-known process, and although acid and basic hydrolysis are relatively fast methods that produce high glucose concentrations, they have several drawbacks, such as the need to neutralize the reaction medium after the process, high corrosion of the equipment and the generation of solid wastes. Due to environmental problems posed by acidic and basic hydrolysis, the enzymatic process is more widely used nowadays [26]. The main advantage of enzymatic hydrolysis is that it does not generate fermentation microorganism inhibitors. In contrast, it is a very slow process, which makes it expensive, and the enzymes are difficult to be recovered and reused. Furthermore, the enzymes cannot break down the lignocellulosic complex, so enzymatic hydrolysis requires a pretreatment that allows for the enzymes to access the cellulose and the hemicellulose [27–28].

Considerable research is directed towards pretreatment to improve enzymatic hydrolysis. Different approaches are being tested: acid, alkali, sulfite, liquid hot water, steam explosion, carbon dioxide explosion, ammonia fiber explosion, ozonolyzis, wet oxidation, ionic liquids, organosolv, microwave and ultrasound [24, 29–31]. Ionic liquids have been used as solvents and reaction mediums for processing biofuels, since their hydrophobic nature, less viscosity, kosmotropic anion and chaotropic cation usually enhance the activity and stability of enzymes [32]. Another option that has been recently studied for the hydrolysis step of lignocellulosic ethanol production is using sub/supercritical water as a reaction medium, in an autocalatylic process [33–35]. Although the hydrolysis processes are not fully developed, they have been given special attention over the past years, and enormous progress is being made towards making them technical and economically feasible.

Unfortunately, no natural organism can metabolize simultaneously both hexoses and pentoses, despite the fact that many naturally occurring yeasts and bacteria have the capability of metabolizing one or the other. Because the industrially adapted strains of yeast that are used for ethanol production cannot use xylose, strains of yeast with this capability have been developed through genetic engineering. To make lignocellulosic ethanol feasible, in addition to using all hexose and pentose sugars, microbial strains must be resistant to the compounds produced or released during biomass degradation [36–37]. *Saccharomyces cerevisiae* and *Zymomonas mobilis* are native ethanol producers that can efficiently convert glucose to ethanol, but cannot use pentose sugars as carbon sources. In contrast, *Escherichia coli* can utilize most carbohydrate components present in lignocellulosics but produce only a small amount of ethanol during fermentation, so *E. coli* strains have been metabolically engineered for enhanced ethanol production. The two scientific approaches to make sure that both hexoses and pentoses are fermented into ethanol include: (i) genetically introducing the missing xylose pathway into the target organism; and (ii) co-culturing xylose and glucose metabolizing organisms [38–39].

Within this context, there is much to be exploited in the field to improve the second generation bioethanol yields and costs. Nonetheless, lignocellulosic ethanol pathways offer a positive (compared to fossil) energy gain and a substantial opportunity to reduce GHG emissions relative to gasoline and corn ethanol [40].

Currently, Canada, Sweden, the US, Denmark, Spain, Italy, France, Japan, India, Australia and Norway are producing ethanol from cellulosic feedstock at different development stages, and several public/private international projects have been developed in the biorenewable sector to promote a bio-based economy [30–31, 41]. However, conversion technologies for producing bioethanol from cellulosic biomass resources such as forest materials, agricultural residues and urban waste have not yet been demonstrated commercially [5].

Despite substantial progress in cellulosic ethanol research and development, many challenges remain to be overcome, such as [13, 15]:

- Development of a suitable and economically viable hydrolysis process step;
- High energy consumption of biomass pretreatment;
- Improvement in the conversion rate and yield of hemicellulose sugars;
- Process scale-up;
- Capital equipment required for commercial demonstrations of some technologies, such as steam explosion, does not exist;
- The recovery of pretreatment chemicals and wastewater;
- Engineering of microorganisms that can produce ethanol from a different sugar mix than that found in first generation fermentation;
- Process integration to minimize energy demands.

Considering this scenario, further research and development is required in this field. Because Brazil is the second major producer of first generation bioethanol in the world, this country has been strongly investing to make second generation bioethanol technically and economically feasible. Other countries, especially the US, are also undertaking efforts to enable the production of second generation ethanol.

2.2.3.2   Syngas

Bio-synthetic gas (Bio-SG), also known as syngas, is biomethane derived from biomass via thermochemical processes, such as gasification. Gasification was developed more than 100 years ago to generate gaseous fuel, town gas, from coal and peat. In principle, biomass gasification is simply the incomplete combustion of biomass, due to low oxygen, to produce CO, CO_2 and H_2 instead of CO_2 and water. The process involves overlapping reactions and stages such as drying, pyrolysis, char gasification, and oxidation. This conversion typically occurs at elevated temperatures (500–1400°C) and pressures ranging from atmospheric pressure to as much as 3 MPa. An oxidant is used: air, pure oxygen, steam or a mixture of these gases. An important process parameter is the operational mode of the gasifier. Generally, gasifiers operate in a fixed bed, fluidized bed, or entrained flow mode. This process is also strongly favored by catalysts; the most used catalysts are dolomite, alkali (metal) and nickel [37].

Syngas uses include direct combustion in boilers, turbines or internal combustion engines to generate electricity, heat or both. Moreover, the gasification allows the production of very clean synthetic biofuels, which can be liquefied into a broad range of liquid fuels, including biodiesel [37].

The deployment of natural gas vehicles has started to grow rapidly, particularly during the last decade, reaching shares of around 25% of the total vehicle fleet in countries like Bangladesh, Armenia and Pakistan [9]. An alternative of clean fuels for these vehicles is biomethane derived from anaerobic digestion (first generation) or gasification (second generation) of biomass [42].

Gasification has been practiced for many years and while there are many examples of demonstration and pre-commercial activities there are still surprisingly few successful operational units [21]. The first demonstration plant producing biomethane thermochemically from solid biomass started operating in late 2008 in Güssing, Austria, and a plant is planned in Gothenburg, Sweden.

2.2.3.3   Advanced Biodiesel from Biomass

The second generation biodiesel, also known as synthetic biodiesel, is a liquid biofuel produced from lignocellulosic biomass via thermochemical processes. Advanced biodiesel includes:

- Hydrotreated vegetable oil (HVO), which is produced by hydrogenating vegetable oils or animal fats;
- Biomass-to-Liquid (BtL) diesel, also referred to as Fischer-Tropsch diesel.

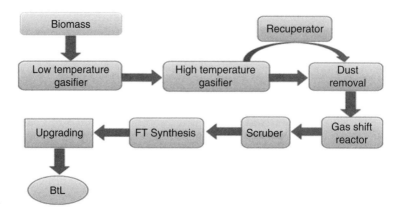

Figure 2.7 Typical process for obtaining BtL biodiesel.

The hydroprocessing of oils into hydrocarbons using an appropriate catalyst and appropriate operating conditions leads to the production of green diesel, which can be blended with diesel fractions of petroleum origin. Besides a much higher cetane number, the advantage of green diesel compared with conventional biodiesel is the higher stability [23].

BtL is the most used process to produce second generation biodiesel. The BtL biodiesel may be derived from any biomass, especially lignocellulosics with low moisture content. A typical process for producing BtL biodiesel is shown in Figure 2.7.

The BtL process consists of two steps. In the first step, gasification, the biomass is converted to a syngas rich in hydrogen and carbon monoxide. The syngas is produced by a three-stage gasification process: i) low temperature gasification; ii) high temperature gasification; and iii) endothermic entrained bed gasification. A purification step follows to eliminate alkylating agents, which are small particles of gaseous contaminants (dust). The syngas must have its H_2/CO ratio adjusted by water-gas-shift before it is used in the next step [23]. After cleaning, the syngas is catalytically converted into a broad range of hydrocarbon liquids, including synthetic diesel and bio-kerosene. The main catalytic process to convert the syngas into a liquid fuel is known as Fischer-Tropsch (FT). This process occurs at high temperatures; it consists of a reaction between hydrogen and carbon monoxide that is explained by the following equation:

$$(2n + 1)H_2 + n(CO) \longrightarrow C_nH_{2n+2} + n(H_2O) \qquad (2.1)$$

The catalysts most commonly used are Mo, W, Co and Ni, and the recent studies have been focusing on new and more efficient catalysts [23]. The FT process results in the synthesis of the BtL biodiesel. While the FT process has been commercially available for decades, some efficiency and environmental issues associated with syngas production from biomass still need to be addressed; therefore, key recent efforts have been focusing on these issues. Moreover, in recent years hydroprocessing in subcritical and supercritical water mediums as well as in supercritical alcohols have been attracting attention [23].

The first large-scale plants producing HVO have been built in Finland and Singapore, but the process has not yet been fully commercialized [22].

Some innovative processes have been evaluated for biodiesel production, such as enzymatic hydrolysis and cracking (pyrolysis), which are frequently pointed out as breakthroughs for vegetable oil conversion to fuels. The H-Bio process patented by Petrobras is a special case of the cracking process, which involves a catalytic hydroconversion of a mixture of diesel fractions and vegetable oil to linear hydrocarbon chains, similar to what already exists in conventional diesel, with a very high conversion yield, at least 95%, without residue generation and a small amount of propane produced as a by-product [1].

2.2.3.4 Bio-oil from Pyrolysis

Pyrolysis has been applied for thousands of years to charcoal production but it is only in the last 30 years that fast pyrolysis at moderate temperatures and with very short reaction times has become of considerable interest. Fast pyrolysis is a thermal decomposition process consisting of rapidly (2–3 s) heating the ground biomass to temperatures between 400°C and 600°C in the absence of air, followed by rapid cooling (Figure 2.8). Through this process, thermally unstable biomass compounds are converted to a heavy liquid fuel. Crude pyrolysis liquid, or bio-oil, is dark brown and resembles biomass in elemental composition. The bio-oil formed has a heating value about half that of conventional fuel oil. Moreover, it offers no significant health, environment or safety risks [21, 37].

The potential attractiveness of bioconversion technologies for liquid fuel production is related to the idea that they may have lower capital costs than thermal conversion methods at scales that do not

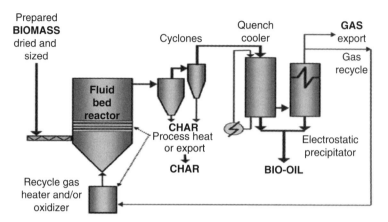

Figure 2.8 Scheme of the fast pyrolysis process. Reprinted from Biomass and Bioenergy, 38, A.V. Bridgwater, Review of fast pyrolysis of biomass and product upgrading, 68–94, 2012, with permission from Elsevier [21].

require the long distance transport of biomass. The oil obtained by pyrolysis is more suitable for long-distance transport than, for instance, straw or wood-chips. Moreover, "coal", or biochar, is produced as a by-product from this process. It can be used as solid fuel, or applied on land as a measure for carbon sequestration and soil fertilization. The pyrolysis oil can be processed in ways similar to crude oil. Alternatively, the bio-oil and the biochar can be treated as raw materials in the production of BtL [21, 36, 43].

Virtually any form of biomass can be considered for fast pyrolysis. While most work has been carried out on wood, over 100 different biomass types have been tested, ranging from agricultural wastes such as straw, olive pits and nut shells to energy crops such as miscanthus and sorghum, forestry wastes such as bark, and solid wastes such as sewage sludge and leather wastes [21]. On the other hand, the information on related to the pyrolysis of animal fats, vegetable oils and sewage sludge is still limited. This may be, at least partly, attributed to the loss of valuable liquid hydrocarbons to gas and coke, generally observed during pyrolysis of biomass. Moreover, because of the high content of guaiacols and phenols, liquids from pyrolysis are not stable. Therefore, in situ catalysts are being used to improve the quality of pyrolysis liquids. The use of suitable catalysts significantly increase the yields and also optimize the composition of the output products [23, 37]. However, there is still much to be done to optimize the production of pyrolysis bio-oil.

Bio-oils are complex colloidal multidispersed systems containing a large amount of water, carboxylic acids, carbohydrates, and lignin-derived substances. As a result, they exhibit some undesired properties, such as acidity, thermal instability, high oxygen content (35–40%), low heating value, high viscosity, corrosiveness, and chemical instability, which limits their application for vehicle fuels. Therefore, several research efforts are currently being undertaken to upgrade pyrolysis oil to advanced biofuels. Upgrading bio-oil to a conventional transport fuel such as diesel, gasoline, kerosene, methane and LPG requires full deoxygenation and conventional refining, which can be accomplished by integrated catalytic pyrolysis, hydrotreatment, esterification (biodiesel), gasification (hydrogen, syngas), blending or other processes. There is also interest in partial upgrading to a product that is compatible with refinery streams in order to take advantage of the economy of scale and experience in a conventional refinery [21, 44–45].

On the other hand, if bio-oils are to become a source of chemical products (as opposed to fuels), separation technology will be important. Phase separation of bio-oils can occur spontaneously or by adding water or salts to form two different phases. When aqueous salt solutions are added, phase separation of the pyrolysis bio-oil can occur to form two different phases: the upper layer with high contents of water, acetic acid, and water-soluble compounds and the bottom layers with low water and high lignin pyrolysis compound contents [44–45].

High ash content is a significant technical barrier for the production of a quality product from poorer quality biomass e.g. agricultural residues. Acid washing, water leaching, hot gas filtration, and post pyrolysis filtration can improve the quality of the resulting bio-oil, but more research in this area is required. Moreover, maximization of the liquid yield from fast pyrolysis, the traditional goal, may need to be reconsidered when the desired properties of the bio-oil as well as downstream processing requirements are taken into account [45].

There are a few industrial units operating with fast pyrolysis in Canada, the UK, Spain, Finland, the Netherlands, the US, Germany, Austria and China. This technology has been successfully demonstrated on a small scale for the production of liquid fuel, and several demonstration and commercial plants are in operation. The process is on the verge of commercial application but it is still relatively expensive compared with fossil-based energy, and thus faces economic and other non-technical barriers when trying to penetrate energy markets. [21, 37, 45].

2.2.3.5   Bio-oil from Hydrothermal Process

Biomass can be processed in a liquid medium (typically water) under pressure at temperatures of 200–450°C. In contrast to fast pyrolysis, residence times of up to 30 min are required. The reaction yields oils and residual solids that have low water content, and lower oxygen content than oils from fast pyrolysis. The yields of hydrothermal processing are also usually higher than those of pyrolysis. Most of the polar components present in the aqueous medium can be converted to hydrocarbons. Then, the separation of hydrophobic hydrocarbon phase from the aqueous phase is rather simple [23].

The hydrothermal liquefaction process can be conducted with or without a catalyst. Under sub/supercritical conditions, water acts as a reactant, catalyst, and solvent for typically acid- or base-catalyzed reactions. Apparently, water may also supply hydrogen for cracking, which is the hydrolysis of triglycerides, followed by decarboxylation. The high-temperature and high-pressure water reduces the formation of gaseous products and minimizes the formation of chars [23].

A recent trend in hydroprocessing is using polar protic solvents (e.g., methanol, ethanol, diethylene glycol, etc.), as the homogeneity of the bio-oil may be improved by diluting it in these solvents. This may be critical for bio-oils derived from municipal solid waste. The homogeneity of the medium can be further improved under supercritical conditions [23].

This technology is being developed for use on algae and waste biomass. Developers of this technology include Changing World Technologies (West Hampstead, NY, US), EnerTech Environmental Inc. (Atlanta, GA, US) and Biofuel B.V. (Heemskerk, Netherlands) [37].

2.2.3.6   Dimethylether

Dimethylether (DME) can be produced from methanol by catalytic dehydration or from syngas. DME is the simplest ether and can replace propane in liquefied petroleum gas used as fuel. It is a promising fuel in diesel engines, due to its high cetane number. The production of DME through the gasification of biomass is in the demonstration stage, and the first plant running with this technology started operating in September 2010 in Sweden [9].

2.2.3.7 Biobutanol

Biobutanol can be used as fuel in internal combustion engines. It has a higher energy density (29.2 MJ/L) and is more similar to gasoline than ethanol, and could thus be distributed through existing gasoline infrastructure. Biobutanol can be produced by fermenting sugar via the acetone-butanol-ethanol (ABE) process using *Clostridium acetobutylicum* and some engineered *Eschericia coli* [38]. The anaerobic growth of selected strains combined with stripping the isobutanol from the broth through continuous flash evaporation, has resulted in more than 90% of the theoretical yield. Yeast strains that can metabolize five-carbon sugars were also proposed [39].

Biobutanol can be produced from the same starch and sugar feedstocks that are used for conventional ethanol. In addition, sugars can be derived from lignocellulosic biomass, using the same biochemical conversion steps required for advanced ethanol production. This fact accentuates the need for research into the biochemical conversion process of biomass to sugar [9].

Demonstration plants are producing biobutanol in Germany and the US and others are currently under construction.

2.2.4 Third Generation Biofuels

Biofuels of the third generation come from algae and hydrogen produced from lignocellulosic biomass. The products resulting from their conversion are described as third generation because they do no longer require the use of land. Their production technologies use catalytic reforming routes to convert sugar, starch and all forms of lignocellulose into targeted short-chain carbon compounds. The technologies for third generation biofuel production are still in the development phase and their large scale production is expected in the medium to long term.

2.2.4.1 Hydrogen from Biomass

Hydrogen, the "fuel of the future", has long been considered a good source of energy due to its high energy density and sustainability, and may finally become an important component of the energy balance. Its demand is not limited to its use as a source of energy; the hydrogen gas is used for the production of chemicals, for the hydrogenation of fats and oils in the food industry for margarine production, for processing steel, and for the desulfurization and reformulation of gasoline in refineries. Disadvantages of hydrogen, on the other hand, include the difficulty of easily and inexpensively storing and dispensing the hydrogen gas, as would be required in widespread commercial applications [46–49].

A variety of photosynthetic and non-photosynthetic microorganisms, including unicellular green algae, cyanobacteria, anoxygenic photosynthetic bacteria, obligate anaerobic, and nitrogen-fixing bacteria are endowed with genes and proteins for H_2 production; however, their efficiency is still low [38, 48]. Other low-cost hydrogen-based fuel cells having different process routes for hydrogen production from biomass can be broadly classified as [49]:

- thermochemical gasification coupled to water gas shift;
- fast pyrolysis followed by reforming of carbohydrate fractions of bio-oil;
- direct solar gasification;
- miscellaneous novel gasification processes;
- conversion of syngas derived from biomass;
- supercritical conversion of biomass;
- microbial conversion of biomass.

Hydrogen can be obtained from both fossil and biomass resources. The hydrogen obtained from renewable biomass is a sustainable source of energy, and is therefore known as biohydrogen. Biohydrogen production technologies are still in the very early stage of research and development. The optimization of bioreactor designs and operational conditions for light, pH, nutrients and microbial flora, testing and validation of biological, chemical and physical pretreatments, rapid removal and purification of gases, immobilization of microorganisms, and genetic modifications of enzymatic metabolic pathways that compete with hydrogen producing enzyme systems offer exciting prospects for improving biohydrogen production systems [48]. Specific areas of research for enhanced biohydrogen production include [12]:

- the reengineering of photosynthetic microorganisms for achieving high hydrogen production capacity;
- the development of techniques for effectively separating and refining the hydrogen formed;

- the design of integrated systems for biohydrogen production followed by technical evaluations and cost-benefit analyses.

Today biomass gasification offers the earliest and most economical route for the production of renewable hydrogen. It is expected that steam reforming of natural gas and gasification of biomass will become the dominant technologies by the end of the 21st century [49].

2.2.4.2 Algae-based Fuels

Algae have been cultivated commercially since the 1950s, mainly for the pharmaceutical industry, but recently they have gained attention as potential sources of biomass because they produce chemicals and substances that have a number of uses, including transport fuels. The production of biofuels from algae has been looked at in some depth over the past 50 years by, in particular, the US and Japan. However, the US and Japanese programs were discontinued without researchers reaching the ultimate goal of finding a cost effective algae biofuel production method. On the other hand, with the increasing cost of conventional fossil fuels, the concern about using edible feedstocks for biofuel production and the development of more sophisticated tools for manipulating the algae, a renewed interest in this area has been observed [25]. Within this context, photosynthetic algae have been considered either as a biomass source or host strain for the production of biofuel [38]. The main advantage of using algae as feedstock for biofuel production is the potential for efficient land and resource use, as the ponds or reactors in which they are grown can be situated on unproductive land [15, 50].

Algae feedstocks can be classified as:

- Macroalgae: Algae such as seaweed, which are cultivated almost exclusively in the sea, or harvested from natural stocks. They are mainly used for food and feed and as an industrial feedstock. Macroalgae can be used to produce biogas and bioethanol, but the yield has so far been low [51];
- Microalgae: They are used for high-end cosmetic and medical applications, and their global production today is just a few tons [52]. They are grown in open

tanks or closed photo-bioreactors, and contain certain percentages of oil that can be used for biodiesel production, while the rest of the biomass can, in principle, be used for ethanol and biogas production. The cultivation of microalgae requires water and energy inputs for circulation and cleaning, which may cause considerable greenhouse gas emissions. The use of genetically modified strains is critical, because even closed systems cannot prevent genetically modified algae from getting into the environment, with its corresponding risks, such as contamination of the ecosystem [53].

Microalgae have been identified as potential sources of third generation biofuels because of their higher photosynthetic efficiency, faster growth rate, and higher area-specific yields compared with other sources of biomass. It has been estimated that 46,000–140,000 liters of ethanol can be produced from 1 ha of microalgae, which is a value several orders of magnitude higher than for ethanol from corn stover (1,000–1,400 L/ha), sugarcane (6,000–7,500 L/ha) and switchgrass (10,700 L/ha). Moreover, some algal species grow well in saline, brackish and wastewater, which makes them more promising feedstock than terrestrial crops that rely exclusively on fresh water. As algae can grow in different aquatic sources (oceans, lakes, estuaries, etc.) containing the wastewaters derived from municipal, agricultural and industrial activities, there is an extra opportunity for combining wastewater treatment by algae via nutrient removal with biofuel production [15, 23, 50].

Figure 2.9 shows the main products and types of fuel that can be obtained from algae, including the technological routes for their production. There are biochemical (e.g., fermentation and anaerobic digestion) and thermochemical (pyrolysis and liquefaction) methods for conversion of algae to biofuels. The former processes can be used to produce first generation biofuels as ethanol and biodiesel, whereas thermochemical processes can be used to produce second generation biodiesel and gas [23].

Using microorganisms such as yeast, heterotrophic algae or cyanobacteria it is possible to convert sugars into alkanes, which are the basic hydrocarbons of gasoline, diesel and jet fuel. The process consists of transforming a variety of water-soluble sugars into hydrogen and chemical intermediaries using aqueous phase reforming, and then into alkanes via a catalytic process. The use of modified

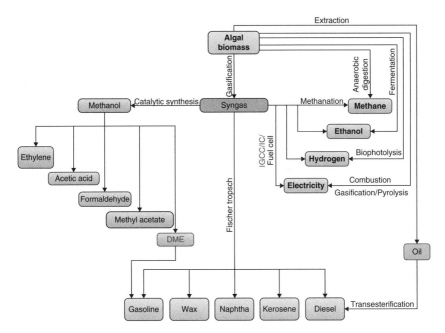

Figure 2.9 Overview of the main pathways for processing algae biomass.

microorganisms allows converting sugars into hydrocarbons that can be hydrogenated to synthetic diesel. However, none of the above processes have been demonstrated at an industrial scale [9].

Production of bioethanol fuel through algal feedstock fermentation starts from strain selection and growth. Two approaches are used: using macroalgae feedstock, often invasive, from the natural habitat; or the growth of an appropriately selected microalgal strain in artificial conditions. Macroalgae require comminution prior to processing. As in cellulosic ethanol from terrestrial raw materials, the algae must undergo pretreatment and saccharification through physical, chemical and/or enzymatic processes before their sugars can be fermented. However, algal biomass is less resistant to conversion into simple sugars than plant biomass due to a lack of lignin. Besides cellulose and hemicelluose, many algal species accumulate a high content of starch as their reserve material [15, 50].

Higher yields of bioethanol can currently be obtained from microalgae than from macroalgae. The main reason is that all major ethanol fermenting organisms have been screened and engineered for an efficient utilization of sugars released from starch hydrolysis, cellulose and hemicellulose (microalgal carbohydrates), but not for macroalgal carbohydrates like agar or carrageenan. As microalgal carbohydrates are similar to those of terrestrial crops, the yields obtained for their processing are higher. It is highly unlikely that a major shift in the productivity of macroalgal ethanol will take place, especially taking into consideration investments in ligno-cellulose ethanol production, which will also favor the microalgal type of biomass. Moreover, with the development of new enzymes for lignocellulose saccharification the gap between conversion efficiencies of macro and micro algal feedstocks is expected to increase. On the other hand, using waste macroalgal biomass for ethanol production deserves special consideration. It shifts algal biofuel production towards a biorefinery concept that is considered a long-term sustainable solution for biomass energy [50].

High oil yielding microalgae can be used as feedstock for bio-diesel. Given the right conditions (temperature, light, sufficient space), there is the potential to overcome the drawbacks that discontinued studies on algae-based fuels. High theoretical productivities and oil accumulation exhibited by certain algal species under conditions of cellular stress made biodiesel production by algal oil transesterification a logical choice for research and development after relative success of biodiesel production from other feedstocks. To date, macroalgal biodiesel has been reported sparingly and yields are much lower than those of microalgae [50].

The microalgae processing for producing biodiesel is comprised of the following steps: microalgae growth, harvest, dewatering and drying. Two methods of algal oil transformation are applied: a two-step method that is divided into oil extraction and oil transesterification; and a single step *in situ* transesterification of algal oils to biodiesel. Three types of conversion methods prevail: chemical, thermochemical and enzymatic. Four major types of catalysts performing this reaction have been applied to algal bio-diesel: alkali, acid, lipase, and heterogenous catalysts. The major drawback of this method is the requirement for methanol to drive the reaction forward; that impacts cost effectiveness, sustainability and the applicability of glycerol, a coproduct, for many industries. Other methods including heterogenous catalysis and supercritical methanol conversion look promising; however, it is too early to say whether they could displace homogenous acid catalysis as the main conversion method. There are few reports on biodiesel synthesis with the biological catalyst lipase. This method has been long seen as a potential breakthrough in biodiesel production; however, high catalyst cost still hinders the development of lipase-catalysed biodiesel production on a large scale [50].

To date, the most effective methods of producing biofuels from algal feedstocks are the fermentation of microalgae to bioethanol and the production of biodiesel via *in situ* transesterification of microalgal biomass. However, both fermentative ethanol production from algal feedstocks and transesterification of algal oils require extensive downstream processing, like dewatering, that is regarded as a major hurdle in algal biofuel commercialization [54]. Therefore, the real breakthrough is expected from the metabolic engineering of photosynthetic organisms to produce and secrete biofuels that promises significant simplification of down-stream processing [50]. In the near future, it is expected that algal strains with improved capability of producing alcohols, diesels, alkanes, and hydrogen will be developed by systems metabolic engineering concepts [38].

There are startup companies attempting to commercialize algae fuels in the US, New Zealand and Israel [55]. Some of them quote high yields for algal oil production and release press statements saying that they will be producing large quantities of algae biodiesel in the coming months or years. However, none of these companies have yet produced algae biofuels at an industrial scale and none have produced evidence of the GHG intensity of their fuel. Algal oil production costs were estimated ranging from $ 2.86/L to $ 3.51/L [56]. Therefore, while this is potentially a promising technology, there is still much to be done in terms of basic research to understand and improve the GHG intensity of the process.

2.2.5 Solar Fuels

There are currently many low-carbon methods to generate electricity, including wind, geothermal, hydroelectric, and solar approaches. When considering the likely contribution of these approaches, it is useful to remember that the Earth receives approximately 7000 times as much energy from the sun as all human energy uses. Energy from the sun can be utilized in the following three ways: via photovoltaic conversion to electricity; by using mirrors to heat liquids that power sterling engines to produce electricity; or by harvesting plant biomass that can be burned as solid or liquid fuels [36].

Biomass can be gasified into syngas using heat generated by a concentrating solar plant, thus potentially improving the conversion efficiency and decreasing GHG emissions. More demonstration plants and further research is needed to make the process more efficient and to allow for commercial-scale operation, though. Another technology that could evolve as a process to produce liquid transport fuels is the splitting of water or CO_2 into hydrogen or carbon monoxide, which can then be converted to liquid fuels via a catalytic process. This process is currently in the laboratory stage and considerable research efforts on large scale are needed to support further development of solar fuels as part of the transport fuel mix in the long term [9].

Moreover, metabolic engineering of photosynthetic organisms seems to be a very promising route to convert solar energy into biofuels [50]. The cyanobacteria, a diverse group capable of carrying out oxygenic photosynthesis, have been proposed as parts of many possible biotechnological processes. Cyanobacteria are appealing since they are capable of direct capture of solar energy and its conversion to useful chemical energy using water as a substrate. This capacity could be used in a number of ways in biofuel production. Their capacity to drive carbon dioxide fixation with photosynthetically derived energy suggests that the newly recycled carbon could be converted to useful biofuels through the introduction of novel (to cyanobacteria) metabolic pathways. This process has been called biophotolysis. This is an inherently appealing and conceptually simple approach in which an abundant substrate, water, and a ubiquitous energy source, solar energy, are used to produce hydrogen, ethanol and isobutyraldehyde, among others. Moreover, useful by-products might be obtained from the biomass produced during use in biofuel production [57].

Development of low-cost photobioreactors suitable for the task (i.e., transparent, durable, hydrogen impermeable, etc.) requires advances in materials science. On the biological side, improvements will come from the application of genetic engineering. Although some effort has been done to adapt cyanobacteria for some liquid biofuels, relatively little has been done until recently to attempt to improve cyanobacterial hydrogen production by metabolic engineering. The near future may bring some real attempts to advance one or more of the cyanobacterial systems to the next level with some real gains in conversion efficiencies [57].

2.2.6   Bio-refineries

The large majority of chemicals are manufactured from petroleum feedstocks. Only a small proportion of the total oil production, around 5%, is used in chemical manufacturing, but the value of these chemicals is high and contributes comparable revenue to fuel and energy products. There is a clear economic advantage in building a similar flexibility into the biofuels market by devoting part of biomass production to the manufacture of chemicals. In fact, this concept makes even more sense in the context of biomass because it is chemically more heterogeneous than crude oil and the conversion to fuels, particularly hydrocarbons, is not so cost effective [21].

The biorefinery concept is analogous to the basic concept of conventional oil refineries: producing a variety of fuels and other products from a certain feedstock. The economic competitiveness of the operation is based on the production of high-value, low-volume co-products in addition to comparably low-value, high-volume biofuels. The key feature of the biorefinery concept is the coproduction of fuels, chemicals and energy from different biomass feedstocks. Therefore, a biorefinery can be defined as an optimized use of biomass for materials, chemicals, fuels and energy applications, where performance relates to costs, economics, markets, yield, environment, impact, carbon balance and social aspects. In other words, there needs to be an optimized use of resources, maximized profitability, maximized benefits and minimized waste [9, 21]. Biorefineries will contribute significantly to the sustainable and efficient use of biomass resources, by providing a variety of products for different markets and sectors [58]. An overview of biorefineries' production technologies and products is shown in Figure 2.10.

Biofuels may be produced in a stand-alone biorefinery providing that the economic parameters are attractive. In this regard, a number of important facts, i.e., source and location of biomass, type of biomass, transportation costs, conversion technology, scale of biorefinery and associated capital costs, overall emissions, etc., must be taken into consideration.

Once in the refinery, biofuels may be blended with petroleum fuels. The existing infrastructure of petroleum refineries may also be suitable for the co-processing of biofuels with petroleum-derived fuels. Apparently, among the biofuels, those derived from vegetable oil seem to be the most suitable for co-processing with petroleum derivatives because of their high energy density, low oxygen content and liquid form [23].

The only stand-alone commercial process for biofuel production, developed by Neste Oil Corp., produces biofuels from vegetable oil. For a similar purpose, raw material of lignocellulosic origin would

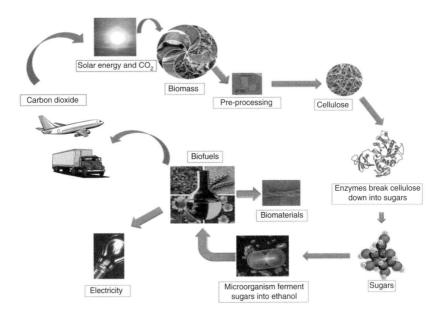

Figure 2.10 The concept of the biorefinery.

require a stabilization step prior to transportation. A commercialization stage is approaching the UOP/Eni Ecofining process developed by Honeywell's UOP. This process is suitable for the production of green diesel and jet fuel. An engineering study for 100-million gal/year was performed with the plant start-up in 2012 [23].

2.3   Future Trends of Biofuels Development

Figure 2.11 shows an overview of advanced biofuels that are currently under study, while Figure 2.12 shows their development status. Depending on the degree of market penetration, the different technologies for the production of biofuels can be associated with distinct steps of the chain of research, development and innovation. While first generation biofuels are currently at the commercialization scale, second generation biofuels are entering the market, and third generation biofuels are still at the laboratory scale. However, it is expected that they will all move to the industrial scale within the next decades. Indeed, currently many countries are fostering or evaluating biofuel programs, in some cases with really good prospects if the proper productivity system is taken into account [1].

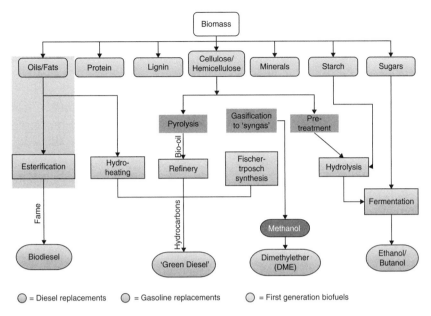

Figure 2.11 Overview of advanced biofuels.

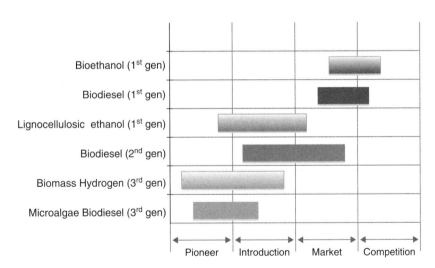

Figure 2.12 Status of the development of biofuels.

The success of implementing a biofuels program results from the proper weighing of the driving forces for their production in each context, taking into account the characteristics of the biofuel alternatives available, in order to maximize the benefits of using them. The decision of which fossil fuel is more convenient to displace in addition to the future of the existing transport infrastructure are also key issues. Last but not least, natural resources availability, mainly land and water, shall be brought into the selection and planning processes. At this stage, the differences among biofuels need to be well understood, by using analytic tools such as energy balance and GHG emissions considering life cycle analysis, as well as considering resource demand, job generation, the availability of necessary technologies, fossil fuels they can replace and costs [1].

Advanced biofuels must be economically competitive with existing products, overcoming the primary economic drivers of feedstock prices, and the overall process productivity and yields. Commercialization requires advancing laboratory-scale processes to yields and productivities that approach the theoretically possible and scaling them up without losing performance [39]. Next, the main fields of research that should contribute to making biofuels economically feasible are presented.

2.3.1   Advances in the Production of Biomass for Biofuels

2.3.1.1   Raw Materials for First Generation Biofuels

There are crops that are not currently widely used for biofuel production or that are not commercially grown, but which could be attractive feedstocks for first generation fuels. These include jatropha for biodiesel production and cassava and sorghum for bioethanol production. They are potentially interesting as first generation biofuel feedstocks because the GHG intensity of the biofuel method in which they would be used is expected to be lower than of other conventional crops. Cassava and sorghum are currently grown at a significant scale worldwide; over 250 million tons of cassava and almost 55 million tons of sorghum were produced in 2011, compared to 1,8 billion tons of sugarcane and 880 million tons of corn [59]. To date, cassava and sorghum have not been used as primary feedstocks for first generation ethanol production; however, some countries and companies are starting to look at these crops as having significant potential as biofuel feedstocks since they can, perhaps, be grown on lower quality land with high yields.

The average yield for cassava is 12.2 t/ha but in some countries, such as India, quote average yields to be as high as 31 t/ha. As a tropical plant, cassava can be grown in warm and humid climates, typically where rainfall is relatively abundant. Significant volumes of cassava are grown in Nigeria, Brazil, Indonesia and Thailand [59]. The cassava root is rich in starch, and its attractive feature for biofuels is the high yield that can be achieved. China, Nigeria, Brazil and Thailand are all interested in expanding cassava production for ethanol fuel [8, 25].

Indigenous to Africa, sorghum is drought and heat tolerant, making it a particularly important crop in arid areas. Its average yield is much lower than of cassava's, 1.4 t/ha. The reason why the yield of sorghum may seem so low is that in the countries where it is grown in largest quantities, including India and Nigeria [59], it is not optimized for commercial operations. Yields could be improved through better management and improved fertilization. Additionally, there has been an increasing interest in the genetic transformation of sorghum. Whilst there are no transgenic crops under cultivation to date, the genetic mapping of sorghum is underway, which should provide the scientific community with tools for improving sorghum production yields. Additionally, some researchers in the US are interested in the fast growing varieties of sorghum that produce large amounts of stover that can be used as lignocellulosic biomass for second generation biofuel production, in addition to the grain that will be destined for first generation bioethanol production [8].

Another source of simple sugars that can be used to produce ethanol is whey. Large quantities of whey are produced as a by-product during the manufacture of cheese. After whey protein has been harvested from whey by ultrafiltration, the remaining permeate is concentrated by reverse osmosis to attain a higher lactose content for efficient fermentation. Lactose in whey permeate is fermented with some special strains of the yeast *Kluyveromyces marxianus* [15].

Biodiesel is produced primarily from triacylglycerol obtained from soy, canola and other oilseeds or from the mesocarp of palm fruits [36]. The average yield of oil crops depends basically on the species, climate, and agricultural management, and varies from hundreds to thousands of liters of vegetable oil per hectare. Soybean, castor, and palm oil are some crops frequently considered for biodiesel production in Brazil. Their biodiesel productivity could reach 0.7 ton/ha, 0.5 ton/ha and 5.0 ton/ha, respectively [1].

Jatropha curcas is a non-edible crop, the seeds of which can be crushed to produce a toxic vegetable oil. It can be grown on semi-arid lands in warm and humid climates. In the past, this crop has been grown for use in soap and candles and also as a medicine. India is particularly interested in growing jatropha for fuel production and has ambitious plans to convert over 11 Mha of wasteland into jatropha fields, since this country has a national policy that does not allow edible oils to be used as fuel. Other regions in which jatropha could be grown include Southern Africa, South East Asia and Latin America.

There are also other crops that are presently not grown extensively but that are being investigated as potential new feedstocks, such as camelina (*Camelina sativa*). Camelina is indigenous to Northern Europe and Central Asia and it currently has an oil yield similar to rapeseed. However, it has not been grown commercially for at least 50 years and consequently there has been little focus on improving yields, which is a field to be explored. Camelina has low tillage and weed control requirements. However, the fact that it has not been grown commercially to date means that it has so far not proven to be a reliable feedstock for biodiesel production.

Because temperate oilseeds have much lower biofuel yields than corn, economic incentives favor ethanol production wherever corn can be grown. Other raw material options are tallow, lard and used cooking oil, which can also be converted to biodiesel, as long as there is enough supply at a competitive price. These prices mainly depend on beef industry production, since there is no sense in fostering cattle ranching aimed at an increase in tallow production; therefore, these raw materials are available in relatively small amounts [1, 36].

2.3.1.2   *Biomass for Second Generation Biofuels*

It would be shortsighted to replace unsustainable petroleum production with unsustainable agricultural production of fuels. Therefore, mainly lignocellulosic residues have been used for second generation biofuel production.

Depending on the conversion technology, 10–25 million tons of dry cellulosic biomass will be required to produce 1 billion gallons of liquid fuel. Partially replacing the consumption of fossil fuels with renewable fuels will require huge quantities of feedstock that will be obtained from both cultivated and collected biomass resources. An annual biorefinery feedstock cycle could consist of agricultural residues in the fall, woody residues or crops in the winter, cover crops, like rye, in the spring, and energy crops, like sorghum or switchgrass, in the summer [37].

In terms of global grain or seed production, maize is the largest crop, producing about 880 million tons of grain [59] and a similar amount of stems and stripped cobs, collectively referred to as stover, which are potentially available for fuel production. Conversion of half of the maize stover in the US to cellulosic ethanol would produce about 13.5 billion gallons of ethanol [36]. In Brazil, sugarcane bagasse is a biomass largely available that represents one third of the plant, and if 50 % of it was used for ethanol production, ethanol yield could increase from 6000 L/ha to 10000 L/ha [60].

The amount of land required to produce enough biofuel to have a significant impact on demand depends entirely on the productivity of a given feedstock on a given parcel of land. The productivity is, in turn, governed by a wide variety of physiological factors, including genetic diversity, agronomic practice, and environmental factors, such as soil quality, water availability, and climate. A challenge for plant biologists is to identify the most highly productive plant species that can be grown on the various types of marginal or abandoned land, to optimize the genetics and production practices, and to evaluate any environmental risks or benefits that may accrue from encouraging the widespread use of such species for energy production. In order to minimize the amount of land diverted from other purposes to energy production it is essential to maximize the "yield" – the amount of biomass produced per unit of land. This can be achieved by optimizing management techniques or genetically modifying the biomasses [36].

Because water is a major limitation of plant productivity, a key goal of developing bioenergy crops will be to maximize water use efficiency and drought tolerance for regions that do not receive excess rainfall. Semi-arid land could be used to grow species with high water-use efficiency and drought resistance, such as *Agave* spp. Salinized soils could be used to grow salt tolerant species such as prairie cordgrass and *Eucalyptus* spp. Moreover, all other environmental stress tolerances, such as flood and cold, are advantageous. A particularly important topic in this respect is to identify species that are not invasive, or to develop technologies, such as conditional sterility, that can prevent an invasive spread [36].

Perennial species, such as sugarcane, energy cane, elephant grass, switchgrass, and Miscanthus, have intrinsically high light, water and nitrogen use efficiency, so these species are potential sources of lignocellulosic biomass for bioethanol production. Moreover, woody biomass can be harvested sustainably for lumber and paper and may, therefore, provide biofuel feedstock for some regions. Globally, large areas of land formerly used for agriculture have reverted to forest, and the continuing trend to electronic media and paper recycling may reduce the demand for pulp woods. This presents an opportunity to reallocate woody biomass for energy [30, 36].

Other important sources of lignocellulosic biomass are residues of the agroindustry, such as sugarcane bagasse, cassava bagasse, corn stover, wheat and rice straw, etc. Even municipal solid waste and residues from the paper industry have been used for the production of second generation biofuels [30]. On the other hand, some potential biomass crops contain secondary metabolites that are toxic to microorganisms. Identification and elimination of such compounds by genetic methods is a priority for research on plant feedstocks if it can be done without creating pest and pathogen problems [36].

2.3.2   Genetic Engineering

Adopting gasoline, diesel, and jet fuel alternatives derived from microbial sources has the potential to significantly limit net GHG emissions, and the main technological bottlenecks for improving overall yield are associated with fermentation, not saccharification [61]. In the effort to establish biofuels, great strides have been made in recent years towards the engineering of microorganisms to produce transportation fuels derived from alcohol, fatty acids, and isoprenoid biosynthesis. In order to develop economical and sustainable processes for biofuel production, the metabolic pathways of biofuel producers need to be optimally redesigned to achieve high performance, which includes improved product yield, higher product concentration and productivity, and product tolerance. Also, the production strains should be designed in a way that the whole process becomes operationally inexpensive by system-wide optimization of midstream and downstream processes [38, 62].

Genetic engineering of ethanol fermentation microorganisms to expand their substrate specificity towards hemicellulosic sugars and macroalgal polysaccharides is probably the most reasonable method to improve the overall yield of bioethanol production from lignocellulosic biomass. To manufacture microbial strains for biofuel production, random mutagenesis and metabolic engineering have been employed as standard strategies over the last couple of decades. The approach of random mutation and selection is difficult for further improvement of cellular performance due to the complexity associated with identifying modified genes as a consequence of random mutation. Metabolic engineering, on the contrary, aims at improving cellular performance by considering metabolic pathways in their entirety and manipulating specific genes. It is expected that this combined strategy will result in the development of microorganisms capable of producing various biofuels cost effectively on an industrial scale [38–39, 50, 62].

Recently, the successful development of microorganisms that are capable of producing several different biofuels including bioethanol, biobutanol, alkane, and biodiesel, and even hydrogen, was achieved. Cyanobacteria have been engineered to produce ethanol, butanol and isoprene but productivities are still low. Traditional strain improvement methods (e.g. mutagenesis and breeding) can also result in improved biofuel productivities from algal strains and these methods have also been recently exploited. However, there are no advances in the molecular engineering of eukaryotic algae for liquid biofuel production. Therefore, despite some successes, challenges still need to be overcome to move advanced biofuels towards commercialization with a competitive price against fossil fuels. Furthermore, the weak link of this technology is carbon capture and sequestration, which continues to elude the coal industry [38–39, 50, 57, 62].

2.3.3   Fourth Generation Biofuels

The concept called "fourth generation algal biofuels" or "photosynthetic biofuels" has recently been introduced. It is anticipated that the breakthrough in algal biofuels will come through the metabolic engineering of photosynthetic microorganisms to produce and secrete biofuel. Fourth-generation technology combines genetically optimized feedstocks, which are designed to capture large amounts of carbon, with genomically synthesized microorganisms that are developed to efficiently make fuels. The key to this process is the capture and sequestration of CO_2, which renders fourth-generation biofuels a carbon negative source of fuel [1, 50, 57].

This approach (providing high biofuel productivity and product separation methods) will significantly decrease the cost of biofuel production by cutting the biomass separation and processing costs associated with traditional approaches. Secreted volatile biofuels (lower alcohols and aldehydes) could be stripped out with gas from growth medium and condensed to liquid biofuel avoiding energy-intensive dewatering steps [50, 57].

2.3.4   Economic Aspects

In Germany and the US, the share of diesel will continue to rise, which makes second generation biodiesel particularly relevant. By contrast, Brazil and the US use mainly gasoline as road transport fuel, so second generation bioethanol is particularly important in these countries. Figure 2.13 shows the potential long term costs of development of second generation biofuels versus gasoline and diesel [51]. The fossil-based fuels are expected to have increasing taxes associated with them, which would ultimately lead to the economic feasibility of biofuels. Moreover, the optimization of technologies for the production of second and third generation biofuels will eventually lead to a decrease in their production costs.

From Figure 2.13, it can be noticed that one of the main challenges today is making biofuels economical. However, there is

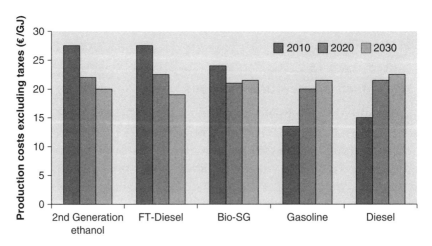

Figure 2.13 Possible cost developments for biofuels up to 2030. Adapted from Fritsch *et al.* [51].

little data on the economics of these new processing technologies because the analyses are very site specific and may vary considerably depending on the current and future regional economic and market situations [5].

For the overall economy of biofuel production, biomass planting, maintaining growth, harvesting, transportation and processing technology, must be considered. The location of the plants for biofuel production and upgrading requires attention as well. For example, upgrading plants may be located either as a stand-alone plant at the point of biomass harvesting or in a separate biorefinery. There is also significant potential for an integration of biofuel production with petroleum refining.

For an overall cost comparison with conventional fuels, the efficiency of conversion and associated emissions involving full cycle must be considered. It is noted that for biofuel production, these issues have not been receiving adequate attention [23].

Some problems still have to be solved for biofuels of lignocellulosic origin before commercialization. The contribution of biomass cost to the overall production cost of lignocellulosic bioethanol proves to be one of the most significant [41]. The pretreatment step is also an intensive energy consuming step, increasing the ethanol cost [63]. Moreover, the contribution of enzyme costs to the economics of lignocellulosic biofuel production continues to be a much debated topic. Whereas some authors argue that the cost of the enzymes is a major barrier for biofuel production, others implicitly assume that it is not, either because they estimate the cost to be relatively low or because they assume that it will decrease with technological innovation or other advances. Other studies simply acknowledge that publicly available detailed information on the cost of enzymes is limited [61].

Literature estimates for the cost contribution of enzymes to the production of lignocellulosic ethanol vary significantly, from $ 0.3/L to $ 0.11/L. The cost contribution of enzymes to ethanol produced by the conversion of corn stover was found to be $ 0.18/L if the sugars in the biomass could be converted at maximum theoretical yields, and $ 0.39/gal if the yields were based on saccharification and fermentation yields that have been previously reported in the scientific literature. On the other hand, the cost of ethanol production can be as low as $ 0.16/L for sugarcane, $ 0.25/L for corn and $ 0.18/L for cassava. Therefore, a significant effort is still required to lower the contribution of enzymes to biofuel production costs [15, 61].

The same problem is found for other biofuels. Despite tremendous investments into isoprenoid and fatty-acid-derived products, commercial success has yet to be demonstrated [39, 62].

As the consumption of biofuels increases to higher levels than today, there will be a natural selection process favoring the more sustainable options, where some aspects are decisive: a good energy balance, high GHG abatement potential, higher yield per cultivated hectare, the possibility of rain fed cultivation or of low water demand, lower production costs and easy end use are requirements to be met [1]. The use of ethanol, for example, is due mainly to its high level of natural production by microorganisms, and not because it is optimal for our existing petroleum-centric transport infrastructure. Ethanol has only 70% of the energy content of gasoline, and it has a high tendency to absorb water from the air, which leads to corrosion in engines and pipes. Moreover, its distillation from the fermentation broth is energy intensive. Biodiesel also has its own limitations: it has only 91% of the energy content of diesel and, because wax can form in the fuel if the temperature is too low, it is difficult to transport with the current distribution infrastructure, so there are geographical limits to its use. [39].

2.3   Conclusions

Biofuels started to be produced in the late 19th century; the interest in the commercial production of biofuels for transport rose again in the mid-1970s and the fastest growth in biofuel production has taken place over the last 10 years, supported by ambitious government policies. Besides energy security and sustainable agriculture concerns, the reduction of CO_2 emissions and the blending mandate policy of road transport promote the importance of biofuels among fossil fuel alternatives.

Biofuels are a complex subject with dimensions that prominently include economics, ecology, environmental sciences, agronomy, plant biology, microbiology, biochemistry, chemistry, genetics, chemical engineering, mechanical engineering, law and policy. The full replacement of fossil fuels by biofuels is not deemed to be achievable but, nonetheless, modern biofuels have a crucial role to fulfill in the long term to help limit the growth of GHG emissions and to lead the transition of the current petroleum-based society towards a more sustainable one. First generation biofuels are now commercially available, and there are crops that are not currently widely used for biofuel production or that are not commercially grown, but which could be attractive feedstocks for first generation fuels. The second generation is starting to enter the commercial stage and mainly lignocellulosic residue has been used for second generation biofuels production. As for third generation biofuels, they are still in the early stages of development, and fourth generation is still embryonic.

The economic feasibility of these processes is a goal to be pursued in the next few years. Commercialization requires advancing laboratory-scale processes to improve yields and productivities. Then, advanced biofuels must be economically competitive with existing products, overcoming the primary economic drivers of feedstock prices, and the overall process of productivity and yield.

Acknowledgements

The authors acknowledge financial support from the São Paulo Research Foundation (FAPESP, projects 2010/08684-8, 2011/19817-1, 2012/11561-0, 2012/11459-1).

References

1. L.A. Nogueira. Does biodiesel make sense? *Energy*, Vol 36(6), p. 3659–3666, 2011.
2. I. Council. *Lighting the way: Toward a sustainable energy future*. InterAcademy Council, FAPESP, 2007.
3. IEA, *Technology Roadmap: Biofuels for Transport*. International Energy Agency, France. p. 56, 2011.
4. Fulton, L. *Biofuels for Transport: An International Perspective*. International Energy Agency: France. 2005.
5. M. Balat and H. Balat. *Applied Energy*, Vol. 86(11), p. 2273–2282, 2009.
6. IEA, *Key world energy statistics*. International Energy Agency. p. 80, 2012.
7. J. Goldemberg. Biomass and energy. *Química nova*, Vol. 32(3), p. 582–587, 2009.
8. FAO, *The state of food and agriculture*. Food and Agriculture Organization of the United Nations: Rome. p. 128, 2008.
9. IEA, Technology Roadmap Biofuels for Transport. International Energy Agency: France. p. 52, 2011.
10. K. Hofsetz and M.A. Silva. *Biomass and Bioenergy*, Vol. 46, p. 564–573, 2012.
11. M.D. Berni and P.C. Manduca. Bioethanol Program in Brazil: Production and Utilization Trade-Offs for CO_2 Abatement, in *3rd International Conference on Future Environment and Energy*: Rome. 2012.

12. A. Mudhoo, TT. Forster-Carneiro and A. Sánchez. *Critical Reviews in Biotechnology*, Vol. 31(3), p. 250–263, 2011.
13. P.S. Nigam and A. Singh. *Progress in Energy and Combustion Science*, Vol. 37(1), p. 52–68, 2011.
14. J.M. Prado and M.A.A. Meireles. *Production of Valuable Compounds by Supercritical Technology Using Residues from Sugarcane Processing*, in *Biorefinery Co-Products: Phytochemicals, Primary Metabolites and Value-Added Biomass Processing*, C. Bergeron, D.J. Carrier, and S. Ramaswamy, Editors. Wiley Online Library: Malaysia. p. 133–151, 2012.
15. S.I. Mussatto, G. Dragonea, P.M.R. Guimarãesa, J.P.A. Silva, L.M. Carneiro, I.C. Roberto, A.Vicentea, L. Domingues and J.A. Teixeira. *Biotechnology Advances*, Vol. 28(6), p. 817–830, 2010.
16. J. Van Gerpen, B. Shanks, R. Pruszko, D. Clements and G. Knothe. Biodiesel production technology. Report for the National Renewable Energy Laboratory. USA: *Department of Energy*, 30: p. 42, 2004.
17. A. Demirbas. *Energy Conversion and Management*, Vol. 50(9), p. 2239–2249, 2009.
18. T. Forster-Carneiro, M. Fernández, M. Pérez García, L.I. Romero García, C.J. Álvarez Gallego and D. Sales. Diseño y optimización de la fase de arranque del proceso SEBAC en el tratamiento de la fracción orgánica de residuos sólidos urbanos. *Residuos*, Vol. 74, p. 54–62, 2003.
19. T. Forster-Carneiro, V. Riau and M. Pérez. *Biomass and Bioenergy*, Vol. 34(12), p. 1805–1812, 2010.
20. R.Rathmann, A. Szklo and R. Schaeffer. *Renewable Energy*,Vol. 35(1), p. 14–22, 2010.
21. A.V. Bridgwater. *Biomass and Bioenergy*, Vol. 38, p. 68–94, 2012.
22. D. Bacovsky, D. Michal, M. Wörgetter. *Status of 2 nd Generation Biofuels Demonstration Facilities in June 2010*. Report T39-P1b, 2010.
23. E. Furimsky. Hydroprocessing challenges in biofuels production. *Catalysis Today*, 2013.
24. P. Langan, S. Gnanakaran, K.D. Rector, N. Pawley, D.T. Fox, D.W. Cho and K.E. Hammel. *Energy & Environmental Science*,Vol., 4(10), p. 3820–3833, 2011.
25. E4TECH, *Biofuels review: advanced technologies overview. Renewable Fuels Agency*, 2008.
26. P. Alvira, E. Tomás-Pejó, M. Ballesteros and M.J. Negro. *Bioresource Technology*, Vol. 101(13), p. 4851–4861, 2010.
27. M.J. Taherzadeh and K. Karimi. *International Journal of Molecular Sciences*, Vol. 9(9): p. 1621–1651, 2008.
28. T. Eggeman and R.T. Elander. *Bioresource Technology*, Vol. 96(18), p. 2019–2025, 2005.
29. H. Tadesse and R. Luque. *Energy & Environmental Science*, Vol. 4(10), p. 3913–3929, 2011.
30. C.R. Soccol, L. Vandenberghe, A. Medeiros, S. Karpa, B. Buckeridge, L.B. Ramos, A. Pitarelo, V. Leitão, L. Gottschalk, M. Ferrara, E. Bon, L. Moraesh, J. Araújo and F. Gonçalves. *Bioresource Technology*, Vol. 101(3), p. 4820–4825, 2010.
31. E. Tomás-Pejó. *Pretreatment Technologies for Lignocellulose-to-Bioethanol Conversion*, in *Biofuels: alternative feedstocks and conversion processes*. In: A. Pandey, C. Larroche, R. Steven, D. Claude-Gilles, G. Edgard. Editors. Elsevier: USA. p. 149–176, 2011.

32. M. Naushada, Z.A. ALOthmana, A.B. Khan and M. Ali. *International Journal of Biological Macromolecules*, Vol. 51(4), p. 555–560, 2012.

33. C. Schacht, C. Zetzl and G. Brunner. *The Journal of Supercritical Fluids*, Vol. 46(3), p. 299–321, 2008.

34. Y Zhao, WJ Lu, HT Wang and D Li. *Environmental science & technology*, Vol. 43(5), p. 1565–1570, 2009.

35. S.E. Jacobsen and C.E. Wyman. *Industrial & engineering chemistry research*, Vol. 41(6), p. 1454–1461, 2009.

36. H. Youngs and C. Somerville. Development of feedstocks for cellulosic biofuels. F1000 *Biology Reports*, 2012.

37. M.A. Sharara, E.C. Clausen and D.J. Carrier. *An Overview of Biorefinery Technology*, in *Biorefinery Co-Products: Phytochemicals, Primary Metabolites and Value-Added Biomass Processing*, Ramaswamy, Editors. Wiley. p. 1–18, 2012.

38. Y.S. Jang, J. Park, S. Choi, Y Choia, D. Seung, J. Cho and S. Lee. *Biotechnology advances*, Vol. 30(5), p. 989–1000, 2012.

39. P.P. Peralta-Yahya, F. Zhang, S. Cardayre and J. Keasling. *Nature*, Vol. 488(7411), p. 320–328, 2012.

40. S. Spatari, D.M. Bagley and H.L. MacLean. *Bioresource Technology*, Vol., 101(2), p. 654–667, 2010.

41. E. Gnansounou and A. Dauriat. *Technoeconomic Analysis of Lignocellulosic Ethanol*, in *Biofuels: alternative feedstocks and conversion processes*, A. Pandey, et al., Editors. Elsevier: USA. p. 123–148. 2011.

42. M. Nijboer. *The Contribution of Natural Gas Vehicles to Sustainable Transport*. International Energy Agency, 2010.

43. L.A.B. Cortez. *Bioetanol de cana-de-açúcar: P&D para produtividade e sustentabilidade*. Editora Blucher, 2010. 992p.

44. H.W. Chen, Q.H. Song, B. Liao and Q. Guo. Further Separation, *Energy & Fuels*, Vol. 25(10), p. 4655–4661, 2011.

45. E. Butler, G. Devlin, D. Meier and K. McDonnell. *Renewable and Sustainable Energy Reviews*, Vol. 15(8), p. 4171–4186, 2011.

46. I.K. Kapdan, F. Kargi. *Enzyme and Microbial Technology*, Vol. 38(5), p. 569–582. 2006.

47. S.M. Kotay and D. Das. *International Journal of Hydrogen Energy*, Vol. 33(1), p. 258–263, 2008.

48. E. Eroglu and A. Melis. *Bioresource Technology*, Vol. 102(18), p. 8403–8413, 2011.

49. E. Kırtay. *Energy Conversion and Management*, Vol. 52(4), p. 1778–1789, 2011.

50. M. Daroch, S. Geng and G. Wang. *Applied Energy*, 2012.

51. U.R. Fritsche, H. Fehrenbach and S. Koppen. *Biofuels - what role in the future energy mix?* Shell Deutschland Oil: Darmstadt, Germany. p. 42. 2012.

52. IEA, *Energy technology perspectives 2012: pathways to a clean energy system*. International Energy Agency. 2012. p. 690.

53. A.A. Snow and V.H. Smith. *BioScience*, Vol. 62(8): p. 765–768. 2012.

54. N. Uduman, Y. Qi, M. Danquah, G. Forde, A. Hoadley. *Journal of renewable and sustainable Energy*, Vol. 2: p. 012701, 2010.

55. Y. Chisti and J. Yan. *Applied Energy*, Vol. 88(10), p. 3277–3279. 2011.

56. A. Sun, R. Davis, M. Starbuck, A. Amotze, R. Patea and P. Pienkos. *Energy*, Vol. 36(8), p. 5169–5179. 2011

57. P. Hallenbeck. *Hydrogen Production by Cyanobacteria*, in *Microbial Technologies in Advanced Biofuels Production*, P.C. Hallenbeck, Editor., Springer US. p. 15–28. 2012

58. M. Berni, S. Bajay and P. Manduca. Biofuels for urban transport: Brazilian potential and implications for sustainable development, in *18th International Conference on Urban Transport and the Environment*. WIT Press: Spain. 2012.
59. FAO. *FAOSTAT*. 2013 [cited 2013 February, 2013]; Available from: http://faostat3.fao.org/home/index.html#VISUALIZE.
60. A. Pandey, C.R. Soccol, P. Nigam, V. Soccol. *Bioresource Technology*, Vol. 74 (1), p. 69–80, 2000.
61. D. Klein-Marcuschamer, P. Oleskowicz-Popiel, B.A. Simmons, H.W. Blanch.. *Biotechnology and Bioengineering*, 2012.
62. Y. Kung, W. Runguphan and J.D. Keasling. *ACS Synthetic Biology*, Vol. 1(11), p. 498–513, 2012.
63. K.A. Ojeda, E. Sánchez, J. Suarez, O. Avila, V. Quintero, M. El-Halwagi, V. Kafarov. *Industrial & Engineering Chemistry Research*, Vol. 50(5), p. 2768–2772. 2010.

Processing of Biofuels

Divya Gupta[1,#], Ajeet Singh[1], Ashwani Sharma[2] and Anshul Nigam[3,*,#]

[1]*Department of Biotechnology, G. B. Pant Engineering College, Pauri Garhwal, Uttrakhand-246001,India.*
[2]*Computer-Chemie-Centrum, Universität Erlangen-Nürnberg Nägelsbachstr. 25, 91052 Erlangen, Germany.*
[3]*IPLS (renamed as BUILDER), Pondicherry University, Puducherry-605014, India.*

Abstract

Biofuels need processing like other fuels. Owing to burgeoning criticism of first generation biofuel as they compete with food resources, therefore emphasis is laid on second generation biofuel as they do not compete with food. Although processing of the first generation biofuels is well defined, processing of second generation biofuel is still not economically viable. The major hurdles in the processing of algal biodiesel, is biomass production, oil extraction and its processing. Similarly, the major obstacle in the processing of cellulosic ethanol is hydrolysis of cellulose to sugars and mass transfer in case of syngas. The processing of biofuels discussed in the chapter is compatible with vehicle engines used at the present time.

Keywords: Biofuel, algal biodiesel, cellulosic ethanol and syngas.

3.1 Introduction

Like crude oil needs processing for the production of fuels, similar processing is required for production of biofuels. The complexities associated with the processing of biofuels depend on the source

#*Equally contributing authors*
Corresponding author: anshulnigam2006@gmail.com

Dr.Vikash Babu, Dr. Ashish Thapliyal & Dr. Girijesh Kumar Patel (eds.) Biofuels Production, (59–84) 2014 © Scrivener Publishing LLC

from which it is obtained. The biofuels are of two types depending on the nature of their source:

3.1.1 First Generation Biofuels

First generation biofuels are derived from food/feed stock like corn, sugar cane [1] or vegetable oil [2]. Processing of these fuels is well defined particularly for bio-ethanol as it is the most widely used biofuel [3]. Since these fuels are produced from agricultural sources they are under severe criticism leading to the fuel versus food debate [4, 5]. These fuels form a major share of biofuels which are available in the market today.

3.1.2 Second Generation Biofuels

Second generation biofuels are derived from non-feed/food sources like cellulose, hemicelluloses and lignin from plants, biodiesel from algal lipids and biomass (Table 3.1). Their production is limited to the laboratory level/pilot scale and is not economically viable [6].

The production of first generation biofuels from feedstock is not being encouraged by many governments as it competes with food. This has lead to the development of a second generation of biofuels that do not compete with food [7]. The Scientific community is eagerly looking forward to these biofuels particularly cellulosic alcohol [8–10] biodiesel from algae [11] and syngas from biomass [12–14]. The major roadblock for second-generation biofuels is the high initial investment leading to higher cost of the end product

Table 3.1 Biofuels with their source and process involved.

S. No.	Biofuel Source	Specific biofuel	Processes involved
1.	Algal Lipids	Biodiesel	Production of algal biomass/ Oil Extraction/ Viscosity reduction/ Trans-esterification / Pyrolysis and Hydroprocessing
2.	Cellulose/ Lignin	Ethanol	Hydrolysis/fermentation
3.	Biomass	Syngas	Gassification/fermentation

in comparison to fossil fuels or other first generation biofuels. The objective of this chapter is to introduce the readers to the processes and caveats involved in the production of second generation biofuels.

3.2 Biodiesel from Algae

Biodiesel from algae has attracted considerable interest because: (1) Productivity of algae is higher than productivities of land plants; (2) They can accumulate very large amounts of triacylglycerides and (3) there is no requirement of agricultural land [15].

Following are the steps of the biodiesel production of algae:

1. Production of algal biomass
2. Oil Extraction
3. Trans-esterification
4. Pyrolysis and
5. Hydroprocessing

3.2.1 Production of Algal Biomass

The production of algal biomass is done in a photo-bioreactor. Efficient design of the photo-bioreactor depends on its ability to provide sunlight to algal cells, assimilation of CO_2, aeration, water and nutrient utilization, temperature and pH [16]. The photo-bioreactor must be designed so that it may also include all possible ergonomic features like low capital investment and operational costs [17]. The parameters, other than light source, are common in other fermentation processes too. Although an open reactor appears to be the most feasible, closed form of reactors are also attractive as it is easy to control the parameters of growth in them. Generally, industrial level microalgae production uses open pond systems consisting of shallow water bodies in which the water is mixed/circulated with a paddlewheel device. These are also known as raceway reactors [18]. The schematic diagram of a raceway reactor is shown in Figure 3.1. Other open algal cultivation systems include large bags, open ponds and circular ponds; these ponds may have paddled wheels for mixing purposes like raceway ponds. The major advantage of these reactors is that they are economical and easy to maintain but are subjected to poor productivity owing to a lower

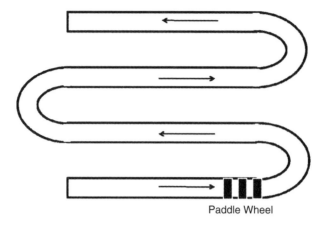

Paddle Wheel

Figure 3.1 Schematic representation of raceway pond. The arrows represent the flow of water while a paddle wheel is provided for mixing purpose.

rate of illumination, a lower mass transfer rate of carbon dioxide as it escapes into the atmosphere and inappropriate mixing. Furthermore, they are highly prone to contamination with the risk that contaminating species may possess an antagonistic relationship with a cultured strain. Moreover, water loss through evaporation increases the salinity [19].

In closed photo-bioreactors, it is far easier to maintain pure culture, a higher rate of illumination, mixing, low water/mineral requirement and aeration. The light source in these reactors may be artificial or natural. Closed photo-bioreactors have been classified into three types in terms of design:

3.2.1.1   Vertical-column Type

Various prototypes of vertical-column photo-bioreactors have been tested from time to time [20–23] and are considered to be the most promising for the purpose of scaling up [24]. These reactors are characterized by high volumetric gas transfer coefficients. This is due to the fact that gas bubbles travel along the column, which also aids in mixing (Figure 3.2). CO_2/O_2 and pH gradients are created due to the same reason [25]. The major limitation of these reactors is a low illumination surface area and an increase in the length of the column, which causes a shading effect [19]. The shadow of the longer column falls over the column in close proximity.

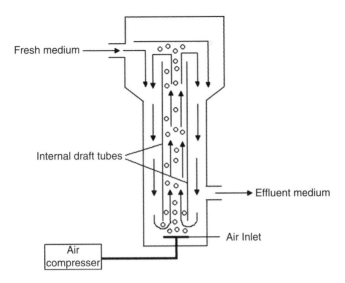

Fresh medium

Internal draft tubes

Effluent medium

Air Inlet

Air compresser

Figure 3.2 Schematic representation of Vertical-column Photo-bioreactor.

3.2.1.2   Flat-plate

The major characteristic of these reactors is the availability of a large surface to volume ratio for illumination making them suitable for the mass cultivation of algae [17–19]. The prototype of this photobioreactor was introduced by Milner [26]. The designs of FBR were improved by several researchers [27–29]. Due to the availability of a large surface area for illumination, higher photosynthetic efficiencies have been achieved [30] (Figure 3.3). The photosynthetic efficiencies may be improved further by tilting the angle of the reactor with reference to the ground [31]. This increases the exposure time of the reactor for sunlight. The major problem in designing an ergonomic reactor is the cost of panels, wall growth that impairs the diffusion of light and moreover may inhibit the growth of some species owing to hydraulic stress.

3.2.1.3   Tubular Photo-bioreactor

Tubular photo-bioreactors are mostly manufactured from hollow glass or plastic tubes (Figure 3.4). Aeration and mixing in these reactors is mostly done by air pump or airlift systems [32]. The arrangement of tubing may vary from horizontal/spiral [33, 34], vertical [35] to conical [36, 37]. The orientation of these reactors may be horizontal, vertical or inclined. The formation of gradients

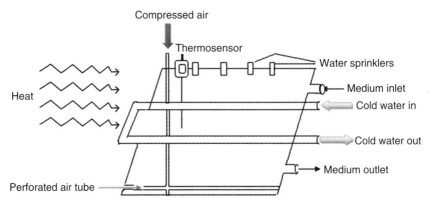

Figure 3.3 Schematic representation of a flat plate Photo-bioreactor.

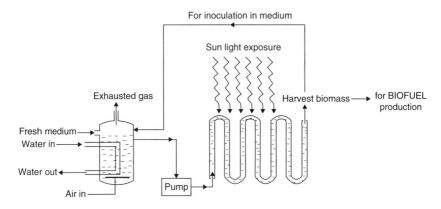

Figure 3.4 Schematic representation of tubular photo-bioreactor.

of oxygen and carbon dioxide is the unique characteristic of these photo-bioreactors. The carbon dioxide gradient also creates pH gradients in them. An increase in the diameter of the tubing results in a decrease of photosynthetic efficiency due to a reduction in the surface to illumination ratio. Increasing tube length has no effect if the tubes are placed horizontally. However, mixing increases the photosynthetic efficiency considerably [38, 39]. Alternatively, photosynthetic efficiency can also be enhanced by tilting the angle of these reactors since it increases the exposure time of the reactor.

The bio-reactors discussed here are being considered for scale up. However, laboratory level PBRs are provided with artificial light shroud and their intensity can be measured /regulated along with other parameters. These reactors are helpful in measuring efficacy of algal cells in capturing light/carbon dioxide and in understanding the process of carbon fixation. Since these reactors can be sterilized; hence pure cultures can be maintained with ease.

3.2.2   Oil Extraction

Lipid content in an algal cell is about 40% of its total mass, which can be extracted and converted into biodiesel. However, to extract the oil from algae is a tedious process because of the small size of an algal cell and thickness of its cell wall. The algal cell wall and cell membrane need to be breached in order to extract this oil [40]. Extraction of oil from algae can be broadly classified in following ways (Figure 3.5):

3.2.2.1   *Mechanical Method*

3.2.2.1.1   Pressing

This technique is based on the use of high pressure, which is employed to remove cell walls and cell membranes for the extraction

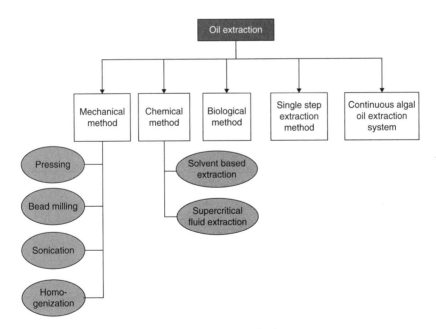

Figure 3.5 Different types of Oil Extraction Methods.

of oil. Pressing dries the algae while still retaining its oil content. The oil can then be collected from the cells using different types of press arrangements such as a viz, screw, expeller, piston etc. [41].

3.2.2.1.2   Bead milling

Bead milling incorporates the agitation of algal biomass in the presence of small steel or glass beads in a closed vessel. This collision between cells and beads generates high liquid shear gradient, which helps to breach algal cell walls and allows the extraction of oil. To enhance the fidelity of this method, nowadays it is coupled with a solvent based extraction method [42].

3.2.2.1.3   Ultrasonic based extraction/ Sonication

Sonication is used to create vapor bubbles of a liquid form in a solvent material that results in cavitation. When these bubbles collapse near algal cells, it results in the breaching of cell walls [43, 44].

3.2.2.1.4   Homogenization

This constitutes rapid pressure drops through an orifice generating high shear gradient and disrupting cell walls.

3.2.2.2   Chemical Method

Chemical based oil extraction methods are based on the use of chemicals for extraction of oil. This method can be further classified into two parts:

3.2.2.2.1   Solvent based extraction

Organic solvents such as benzene, ether, hexane, cyclohexane, acetone and chloroform can be used for oil extraction from an algal biomass by breaching its cell wall [43]. Out of those, hexane is popularly used for the extraction of oil. Again, this method also requires drying of the algal biomass, which is an energy intensive process [40]. Combinations of the hexane solvent method with pressing can release about 95% of oil from algae. Concurrent use of solvent extraction with transesterification enhances its efficacy, involving the direct conversion of triacylglycerides into methyl esters by the addition of alcohol and a catalyst [45].

3.2.2.2.2   Supercritical fluid extraction

Carbon dioxide, which acts as a supercritical fluid, works as an organic solvent when allowed to mix with algal biomass. It penetrates the cell wall and removes oil from algal cell efficiently [40].

In 2010, Couto *et al.* described the extraction of lipids using supercritical CO_2 followed by its conversion into biodiesel [46]. This method does not require any additional organic solvent and hence produces highly purified extracts. Moreover, supercritical CO_2 can be easily removed from the reaction mixture since it is a gas at room temperature [45–49]. Also this method helps to maintain algal lipid characteristics at the time of oil extraction since supercritical CO_2 requires low temperatures for reaction [40].

The efficiency of this method is about 100%. However, the requirement of additional costly instruments reduces its use. Use of supercritical CO_2 in conjunction with methanol modifies its polarity and viscosity, which subsequently enhances solvating power and ultimately produces a more efficient extraction method [48–51].

3.2.2.3   Biological

Enzymes can be used to breach the algal cell wall (either completely or partially) and to release its intracellular content into a medium. Enzymatic extraction is a costly method and also tedious because of its degradation selectivity. However, some biopolymer viz algaenan present in the tri-laminar sheath of an algal cell wall during mechanical degradation demands extensive energy and numerous passage over disruption equipment. In such cases, the use of specific enzymes to breach the algal cell wall becomes an advantageous method [52]. Brune and Beecher in 2007 have already showed the use of brine shrimp for the extraction of algal oil [53].

3.2.2.4   Single Step Extraction Method

OriginOil™ has recently produced a single step extraction process which involves the exposure of algal biomass to quantum fracturing, pulsed electromagnetic field and CO_2 (to alter pH) for the breaching of the cell wall while simultaneously separating algal biomass from oil using a gravity clarifier. This method is advantageous because it is simpler, more efficient, provides continuous extraction, is inexpensive, versatile and is not dependent on using chemical or heavy instruments [41–54].

3.2.2.5   Continuous Algal Oil Extraction System

Recently, Cavitation Technologies, Inc. (CTi) has produced CTi's nanoreactor, which is used to develop cavitation bubbles in a medium having algal biomass. Shock waves are generated as these cavitation bubbles collapse, breaching cell walls and in turn helping to release intracellular content [41].

3.2.3   Transesterification / Base Catalysis

Extracted oil from algal biomass can be used to produce biodiesel using a widely known method of transforming lipids into biodiesel viz transesterification [40]. Basically, transesterification involves the conversion of an ester form into another by reciprocate implying alkoxy moiety [55] (Figure 3.6). This reaction is performed by using a base catalyst that ends by forming an ester product called biodiesel, hence named base catalyzed transesterification [56]. This process is initiated by providing a suitable temperature to extract oil, followed by the addition of an alcohol (usually methanol) and a catalyst (usually Sodium ethanolate). The catalyst will deprotonate methanol and make it suitable for nucleophilic attack on a triglyceride, finally forming biodiesel (methyl ester) and glycerol as co-product [55–57]. Glycerol, being heavier, settles down with soapy material in the middle while biodiesel remains floating on the top layer [56] (Figure 3.7)

Transesterification is usually carried out at about 60–66°C, 20 psi pressure. An increase in temperature can affect the reaction rate since temperature above the boiling point of methanol (68°C) vapourizes it, resulting in the formation of bubbles which ultimately reduces the rate of the reaction [55–56].

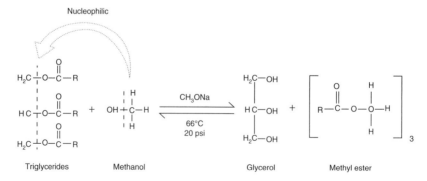

Figure 3.6 Mechanism of transesterification reaction between triglycerides and methanol.

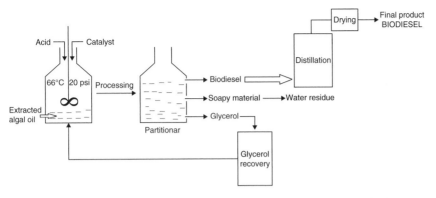

Figure 3.7 Schematic representation of biodiesel production from algal oil using transesterification process.

There are various types of catalysts that can be incorporated into the transesterification process. Viz sodium hydroxide (NaOH), potassium hydroxide (KOH), sodium methoxide (CH_3ONa), potassium methoxide (CH_3OK) are possible options, each with their own advantages and disadvantages. In 2007, Encinar *et al.* demonstrated higher productivity (about 70%) of methyl ester in the presence of potassium hydroxide as a catalyst [58]. Demirbus in 2009 and Gupta *et al.* in 2007demonstrated better productivity of biodiesel using KOH as a catalyst [59, 60]. Furthermore, KOH and NaOH are also cheap in cost and easy to handle during experiments by workers [55]. However, KOH and NaOH require high concentrations (about 0.5%-1.5% of the weight of the oil) in comparison to sodium methoxide (about 0.3%-0.5% of the weight of the oil) for the same yield of biodiesel [61]. Hence, sodium methoxide is used on a pilot scale, while NaOH or KOH are used on laboratory scale [62].

Methanol is a frequently used alcohol in transesterification reaction. However, ethanol or butanol can also be used in this process [63, 64]. Methanol being more accessible, inexpensive and having less steric resistance than methanol, is mostly used for transesterification reaction [65]. However, ethanol enhances cetane number, can easily mix with oil compared to methanol and has a renewable source of origin [55].

During the initial phase of the reaction, there is a requirement of vigorous mixing so that oil and methanol get evenly mixed with each other, as they are poorly miscible. However, as the process marks the formation of the product, slight mixing is required to facilitate the settling down of glycerol, so that reactions can continues in the upper layer of methanol and oil [61]. Another hurdle during the process is the production of emulsion, which complicates the process of partitioning glycerol and biodiesel. These emulsions are more enduring in the case of ethanol than in the case of methanol [66].

One important factor when using alcohol is to use a suitable amount of alcohol since this reaction is reversible; a high amount of alcohol can shift the reaction equilibrium to the right and can decline productivity. Menget *et al.* in 2008 and Refaat *et al.* 2008b have explained the use of a 6:1 ratio for better productivity [67, 68].

Extracted algal oil is comprised of triacylglycerides, free fatty acids, phospholipids, glycolipids and sulfolipids. For base catalyzed transesterification to work out, free fatty acid content should be less than 0.1%, otherwise it leads to saponification [40–61].

3.2.4   Pyrolysis

Pyrolysis is a combination of two words where pyro specifies "fire" and lysis specifies "breakage". Pyrolysis defines the process of using heat to break chemical bonds. Pyrolysis, an alternative tool to relegate viscosity, works by breaking chemical bonds at a high temperature to produce biodiesel in the absence of air [56]. Extracted algal oil is heated to 450–600°C in the absence of air resulting in the crumbling of chemical bonds and the formation of vapors, gases and char. These vapors undergo the condensation process forming liquid which can be used as biodiesel after passing through a catalyst viz, zeolites [56, 69, 70] (Figure 3.8).

Pyrolysis is advantageous over transesterification in following ways:

- Produces a higher yield than transesterification, which is about 70–75%,
- No need to extract oil in the initial steps.
- After the condensation of vapours into liquid and the passage of it over the catalyst, there is no need for additional washing, drying or distillation.
- Co-products like char and gases can also be reused. Char can be used as a fertilizer and gases, being combustible, can be used in the heating step.
- CO_2 produced while the fuels are burning in the heating step can be used to enhance algal growth.

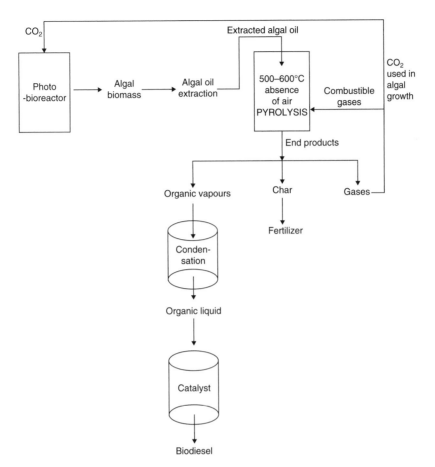

Figure 3.8 Schematic representation of biodiesel production from algal oil using pyrolysis process.

However, this method requires high temperatures and vacuum conditions to process biodiesel, which is an energy intensive step.

3.2.5   Hydroprocessing

Hydroprocessing, as its name defines, is a technique of processing oil following pyrolysis into fuel or biodiesel in the presence of hydrogen. The major objective of this process is the removal of oxygen, which is removed as water. The process is carried out in the presence of a catalyst and the oil obtained is of fuel grade as is rich in hydrocarbons. During this process glycerol is formed from triacyl glycerides and is transformed into propane. The alkane mixture obtained in this step is subjected to fractionation. The fractions obtained include synthetic kerosene jet fuel and hydrogenation derived renewable diesel (HDRD). HDRD has a high cetane number and can also be obtained from the processing of any other animal or plant derived oil. Also, it is environmental friendly as it is low in sulphur content [40]. In January 2009, Continental Airlines tested its Continental jet 516, a two engine Boeing 737–800 for a two hour flight. One of the Continental jet's two engines was powered by 50% of synthetic kerosene obtained from jatropha and algal oil, and the remaining 50% was from a petroleum based jet fuel [71].

3.3   Cellulosic Ethanol

Brazil and the US account for the production of 5 billion gallons of ethanol from cane sugar and starch from corn. The amount produced is not sufficient to make an impact in world petroleum consumption. From the estimates made by different groups of researchers it's clear that ethanol from conventional agriculture products cannot make an impact on transport fuel usage. The major limiting factor is the productivity of agricultural land. Moreover, the overall energy inputs for the production of ethanol account for a substantial fraction of energy output. Many experts agree that cellulosic biomass can support the production of biofuels at a scale that can match global petroleum use. The ethanol synthesis from cellulose through fermentation route requires the unlocking of sugars from cellulose and hemicelluloses.

Lignocellulose is comprised mainly of cellulose, hemicellulose and lignin. Cellulose is a linear polymer of D-glucose units linked by β-1, 4 glycosidic linkages with a high degree of polymerization [72], that varies from plant to plant. Cellulose being chemically stable and structurally inherent provides resistance against microbial attack and a high tensile strength to the cell wall [73]. Hemicellulose is hetero polysaccharide consists of pentoses (xylose, arabinose), hexoses (mannose, glucose, galactose) and sugar acids [74, 75]. Hemicellulose being branched and having an amorphous nature can be easily hydrolyzed into its monomer components [76]. Lignin is a highly complex network of phenyl propane units linked by ether bonds [77]. Lignin provides a high degree of mechanical strength to the cell wall. Lignin presents difficulty in hydrolysis by preventing the swelling of the cell wall, shielding cellulose surfaces, therefore affecting enzyme accessibility to cellulose. Another hurdle in the hydrolysis of lignin is that it is a non-carbohydrate in nature, composed mainly from 3 hydroxycinnamyl alcohols methoxylated to various degree viz. *p*-coumaryl, coniferyl, and sinapyl alcohol [78–80].

A basic representation of the conversion of lignocellulosic biomass into ethanol by enzymatic and chemical hydrolysis is shown below (Figure 3.9):

3.3.1 Size Reduction

The very first step in cellulosic ethanol production is the processing of lignocellulosic biomass into a small size using physical methods viz milling, grinding etc [71]. Following this step, the whole process

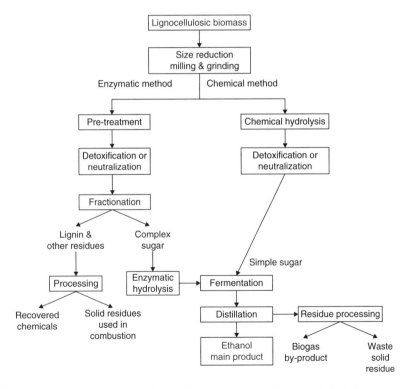

Figure 3.9 A basic representation of the conversion of ligno-cellulosic biomass into ethanol.

gets bifurcate into two parts, one is enzymatic hydrolysis and the other is chemical hydrolysis.

3.3.2 Enzymatic Hydrolysis

This method is based on using enzymes for the hydrolysis of complex sugars into simple sugars.

3.3.2.1 Pre-Treatment

Enzymatic hydrolysis starts from the pre-treatment of milled substances to effectively hydrolyse cellulose and hemicellulose into their corresponding monomeric units. Pre-treatment before enzymatic hydrolysis is necessary to enhance enzyme accessibility to cellulose or hemicellulose, raise surface area, remove lignin, degrade crystallinity of cellulose, and to solubilize cellulose or hemicellulose. There are various pre-treatment methods available, each one based on a different principal. Some are described below:

3.3.2.2 Steam Explosion Method

The steam explosion method is based on the processing of a milled product with high-pressure saturated steam at 180–210°C for about 1–10 min. This is then combined with an instantaneous decrease of steam pressure to obtain an explosive discharge [81]. It significantly enhances the enzymatic hydrolysis yield. However, it is unfavorable because it is energy intensive, costly and there is a loss of free sugar in the washing and purification step [82].

Today, steam explosion is used in combination with acid as a catalyst to enhance fidelity. In 2004, Varga et al. demonstrated that the use of a combination of steam explosion with 2% H_2SO_4 enhances enzymatic hydrolysis four fold in comparison to non-treated biomass [83].

3.3.2.3 Dilute Acid Method

The dilute acid pre-treatment method involves the addition of dilute acid (typically 0.2–3.0% H_2SO_4) and water into the milled product to form a slurry after providing requisite temperature (>160°C) through steaming. The two main differences between this method and the steam explosion method are the use of a smaller particle size in the dilute acid method and the use of a different starting material (the steam explosion method uses a moist starting material while the dilute acid method requires slurry with a lower dry matter content, i.e., <5%) [84].

3.3.2.4 Lime Pre-treatment Method

This method incorporates the mixing of calcium carbonate or sodium hydroxide with the milled product at 85–150°C for 1 hour, for several days, depending upon the starting material used. However, this method only helps to remove lignin [85, 86].

3.3.2.5 Ammonia Fiber Explosion (AFEX)

As its name implies, it is based on the use of liquid ammonia at 90–100°C and 17–20 bar pressure for 5–10 min. It is a rapid technique and can process material with 60% dry matter without any loss of free sugar [87].

3.3.2.6 Detoxification/Neutralization

The pre-treatment method results in the simultaneous production of some toxic components such as phenolic compounds, furan derivatives and aliphatic acids from sugar and lignin degradation. To detoxify such toxic components from the reaction mixture, some detoxification and neutralization methods are employed such as ion-exchange resins, treatment with lime, treatment with sulphite etc. [88].

3.3.2.7 Enzymatic Hydrolysis

After passing through the pre-treatment methods, the reaction mixture mainly comprises of cellulose and hemicellulose with a limited amount of lignin. Enzymes need suitable environmental conditions in order to work efficiently.

3.3.2.7.1 Cellulose degrading enzymes

1. Endo-1, 4-β- glucanase (EC 3.2.1.4): randomly hydrolyzing internal β-1, 4-D-glucosidic linkage.
2. exo-1,4-β-D-glucanases / Cellobiohydrolases (EC3.2.1.91): hydrolyzing cellobiose units from the ends.
3. β-glucosidase (EC 3.2.1.21): cleaving cellobiose units and producing glucose.

Trichoderma reesei, Humicola, Acremonium are some examples of cellulases from fungal origin. *Clostridium thermocellum* represents an example of bacterial cellulases.

3.3.2.7.2   Hemicellulose degrading enzyme

Xylan Chain degrading enzymes:

1. Endo-1, 4-β-D- Xylanase (EC 3.2.1.8), backbone degrading enzyme
2. 1, 4- β-D-Xylosidase (EC 3.2.1.37), hydrolyzing xylo-oligosaccharides to xylose.

Mannase degrading enzymes:

1. Endo-1, 4-β-D-mannanases (EC 3.2.1.78), randomly hydrolyzing internal bonds
2. 1, 4- β-D-mannosidases (EC 3.2.1.25), hydrolyzing mannooligosaccharides to mannose.

Side group removing enzymes:

1. α-D-galactosidases (EC 3.2.1.22)
2. α-L-arabinofuranosidases (EC 3.2.1.55)
3. α-glucuronidases (EC 3.2.1.139)
4. acetylxylanesterases (EC 3.1.1.72)
5. feruloyl and *p*-cumaric acid esterases (EC 3.1.1.73) [89–90]

Lignin modifying enzymes:

1. Laccase (EC 1.10.3.2): having copper as cofactor involved in lignin degradation
2. Lignin Peroxidase (EC 1.11.1.14): having heme as a cofactor involved in the oxidative breakdown of lignin

3.3.3   Chemical Method

This method is based on using chemicals for the hydrolysis of complex sugars into simple sugars.

3.3.3.1   *Chemical Hydrolysis*

The milled product is allowed to undergo chemical treatment at a suitable temperature for a specific time period, resulting in the hydrolysis of cellulose and hemicellulose into their corresponding monomeric units. HCl and H_2SO_4 can both be used for chemical hydrolysis, although H_2SO_4 is the mainly used chemical [72]. There are two ways of performing chemical hydrolysis:

1. Concentrated acid hydrolysis
2. Diluted acid hydrolysis

3.3.3.1.1 Concentrated acid hydrolysis
Concentrated acid hydrolysis, a comparatively obsolete procedure that requires a low temperature (40°C) to provide a high sugar yield. However, the use of high acid consumption in this process leads to equipment corrosion. Also, it is an energy intensive and time consuming process [72].

3.3.3.1.2 Dilute acid hydrolysis
Dilute acid hydrolysis is presumably the most generally used process. It requires low acid consumption and less processing time comparatively for the hydrolysis of pentoses and hexoses into simple sugars. However, it is based on high temperature usage, which leads to the formation of some undesirable by-products. Reactors like plug-flow, percolation, counter-current and shrinking bed reactors are used for the dilute acid hydrolysis of hexoses and pentoses in batch mode [72].

3.3.3.1.3 Detoxification / Neutralization
After finishing the chemical hydrolysis process, the reaction mixture needs to pass through a detoxification or neutralization step to recover acid or to detoxify some components. The acid component can be recovered by passing through an anion membrane or by converting it into H_2S. Dilute acid hydrolysis sometimes results in the formation of some toxic products like 5-hydroxymethyl furfural (HMF), acetic acid, uronic acid, 4-hydroxy benzoic acid, phenol, formaldehyde etc. that can be recovered by passing a reaction mixture through an ion-exchange resin or treating it with lime or sulphite.

3.3.4   Fermentation Process

After passing through the detoxification, neutralization or fraction-ation steps, resulting hydrolyzate still represents a mixture of many compounds, including some inhibitory compounds. The type, composition and concentration of these inhibitory compounds are based on the kind of lignocellulosic material used, the hydrolysis method, either enzymatic or chemical, and on the pre-treatment method involved) [72]. Also, hydrolysis of cellulose produces glucose but hydrolysis of hemicellulose results in the production of some pentoses, mainly xylose, with some amount of arabinose, galactose, mannose and fucose based on the starting material [91]. Naturally occurring microorganisms like *Saccharomyces cerevisiae* and *Zymomonas mobilis*, commonly used fermenting microorganisms, can easily and efficiently ferment glucose but fail to ferment other sugars like xylose or arabinose [92].

All these things make it difficult to ferment hydrolyzates compared to fermenting sugarcane juice or grains.

Some examples of microorganisms that can ferment xylose into ethanol are given below:

Yeast Species: *Candida, Kluveromyces, Pichia, Schizosaccharomyces*
Fungal Species: *Fusarium, Rhizopus, Monilia, Mucor*
Bacterial Species: *Clostridium, Eruinia, Bacillus, Thermoanaerobacter* [93–96]

Today, some genetically modified organisms that are able to ferment both glucose and xylose are beginning to be used. However, since this is only operating at the laboratory scale there is a need to transfer them to the pilot scale. In 1996, Padukone *et al.* showed the use of the recombinant *E.Coli* ATCC 11303 (pLOI 297) to ferment glucose and xylose into ethanol with a theoretical yield of about 84% [97]. In 1999, Ingram *et al.* described the use of a recombinant strain of *Klebsiella oxyticato* to directly ferment cellobiose and cellotriose [98]. In 2004, Abbas *et al.* used a genetically modified strain, *Saccharomyces cerevisiae* r424A, to ferment corn fiber hydrolyzate with a yield of 58g/l ethanol in fermented medium [99].

After completing this fermentation process, the fermented medium undergoes the distillation process in order to separate the ethanol. The residual processing of lignin, non-reacted cellulose and hemicellulose, micro biomass, chemicals and enzymes can result in the production of biogas as a by-product and solid waste residue [100, 101].

3.4 Syngas

Synthetic gas or syngas is derived from biomass through its partial oxidation in a gasifier or through fermentation (Figure 3.10). Syngas is a mixture mainly composed of carbon monoxide (CO),

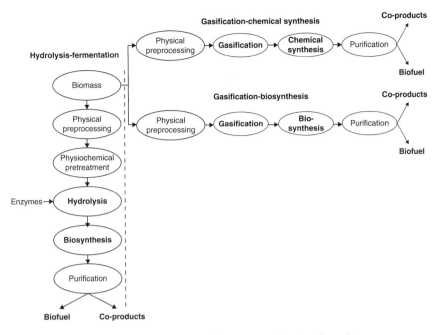

Figure 3.10 Schematic representation of syngas production from biomass through hydrolysis and gasification methods.

carbon dioxide (CO_2), and hydrogen (H_2) [102]. Here, biomass is gasified through fermentation resulting in formation of CO, CO_2, and H_2 and other by-products like tar, hydrocarbon, nitric oxide (NO), and ammonia. It can be further converted into liquid bio-fuel through fermentation. The microbes used for this conversion include *Clostridium ljungdahlii* [103], *Clostridium autoethanogenum* [104] and *Butyribacterium methylotrophicum* [105]. These organisms use the acetyl-CoA pathway, also known as the "Wood/Ljungdahl" pathway for energy and growth [106]. Acetogens derive energy by reducing carbon dioxide as the terminal electron acceptor to acetyl-CoA [103]. The major hurdle in the production of biofuels from syn-gas is attributed to the low gas-to-liquid mass transfer rate of CO, CO_2 and H_2. Current research focuses on improving this challenge. Generally, mass transfer of these gases is enhanced by increasing the gas flow rate per unit of liquid volume [107] or by the use of micro spargers to produce micro-bubbles that increases the surface area of the gas liquid interface [108]. Maintaining high cell densities also improve the mass transfer rate of the gases resulting in their efficient conversion into fuel [109]. This is presumably due to the fact that the gas entering the reactor is immediately consumed.

3.5 Conclusion

Development of second generation biofuels is necessary since food resources cannot be jeopardized for fuel. To ensure food safety, in many countries the production of biofuel has been restricted to plant products that can be grown on wastelands or generated through waste. Since many issues related to the scaling up of second generation biofuels are common to those of the first generation, lessons learned from the past may be applied. The process of refinement of algal lipids to biofuel involves similar steps to that of vegetable oil. It can be concluded that the future of second generation biofuels depends not only on biotechnology but also on an improvement in production methods or an amalgamation of both.

Note: Processing of bio-hydrogen is not discussed here as it is not able to power present day vehicle engines as fuel. Our objective is to introduce readers to the processing technologies of second generation biofuels that can be used as transport fuel. It is important to note that both cellulose (after being hydrolyzed to sugars) and syngas can also be diverted to butanol synthesis by altering the biocatalyst.

References

1. C. Schubert, *Natural Biotechnology*, Vol. 24, pp. 777–784, 2006.
2. A. Demirbas, *Progress in Energy and Combustion Science*, Vol. 33, pp. 1–18, 2007.
3. S.N. Naik, V.V. Goud, *et al.*, *Renewable and Sustainable Energy Reviews*, Vol. 14, pp. 578–597, 2010.
4. J. Hill, E. Nelson *et al.*, *Proceedings of the National Academy of Sciences of the United States of America*, Vol. 103, pp. 11206–11210, 2006.
5. M.J. Taherzadeh and K. K, *BioResources*, Vol. 2, pp. 707–738, 2007.
6. A. Eisentraut, Sustainable Production of Second - Generation Bio-fuels: Potential and perspectives in major economies and developing countries, Paris, France, International Energy Agency, 2010.
7. D. Kumar and G.S. Murthy, *Biotechnology for Bio-fuels*, Vol. 4, pp. 1–19, 2011.
8. Y. Sun and C. J, *Bioresource technology*, Vol. 83, pp. 1–11, 2002.
9. F. Teymouri, L. Laureano-Pérez, *et al.*, *Applied Biochemistry and Biotechnology*, Vol. 115, pp. 951–963, 2004.
10. P. Bansal, M. Hall, *et al.*, *Biotechnology Advances*, Vol. 27, pp. 833–848, 2009.
11. M. Hannon, J. Gimpel, *et al.*, *Bio-fuels*, Vol. 1, pp. 763–784, 2010.
12. J.R. Rostrup-Nielsen, *Science*, Vol. 308, pp. 1421–1422, 2005.
13. P.C. Munasinghe and S. K. Khanal, *Bioresource Technology*, Vol. 101, pp. 5013– 5022, 2010.

14. M. Mohammadi, G.D. Najafpour, *et al., Renewable and Sustainable Energy Reviews*, Vol. 15, pp. 4255–4273, 2011.
15. S.A. Scott, M.P. Davey, *et al., Current Opinion in Biotechnology*, Vol. 21, pp. 277–286, 2010.
16. A.M. Kunjapur and R.B. Eldridge, *Industrial and Engineering Chemistery Research*, Vol. 49, pp. 3516–3526, 2010.
17. B. Wang, C.Q. Lan, *et al., Biotechnology Advances*, Vol. 30, pp. 904–912, 2012.
18. M.A. Borowitzka, *Journal of Biotechnology*, Vol. 70, pp. 313–321, 1999.
19. C.U. Ugwu, H. Aoyagi, *et al., Bioresource Technology*, Vol. 99, pp. 4021–4028, 2008.
20. F.G. Camacho, A.C. Gomez, *et al., Enzyme and Microbial Technology*, Vol. 24, pp. 164–172, 1999.
21. S. L. Choi, I. S. Suh, *et al., Enzyme and Microbial Technology*, Vol. 33, pp. 403–409, 2003.
22. J. Vega-Estrada, M.C. Montes-Horcasitas , *et al., Applied Microbiology and Biotechnology*, Vol. 68, pp. 31–35, 2005.
23. S.Y. Chiu, M.T. Tsai, *et al., Engineering in Life Sciences*, Vol. 9, pp. 254–260, 2009.
24. C.U. Ugwu, H. Aoyagi, *et al., Bioresource Technology*, Vol. 99, pp. 4021–4028, 2008.
25. L. Xu, P.J. Weathers, *et al., Engineering in Life Sciences*, Vol. 9, pp. 178–189, 2009.
26. Burlew, J. S., Ed. (1953). Milner Rocking tray Algal Culture from Laboratory to Pilot Plant. Washington, Carnegie Institution.
27. R.D. Ortega and J.C. Roux, *Biomass*, Vol. 10, pp. 141–156, 1986.
28. M.R. Tredici and R. Materassi, *Journal of Applied Phyco*logy, Vol. 4, pp. 221–231, 1992.
29. Q. Hu, H. Guterman, *et al., Biotechnology and Bioeng*ineering, Vol. 51, pp. 51–60, 1996.
30. A. Richmond, *Journal of Applied Phyco*logy, Vol. 12, pp. 441–451, 2000.
31. Q. Hu, D. Faiman, *et al., Journal of Fermentation and Bioengineering*, Vol. 85, pp. 230–236, 1998.
32. L. Travieso, D. O. Hall, *et al., International Biodeterioration and Biodegradation*, Vol. 47, pp. 151–155, 2001.
33. D. Chaumont, C. Thepenier, *et al.*, "Scaling up a tubular photoreactor for continuous culture of Porphyridium cruentum - From laboratory to pilot plant," in T. Stadler, J. Morillon, M.C. Verdus, W. Karamanos, H. Morvan, D. Christiaen, eds., *Algal Biotechnology*, Elsevier Applied Science, London, pp. 199–208, 1988.
34. E. Molina, J. Fernandez, *et al., Journal of Biotechno*logy, Vol. 92, pp. 113–131, 2001.
35. A.A. Henrard, M. G. de Morais, *et al., Bioresource technology*, Vol. 102(7), pp. 4897–4900, 2011.
36. M. Morita, Y. Watanabe, *et al., Biotechnology and Bioengineering*, Vol. 74(2), pp. 136–144, 2001.
37. M. Morita, Y. Watanabe, *et al., Biotechnology and Bioengineering*, Vol. 77(2), pp. 155–162, 2002.
38. C.U. Ugwu, J. C. Ogbonna, *et al., Journal of Applied Phycology*, Vol. 15, pp. 217–223, 2003.
39. C.U. Ugwu, H. Aoyagi, *et al., Bioresource Technology*, Vol. 99, pp. 4021–4028, 2008.

40. Current Status and Potential for Algal BiofuelsProduction, A report to IEA bioenergy task 39, National renewable energy laboratory, Report T39-T2 6, 2010

41. Oilgae, Extraction of algal oil by mechanical methods, http://www.oilgae.com/algae/oil/extract/mec/mec.html, 2013.

42. P. Mercer and R.E. Armenta, *European Journal of Lipid Science and Technology*, Vol. 113(5), pp. 539–547, 2011.

43. R. Harun, M. Singh, G.M. Forde, M.K. Danquah, *Renewable and Sustainable Energy Reviwes*, Vol. 14, pp. 1037–1047, 2010.

44. F. Wei, G.Z. Gao, X.F. Wang, X.Y. Dong, *et al.*, *Ultrasonics Sonochemistry*, Vol. 15, pp. 938–942, 2008.

45. R.L. Mendes, B.P. Nobre, M.T. Cardoso, A.P. Pereira, A.F. Palavra, *Inorganica Chimica Acta*, Vol. 356 (3), pp. 328–334, 2003.

46. R.M. Couto, P.C. Simões, A. Reis, T.L. Da Silva, V.H. Martins, Y. Sánchez-Vicente, *Engineering in Life Science*, Vol. 10, pp. 158–164, 2010.

47. F. Sahena, I.S.M. Zaidul, S. Jinap, A.A. Karim, *et al.*, *Journal of Food Engineering*, Vol. 95, pp. 240–253, 2009.

48. J.A. Mendiola, L. Jaime, S. Santoyo, G. Reglero, *et al.*, *Food Chemistry*, Vol. 102, pp. 1357–1367, 2007.

49. R.L. Mendes, A.D. Reis, A.F. Palavra, *Food Chemistry*, Vol. 99, pp. 57–63, 2006.

50. R.L. Mendes, A.D. Reis, A.P. Pereira, M.T. Cardoso, *et al.*, *Chemical Engineering Journal*, Vol. 105, pp. 147–151, 2005.

51. M. Herrero, A. Cifuentes, E. Iban˜ ez, *Food Chemistry*, Vol. 98, pp. 136–148, 2006.

52. K. Sander, G.S. Murthy, Enzymatic Degradation of Microalgal Cell Walls, An ASABE Meeting Presentation, Paper Number: 1035636, 2009.

53. D.E. Brune, L.E. Beecher, Proceedings of the 29th Annual Symposium on Biotechnology for Fuels and Chemicals, www.simhq.org/meetings/29symp/index.html, 2007.

54. Origin oil, http://www.originoil.com/technology/overview.html.

55. A.A. Refaat, *International Journal of Environment Science and Technology*, Vol. 7(1), pp. 183–213, 2010.

56. White Paper, Foxboro, Invensys, Instrumentation for Biodiesel fuel Production, 2012.

57. P.E. Wiley, K.J. Brenneman, *Water Environment Research*, Vol. 81(7), pp. 702– 708, 2009.

58. J.M. Encinar, F. Juan, J.F. Gonzalez, A. Rodriguez-Reinares, *Fuel Processing Technology*, Vol. 88 (5), pp. 513–522, 2007.

59. A. Demirbas, *Energy Conversion and Management*, Vol. 50 (4), pp. 923- 927, 2009.

60. A. Gupta, S.K. Sharma, A. Pal Toor, *Journal of Petrotech Society*, Vol. IV (1), pp. 40–45, 2007.

61. Biodiesel Production Techniques, Food technology fact sheet, 405–744-6071, www. fapc.biz.

62. A. Singh, B. He, J. Thompson, J. van Gerpen, *Applied Engineering in Agriculture*, Vol. 22(4), pp. 597–600, 2006.

63. X. Yuan, J. Liu, G. Zeng, J. Shi, J. Tong, G. Huang, *Renewable Energy*, Vol. 33 (7), pp. 1678–1684, 2008.

64. P. Felizardo, M.J. Correia, I. Raposo, J.F. Mendes, R. Berkemeier, J.M. Bordado, *Waste Management*, Vol. 26 (5), pp. 487–494, 2006.

65. A.C. Pinto, L.N. Guarieiro, M.J. Rezende, N.M. Ribeiro, N. M., E.A. Torres, W.A. Lopes, P.A. Pereira, J.B. Andrade, *Journal of the Brazilian Chemical Society*, Vol. 16(6B), pp.1313–1330, 2005.

66. W. Zhou, D.G.B. Boocock, *Journal of the American Oil Chemists' Society*, Vol. 83(12), pp. 1041–1045, 2006.

67. X. Meng, G. Chen, Y. Wang, *Fuel Processing Technology*, Vol. 89(9), pp. 851–857, 2008.

68. A.A. Refaat, S.T. El Sheltawy, K.U. Sadek, *International Journal of Environment Science and Technology*, Vol. 5(3), pp. 315–322, 2008b.

69. T.A. Milne, R.J. Evans, and N. Nagle, *Biomass*, Vol. 21(3), pp. 219–232, 1990.

70. M. Ringer, V. Putsche, and J. Scahill, Large-Scale Pyrolysis Oil Production: A Technology Assessment and Economic Analysis, *Technical Report* NREL/TP-510–37779, 2006.

71. Scientific American, http://www.scientificamerican.com/article.cfm?id=air-algae-us-bio-fuel-flight-on-weeds-and-pond-scum, 2009.

72. M.J. Taherzadeh, K. Karimi, *Bioresources*, Vol. 2(3), 472–499, 2007.

73. J.B. Kristensen, Enzymatic hydrolysis of lignocellulose. Substrate interactions and high solids loadings, Forest & Landscape Research No. 42–2008, Forest & Landscape Denmark, Frederiksberg, 130 pp., 2008.

74. E. Sjöström, Wood chemistry: fundamentals and applications, Academic Press, New York, 1981.

75. C.E. Wyman, B.E. Dale, R.T. Elander, M. Holtzapple, M.R. Ladisch, and Y.Y. Lee, *Bioresource Technology*, Vol. 96, pp. 1959–1966, 2005.

76. C.N. Hamelinck, G. Van Hooijdonk, and A.P.C. Faaij, *Biomass Bioenergy*, Vol. 28, pp. 384–410, 2005.

77. B.C. Saha, and R.J. Bothast, *ACS Symposium Series*, Vol. 666, pp. 46–56, 1997.

78. D.E. Akin, *Applied Biochemistry and Biotechnology*, Vol. 137, pp. 3–15, 2007.

79. D.E. Akin, *Bio-fuels, Bioproducts and Biorefining*, Vol. 2, pp. 228–303, 2008.

80. C.A. Mooney, S.D. Mansfield, M.G. Touhy, J.N. Saddler, *Bioresource Technology*, Vol. 64, pp. 113–119, 1998.

81. M.J. Taherzadeh, and C. Niklasson, *ACS Symposium Series*, Vol. 889, pp. 49–68, 2004.

82. B.S. Dien, N. Nagle, K.B. Hicks, V. Singh, R.A. Moreau, M.P. Tucker, N.N. Nichols, D.B. Johnston, M.A. Cotta, Q. Nguyen, and R.J. Bothast, *Applied Biochemistry and Biotechnology*, Vol. 113–116, pp. 937–949, 2004.

83. E. Varga, K. Reczey, and G. Zacchi, *Applied Biochemistry and Biotechnology*, Vol. 113–116, pp. 509–523, 2004.

84. H. Jørgensen, J.B. Kristensen and C. Felby, *Bio-fuels, Bioproducts and Biorefining*, Vol. 1, pp. 119–134, 2007.

85. C.E. Wyman, B.E. Dale, R.T. Elander, M. Holtzapple, M.R. Ladisch and Y.Y. Lee, *Bioresource Technology*, Vol. 96, pp. 1959–1966, 2005.

86. N. Mosier, C. Wyman, B. Dale, R. Elander, Y.Y. Lee, M. Holtzapple *et al.*, *Bioresource Technology*, Vol. 96, pp. 673–686, 2005.

87. F. Tymouri, L. Laureano-Perez, H. Alizadeh and B.E. Dale, *Bioresource Technology*, Vol. 96, pp. 2014–2018, 2005.

88. B.C. Saha, *ACS Symposium Series*, Vol. 889, pp. 2–34, 2004.

89. Q.K. Beg, M. Kapoor, L. Mahajan and G.S. Hoondal, *Applied Microbiology and Biotechnology*, Vol. 56, pp. 326–338, 2001.

90. D. Shallom and Y. Shoham, *Current Opinion in Microbiology*, Vol. 6, pp. 219–228, 2003.

91. N.N. Nichols, B.S. Dien, and R.J. Bothast, *Applied Microbiology and Biotechnology*, Vol. 56, pp. 120–125, 2001.

92. B.C. Saha, *ACS Symposium Series*, Vol. 889, pp. 2–34, 2004.

93. C.L. Flores, C. Rodriguez, T. Petit and C. Gancedo, *FEMS Microbiology Reviews*, Vol. 24(4), pp. 507–529, 2000.

94. J. Zaldivar, J. Nielsen and L. Olsson, *Applied Microbiology and Biotechnology*, Vol. 56 (1–2), pp. 17–34, 2001.

95. H.K. Sreenaath and T.W. Jeffries, *Bioresource Technology*, Vol. 72 (3), pp. 253–260, 2000.

96. R. Millati, L. Edebo, M.J. Taherzadeh, *Enzyme and Microbial Technology*, Vol. 36 (2–3), pp. 294–300, 2005.

97. N. Padukone, "Advanced process options for bioethanol production", in C.E. Wyman, eds., Handbook on Bioethanol: Production and Utilization, Taylor, Francis, 1996.

98. L.O. Ingram, H.C. Aldrich, A.C.C. Borges, T.B. Causey, A. Martinez, F. Morales, A. Saleh, S.A. Underwood. L.P. Yomano, S.W. York, J. Zaldivar and S.D. Zhou, *Biotechnology Progress*, Vol. 15(5), pp. 855–866, 1999.

99. C. Abbas, K. Beery, E. Dennison, and P. Corrington, *ACS Symposium Series*, Vol. 889, pp. 84–97, 2004.

100. N. Mosier, C. Wyman, B. Dale, R. Elander, Y.Y. Lee, M. Holtzapple and M. Ladisch, *Bioresource Technology*, Vol. 96, pp. 673–686, 2005.

101. Y. Lin, and S. Tanaka, *Applied Microbiology and Biotechnology*, Vol. 69, pp. 627–642, 2006.

102. J.R. Rostrup-Nielsen, *Science*, Vol. 308, pp. 1421–1422, 2005.

103. J. Phillips, E. Clausen, *et al.*, *Applied Biochemistry and Biotechnology*, Vol. 45–46, pp. 145–157, 1994.

104. J. Abrini, H. Naveau, *et al.*, *Archives of Microbiology*, Vol. 161, pp. 345–351, 1994.

105. M.D. Bredwell, P. Srivastava, *et al.*, *Biotechnology Progress*, Vol. 15, pp. 834–844, 1999.

106. H.G. Wood, *FASEB Journal*, Vol. 5, pp. 156–163, 1991.

107. A.J. Ungerman, and T. J. Heindel, *Biotechnology Progress*, Vol. 23, pp. 613–620, 2007.

108. D. Birch, and N. Ahmed, *Powder Technology*, Vol. 88, pp. 33–38, 1996.

109. D.K. Kundiyana, R.L. Huhnke *et al.*, *Journal of Bioscience and Bioengineering*, Vol. 109, pp. 492–498, 2010.

4

Bioconversion of Lignocellulosic Biomass for Bioethanol Production

Virendra Kumar[1], Purnima Dhall[1], Rita Kumar[1,*] and Anil Kumar[2]

[1]Institute of Genomics and Integrative Biology, Mall Road,
Delhi-110007, India
[2]National Institute of Immunology, Aruna asif ali marg,
New Delhi-110067, India

Abstract

Lignocellulose is the most abundant source of carbohydrate and phenol polymer on earth. It refers to the dry biomass of plants that are composed of cellulose, hemicelluloses and lignin. Currently lignocelluloses are obtained from trees that are used in paper and wood industries whereas those from agricultural residues are used as animal feed and manure. The lignocelluloses produced from industrial and municipal solid waste are the source of major environmental pollution. The use of lignocellulosic biomass for biofuel production is of great interest nowadays because it has a number of advantages over conventional fuel. Bioethanol production from lignocellulosic biomass has the potential to meet current energy demand as well as reduce greenhouse gas emission and support the rural economy. This production of bioethanol involves five major steps i.e. pre-treatment, hydrolysis, detoxification fermentation, and distillation. Pre-treatment involves dissolving hemicellulose and altering the lignin structure, to provide improved enzyme accessibility for hydrolysis. In hydrolysis, cellulose and hemicelluloses are broken down into simple sugars by an enzymatic or acid treatment. Several physical, chemical and biological alternatives can be used to reduce or detoxify the negative effects of inhibitors produced during the hydrolysis process. The sugar hexose and pentose produced during the hydrolysis of lignocellulosic biomasses are converted into ethnol using various bacterial and yeast species.

Corresponding author: rita@igib.res.in

Dr.Vikash Babu, Dr. Ashish Thapliyal & Dr. Girijesh Kumar Patel (eds.) Biofuels Production, (85–118) 2014 © Scrivener Publishing LLC

85

Sacchromyces cerevisiae is the most commonly used glucose fermenting organism, but it is unable to ferment pentose sugar. Therefore, recently, researchers are focusing on the development of genetically engineered microorganisms for the complete utilization of lignocellulosic biomass for ethanol production.

Keywords: Biofuel, cellulose, hemicellulose, lignin, lignocellulose

4.1　Introduction

Lignocellulose is typically a non-edible plant material and the most abundant renewable biomass, including dedicated crops of wood and grass, and agro-forest residues. It is composed of cellulose, hemicellulose and lignin (Fig. 4.1). Furthermore, minor amounts of proteins, minerals and other components can be found in the lignocellulose composition [1]. The ratio of the three main components varies among different plant species. Dry wood typically consists of 40–50% cellulose, 15–20% hemicellulose, and 20–30% lignin. Other constituents of lignocelluloses are extractives, which usually represent a fraction of less than 10%, and ash, the content

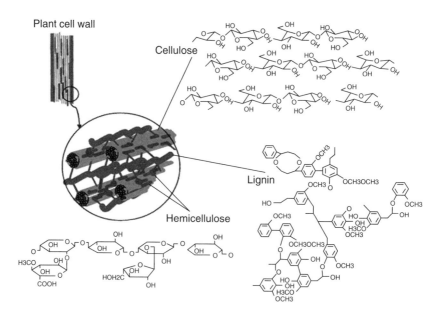

Figure 4.1 Structure of Lignocellulose.

of which is usually less than 1%. Inside the lignocellulose complex, cellulose retains a crystalline fibrous structure and it appears to be the core of the complex. Hemicellulose is positioned both between the micro and the macro-fibrils of cellulose. Lignin provides the structural role of the matrix in which cellulose and hemicellulose are embedded [2].

Cellulose is a long chain polymer, made up of repeating units of D-glucose, a simple sugar. It is the main component of plant cell walls. In the cellulose chain, the glucose units are in 6-membered rings, called pyranoses. They are joined by single oxygen atoms (acetal linkages) between the C-1 of one pyranose ring and the C-4 of the next ring. Since a molecule of water is lost when an alcohol and a hemiacetal react to form an acetal, the glucose units in the cellulose polymer are referred to as anhydroglucose units. The average cellulose chain has a degree of polymerization of about 9,000 to 10,000 units. There are several types of cellulose in the cell wall. Approximately 65 percent of the cellulose is highly oriented, crystalline, and not accessible to water or other solvents [3]. The remaining cellulose, composed of less oriented chains, is only partially accessible to water and other solvents as a result of its association with hemicellulose and lignin.

Hemicellulose is the second most abundant component of lignocelluloses. It is heteropolymer of pentoses (xylose and arabinose), hexoses (glucose, galactose, mannose) and sugar acids (acetic). Sugars are linked together by β-1,4- and occasionally β-1,3-glycosidic bonds. Hemicellulose is also a polysaccharide, consists of shorter chains of 500–3,000 sugar units softwood hemicellulose mainly contain mannose as a major constituent whereas hardwoods mainly contain Xylans [4].

Lignin is composed of three major phenolic components, namely p-coumaryl alcohol (H), coniferyl alcohol (G) and sinapyl alcohol (S). Lignin is synthesized by the polymerization of these components. The polymer is synthesized by the generation of free radicals, which are released in the peroxidase-mediated dehydrogenation of three phenyl propionic alcohols. Coniferyl alcohol is the principal component of softwood lignin, whereas guaiacyl and syringyl alcohols are the main constituents of hardwood lignin. After cellulose, it is the most abundant renewable carbon source on earth. Lignin resists attack by most microorganisms, and anaerobic processes tend not to attack the aromatic rings at all. Lignin is nature's cement along with hemicellulose to exploit the strength of cellulose while conferring flexibility [5].

4.1.1   Sources of Lignocellulosic Biomass

Lignocellulose is a generic term for plant matter and "wastes or biomass" are generated through forestry, agricultural practices, various industries and municipalities. [6–8]. The main sources of lignocellulosic compounds are:

i. **Agricultural Residues:** corn stover, wheat straw, rice straw, sugarcane bagasse grasses, rice hulls, barley crop residues, olive pulp and other agricultural residues (including straw, stover, peelings, cobs, stalks, nutshells and non food seeds).
ii. **Forestry Residues:** dedicated energy crops, residues from harvest operations that are left in the forest after stem wood removal, such as branches, foliage, roots, etc.
iii. **Industrial Waste:** waste generated from pulp and paper industry wood residues (including saw mill and paper mill discards) and food industry residues.
iv. **Biowaste Streams:** municipal solid waste (mainly kitchen and garden waste lignocelluloses garbage and sewage), packaging waste wood, household waste wood, market waste, food processing wastes, and cellulose wastes (newsprint, waste paper, recycled paper)

The composition of lignocellulose highly depends on its source. There is a significant variation of the lignin and (hemi) cellulose content of lignocellulose depending on whether it is derived from hard-wood, softwood, or grasses.

Depending on the source of the lignocellulosic biomass, the percentage of the celluose, hemicellulose and lignin varies. Table 4.1 shows the composition of the celluose, hemicellulose and lignin depending on the source available [9].

The lignocellulosic biomass, which represents the largest renewable reservoir of potentially fermentable carbohydrates on earth [10], is mostly wasted in the form of pre-harvest and post-harvest agricultural losses and the waste of food processing industries. It is used as pulp in the paper industry and burned as fuel, animal feed and manure formation.

Table 4.1 Composition of lignocellulose in several sources on dry basis [9].

Lignocellulosic materials	Cellulose (%)	Hemicellulose (%)	Lignin (%)
Hardwoods stems	40–55	24–40	18–25
Softwood stems	45–50	25–35	25–35
Nut shells	25–30	25–30	30–40
Corn cobs	45	35	15
Grasses	25–40	35–50	10–30
Paper	85–99	0	0–15
Wheat straw	30	50	15
Leaves	15–20	80–85	0
Cotton seed hairs	80–95	5–20	0
Newspaper	40–55	25–40	18–30
Waste papers	60–70	10–20	5–10
Primary wastewater solids	8–15	NA	24–29
Solid cattle manure	1.6–4.7	1.4–3.3	2.7–5.7
Switch grass	45	31.4	12.0

4.1.2 Lignocellulosic Biomass can be used for Bioethanol Production

Lignocellulose is a renewable and widely available biomass that can be converted into valuable chemicals including fuel and polymer precursors of strategic importance for the sustainability, advancement of energy and chemical industries [11–21]. Lignocellulosic biomass based biofuel has the potential to not only significantly displace fossil fuels, but also to add value to agricultural byproducts, forestry residues, or municipal wastes.

Lignocellulosic biomass is a low cost feedstock with the advantage that it is either available in large amounts as agricultural residues or can be cultivated with high yield per hectare and low energy inputs. Therefore, it is considered as foreseeable, feasible and sustainable resource for renewable fuel. Agricultural wastes such as crop residues, food processing wastes and forestry residues are potential sources of lignocellulose. The annual production of biomass is estimated to be 191010 metric tons worldwide [22]. Indeed fuels produced from lignocellulosic biomasses are of particular interest since the amount of carbon dioxide resulting from biomass use is equal to those absorbed during the photosynthesis process. Moreover, lignocellulosic biomass bioconversion for biofuel production is of environmental interest while it adds a solution to the problem of waste management. Furthermore, biofuels, which can be used as additives with gas oil or gasoline, contribute to reduce energy dependence.

There is significant potential for biofuels from cellulosic biomass to become the future alternative fuel resource. Several positive points that are in support of bioethanol production from lignocellulosic biomasses are that cellulose biomass is more abundant and less expensive than other food crops conventionally used for biofuel production. The sugar potential of most forms of cellulosic biomass on a dry ton basis is equal or greater to the amount of sugar obtained by conventional processing from corn or sugarcane. The greenhouse gas emissions from cellulosic biomass ethanol are much less than corn-based ethanol.

4.2   Bioethanol Production Process

Ligocellolusic biomass contains a complex mixture of carbohydrate polymers from the plant cell walls known as cellulose, hemi cellulose and lignin. It is the most promising renewable resource for bioethanol production. The bioconversion of lignocellulosic biomass into bioethanol is a very complex process. Overall fuel ethanol production includes five main steps, biomass pre-treatment, cellulose hydrolysis, fermentation of sugar, and distillation. Furthermore, the detoxification of toxic compounds released during the pretreatment and hydrolysis step can be carried out. (Figure 4.2).

4.2.1   Pretreatment

Lignocelluolosic biomass pretreatment is a very important step for getting fermentable sugar in the hydrolysis process. Pre-treatment

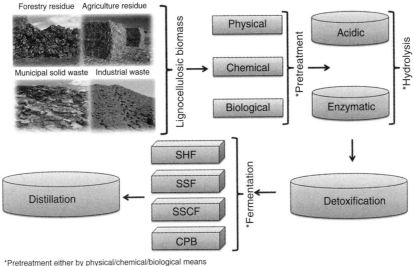

Figure 4.2 Flow diagram of bioethanol production using various lignocellulosic biomass.

aims to decrease the crystallinity of cellulose, increase biomass surface area, remove hemicellulose, break the lignin barrier and increase the porosity of lignocellulosic material. It makes cellulose more accessible to hydrolytic enzymes that facilitate the conversion of carbohydrate polymers into fermentable sugars in a rapid way with the concomitant increase in the yields. Since many lignocellulosics have different physicochemical characteristics, it is necessary to use suitable pre-treatment technology based on their properties. Pretreatment must meet the following requirements: improve the formation of sugars or the ability to subsequently form sugars by hydrolysis, avoid the degradation or loss of carbohydrates, avoid the formation of by-products that are inhibitory to the subsequent hydrolysis and fermentation processes and be cost-effective. Pretreatments include physical, physicochemical, chemical and biological methods, and their combinations.

4.2.1.1 Physical Pretreatment

Physical treatments such as chipping grinding, milling, high temperatures, freeze/thaw cycles, and radiation are aimed at size reduction and mechanical decrystallization. It reduces the particle size and crystallinity and increases the surface area and the bulk density. The following are methods for physical pretreatment.

i. Milling

Chipping and grinding can be applied to reduce cellulose crystallinity. The size of the materials are usually 10–30 mm after chipping and 0.2–2 mm after milling or grinding. Size reduction may provide better results [23, 24] but very fine particle sizes may impose negative effects on the subsequent processing such as pretreatment and enzymatic hydrolysis. It may generate clumps during the subsequent steps involving liquid and may lead to channelling. It is a very energy intensive process and in many cases the energy consumption is higher than the theoretical energy content available in the biomass.

ii. Pyrolysis

Pyrolysis is an endothermic process where less input of energy is required. Pyrolysis has also been used for the pretreatment of lignocellulosic materials, since biomass can be used as a substrate for a fast pyrolysis for thermal conversion of cellulose and hemicelluloses into fermentable sugar with a good yield [25]. In this process, the materials are treated at a temperature greater than 300°C, whereby cellulose rapidly decomposes to produce gaseous products such as H_2, CO and residual char [26, 27]. Pyrolysis in the presence of oxygen results in depolymerization, oxidation and dehydration.

iii. Radiation

Irradiation is reported to cause an increase in the surface area, while its effect on the crystallinity of the cellulose is controversial. The radiation treatments are effective in breaking the lignin-cellulose complex as evidenced by the increased presence of phenolics in the irradiated samples. The irradiation of cellulose by γ-rays leads to a cleavage of β-1,4-glycosidic bonds and gives a larger surface area and a lower crystallinity [28]. This method is far too expensive, however, to be used in a full-scale process.

4.2.1.2   Physico-chemical

Steam explosion is the most widely used physico-chemical pretreatment for lignocellulosic biomass. Besides this, liquid hot water and ammonia fiber explosions have also been reported as good physicochemical pretreatments for linocellolosic biomass. The following are various physico-chemical methods for pretreatment.

i. Steam Explosion (SE)

Steam explosion is a promising method of pretreatment which makes biomass more accessible to cellulase attack [29]. In this method, the biomass is heated using high pressure steam for a few minutes, the reaction is then stopped by a sudden decompression in atmospheric pressure [29, 30]. When steam is allowed to expand within the lignocellulosic matrix it separates the individual fibres [31]. The process causes hemicelluloses degradation and lignin transformation due to the high temperature, thus increasing the potential of cellulose hydrolysis. Hemicellulose is thought to be hydrolyzed by acetic and other acids released during steam-explosion pre-treatment. Lignin is removed only to a limited extent during the pre-treatment but is redistributed on the fibre surfaces as a result of melting and depolymerisation/repolymerization reactions [32].

The advantages of steam-explosion pretreatment include the low energy requirement compared to mechanical grinding and no recycling or environmental costs. The conventional mechanical methods require 70% more energy than steam explosion to achieve the same particle size reduction [33].

Steam explosion is recognized as one of the most cost-effective pretreatment processes for hardwoods and agricultural residues, but it is less effective for softwoods. [9]. Limitations of steam explosion include the destruction of a portion of the xylan fraction, incomplete disruption of the lignin-carbohydrate matrix, and the generation of compounds that might be inhibitory to microorganisms used in downstream processes [34]. Because of the formation of degradation products which are inhibitory to microbial growth, enzymatic hydrolysis, and fermentation, pre-treated biomass needs to be washed with water to remove the inhibitory materials along with water soluble hemicelluloses [35].

ii. Ammonia fiber explosion

Ammonia fiber explosion pretreatment involves liquid ammonia and steam explosion [31]. It is an alkaline thermal pre-treatment which exposes the lignocellulosic materials by high temperature and pressure treatment followed by rapid pressure release. This method does not produce inhibitors of the downstream processes [9, 36]. This pre-treatment has the drawbacks of being less efficient for biomass containing higher lignin contents (e.g. softwood newspaper) as well as of causing solubilization of only a very small fraction of solid material particularly hemicelluloses [9, 37]. It is more effective for the treatment of substrates with less content of lignin compared to sugarcane. This system does not directly liberate any sugars, but allows the polymers (hemicellulose and cellulose) to be attacked enzymatically which break down to sugars.

iii. CO_2 explosion

CO_2 explosion acts in a manner similar to that of the steam and ammonia explosion techniques. However, CO_2 explosion is more cost effective than ammonia explosion and does not cause the formation of inhibitors as in steam explosion [26, 38]. Conversion yields are higher compared to the steam explosion method [38].

4.2.1.3   Chemical Pre-treatment

In crystalline cellulose requires a chemical agent that is capable of breaking the hydrogen bonds of the cellulose. Aqueous solutions of acid and alkali belong to this group of chemical agents. The following are chemical pretreatment methods for lignocellulosic biomasses.

i. Ozonolysis

Ozone treatment is one way of reducing the lignin content of lignocellulosic wastes and unlike other chemical treatments, it does not produce toxic residues. Ozone can be used to degrade lignin and hemicellulose in many lignocellulosic materials such as wheat straw [39], bagasse, green hay, peanut, pine cotton straw and poplar sawdust [40–42]. The degradation is mainly limited to lignin. Hemicellulose is slightly affected, but cellulose is not. Ozonolysis pre-treatment has an advantage that the reactions are carried out at room temperature and normal pressure. Furthermore, the fact that ozone can be easily decomposed by using a catalytic bed or increasing the temperature means that processes can be designed to minimize environmental pollution [43]. A drawback of ozonolysis is that a large amount of ozone is required, which can make the process expensive.

ii. Acid pre-treatment

Pre-treatment with acid hydrolysis can result in the improvement of enzymatic hydrolysis of lignocellulosic biomasses to release fermentable sugars. Acid pre-treatment is widely used from hardwoods to grasses and agricultural residues. [44] In the acid pre-treatment, different types of acids (sulphuric acid, hydrochloric acid, nitric acid and phosphoric acid) are used. Among them sulphuric acid is mostly used for acid pre-treatment [45]. The acid medium attacks the polysaccharides, especially hemicelluloses which are easier to hydrolyze than cellulose [45]. In such a way, dilute or concentrated acids improve cellulose hydrolysis [31]. Concentrated acids are toxic, corrosive, hazardous, and thus require reactors that are resistant to corrosion, which makes the pre-treatment process very expensive. In addition, the concentrated acid must be recovered after hydrolysis to make the process economically feasible [9, 46]. Dilute-acid hydrolysis has been successfully developed for the pre-treatment of lignocellulosic materials. High temperatures in the dilute-acid treatment are favourable for cellulose hydrolysis. However, acid pre-treatment results in the production of various inhibitors like acetic acid, furfural and 5-hydroxymethylfurfural. These products inhibit the growth of microorganisms.

iii. Alkaline pre-treatment

Alkaline pre-treatment of lignocellulosic digests the lignin matrix and makes cellulose and hemicellulose available for enzymatic degradation [47]. Alkali treatment of lignocellulose disrupts the cell wall by dissolving hemicelluloses, and lignin. Effect of alkaline pre-treatment depends on the lignin content of the materials [35, 48]. Alkali pre-treatment processes utilize lower temperatures and pressures than other pre-treatment technologies [36]. Compared with acid processes, alkaline processes cause less sugar degradation. In the alkaline pre-treatment process, hydroxides of sodium, potassium, calcium and ammonium are used. Lignin removal by alkaline pretreatment increases enzyme effectiveness by eliminating non-productive absorption sites and by increasing access to cellulose and hemicellulose. Alkaline pretreatment removes acetyl groups from hemicellulose thereby reducing the steric hindrance of hydrolytic enzymes and greatly enhancing carbohydrate digestibility.

iv. Organosolv pre-treatment
The organosolvation method is a promising pre-treatment strategy, and it has attracted much attention and demonstrated the potential for utilization in lignocellulosic pre-treatment [49]. The utilization of organic solvent/water mixtures eliminates the need to burn the liquor and allows the isolation of the lignins. In the organosolvation process, an organic or aqueous organic solvent mixture with inorganic acid catalysts (HCl or H_2SO_4) is used to break the internal lignin and hemicellulose bonds. Other various organic solvents which can be used for delignification are methanol, ethanol, acetic acid, performic acid, peracetic acid and acetone [50].

4.2.1.4   Biological Pre-treatment

Physical, physicochemical and chemical, pre-treatment of lignocelluloscic biomasses are expensive and negatively affect our environment. In particular, physical and thermochemical processes require abundant energy for biomass conversion. Biological methods that rely on enzymes are particularly attractive. Natural lignocellulose degradation and utilization are carried out by specific enzymes from lignocellulolytic organisms (especially wood-degrading fungi and bacteria). Biomass converting enzymes might provide high specificity, low energy or chemical consumption, or low environmental pollution. In the biological pre-treatment process of lignocelluloses, biomass-converting enzymes degrade hemicelluloses and lignin and increase the accessibility on cellulose for hydrolysis into simple sugars, which can then be fermented by microorganisms into valuable fuel.

Hemicellulose is a group of complex polysaccharides made by different glyco-units and glycosidic bonds. Degradation of hemicellulose, not only "liberates" cellulose for cellulases but also converts hemicellulose into valuable saccharides. Hemicelluloses are heteropolymers and include β-glucan, xylan, xyloglucan, arabinoxylan, mannan, galactomannan, arabinan, galactan, polygalacturonan, etc., which are targets of β-glucanase, xylanase, xyloglucanase, mannanase, arabinase, galactanase, polygalacturonase, glucuronidase, acetyl xylan esterase, and other enzymes [51, 52]. Hemicellulose is a group of enzymes which work in synergy for biomass conversion. Functions of different hemicellulase are given in table 4. 2.

Lignin degrading microbes, mainly white rot fungi, contain oxidoreductases enzymes for lignocellulose degradation [53]. The main tasks of these oxidoreductases are likely aimed at the degradation of lignin, a highly heterogeneous and recalcitrant aromatic polymer entangled with hemicellulose or cellulose. Lignin degradation is an imperative for industrial enzymatic biomass-conversion, because it not only increases (hemi) cellulose accessibility for (hemi) cellulase but it also diminishes (hemi) cellulase inactivation caused by lignin absorption.

Lignin peroxidase, Mn peroxidase and versatile peroxidase are extracellular heme peroxidases with high potency to oxidatively degrade lignin. Upon interaction with H_2O_2, these enzymes form highly reactive Fe (V) or Fe (IV)-oxo species, which oxidize lignin either directly or via Mn (III) species. Laccase is a multi-copper oxidase secreted by numerous lignocellulolytic fungi. This enzyme can directly oxidize phenolic parts of lignin, or indirectly oxidize non-phenolic lignin parts with the aid of suitable redox-active mediator. Aryl-alcohol oxidase, glyoxal oxidase and various carbohydrate oxidases are also involved in natural lignocellulose degradation. They need no chemicals but low hydrolysis rates and low yields impede their implementation [31, 38,]. A drawback of biological pre-treatment is that in most cases of biological pre-treatment and hydrolysis, the process is very slow and bio-delignification generally needs long periods of time.

4.2.2 Hydrolysis

The goal of this process is to generate fermentable monomeric sugars from hemicellulose and the cellulose content of lignocellulosic biomass. This can be accomplished by two different processes, namely, acid hydrolysis and enzymatic hydrolysis.

Table 4.2 Enzymes for degradation of lignocelulosic biomass.

Biomass	Enzyme	Function	Reference
Cellulose	Cellobiohydrolase	Degradation of crystalline cellulose	[54,55]
	endo-1,4-β-Glucanase	Degradation of amorphous cellulose ,	[56]
	β-Glucosidases	Degradation of cellobiose, as well as other cellodextrins act on the non-reducing Glc unit from cellobiose or cellodextrin	[57]
Hemicellulose	endo-β-Xylanases	Hydrolyzes backbone glycosidic bonds in xylan.	[58]
	β-Xylosidase	hydrolyze xylobiose or other xylooligosaccharides, after their production from xylan by endo-β-Xylanases	[59]
	Acetyl Xylan Esterase	Deacetylates substituted O_2 or O_3 sites of backbone glycosyl units in xylan	[60]
	Feruloyl Esterase	Hydrolyzes feruloyl esters at α-L-Ara (O_2 or O_5 site), β-D-galactosyl (Gal, O_6 site), or α-D-Xyl side chains of arabinan/arabinoxylan, rhamnogalacturonan, or xyloglucan	[61]
	Glucuronoyl Esterase	Demethylates O_7-methyl glucuronoyl (GlcU) α(1→2) linked to backbone Xyl in glucuronoarabinoxylan	[62]
	α-L-Arabinofuranosidase,	Removal of Ara substituent	[63]

Table 4.2 (Cont.)

Biomass	Enzyme	Function	Reference
Hemicellulose	α-Galactosidase	Removal of Gal substituent linked via α-glycosidic bonds to galactomannan, pectin, or other hemicelluloses	[64]
	α-Glucuronidase	Removal of α(1→2) linked glucuronoyl or its methyl ester in xylan (often at the O_2 site) or other hemicelluloses	[65]
	Glucanase,	Degradation of β(1→3), (1→4), or (1→6) glucan, endo- or exo-acting, glycosidic bond type-specific	[66]
	Mannanase	Degradation of (galacto)(gluco)mannans, β(1→4)-D-mannosyl or manno/glucopyranosyl polymers with variable α(1→6) D-Gal side chain as well as O_2 and/or O_3 acetylation	[67]
	Xyloglucan Hydrolase	Degradation of xyloglucan, β(1→4) glucan with α(1→6) linked Xyl substituted by either α(1→2) L-Ara or β(1→2) D-Gal units (partially acetylated or substituted by α(1→2) L-fucopyranosyl (Fuc)),	[68]
	Pectinase	degradation of pectic polysaccharides, consisted of α(1→4) poly-α-(rhamno)galacturonic acids with variable backbone methylation/acetylation and Ara and Gal side chains branching	[69,70]

(Continued)

Table 4.2 (*Cont.*)

Biomass	Enzyme	Function	Reference
lignin	Lignin peroxidase	strong oxidants with high-redox potential that oxidize the major non-phenolic structures of lignin.	[71]
	Mn peroxidase	oxidizes a variety of organic compounds in the presence of Mn(II). It oxidizes Mn(II) to Mn(III) In the presence of H_2O_2	[72]
	Laccase	can directly oxidize phenolic parts of lignin, or indirectly oxidize non-phenolic lignin parts with the aid of suitable redox-active mediator	[73]
	versatile peroxidase	combine the catalytic activities of both MnP and LiP and are able to oxidize Mn(2+)	[74]
	Aryl-alcohol oxidase	an extracellular flavoprotein providing the H_2O_2 required by ligninolytic peroxidases for degradation of lignin	[75]
	glyoxal oxidase	extracellular oxidases able to use simple aldehyde, alpha-hydroxycarbonyl, or alpha-dicarbonyl compounds as substrates.	[76]
	Dye decolorizing peroxidase	oxidize high-redox potential anthraquinone dyes and were recently reported to oxidize lignin model compounds.	[77]

4.2.2.1   Acid Hydrolysis

It is one of the oldest methods used in the saccharification of lignocellulosic biomass in which mineral acids such as sulfuric acid, hydrochloric acid, hydrofluoric acid and nitric acid are widely employed for the hydrolysis of lignocellulosic biomass. Among these, the oldest and best understood process utilizes sulphuric acid. The sulphuric acid-based hydrolysis process is operated under two different conditions; (1) a process which uses concentrated acid and operates at a lower temperature and (2) a process that uses dilute acid and operates at a higher temperature. Lignocellullosic biomass can be hydrolyzed using sulphuric acids to produce xylose, arabinose, glucose, and acetic acid by cleavage of the β-1, 4 linkage of cellulose and hemicellulose components.

Some researchers used nitric acid and phosphoric acid at varying concentrations and temperature to hydrolyze the agriculture residue [78, 79]. The hydrolysis of lignocellulosic biomass using acid has several disadvantages due to the formation of various toxic compounds, such as furfural, hydroxylmethylfurfural, acetic acid, formic acid, levulinic acid etc. These compounds are potent inhibitors and negatively affect the fermentation process. It is necessary to remove these compounds before the fermentation process, which increases the production costs. To solve this problem, lime is used to neutralize these acids but it has a disadvantage. Lime treatment causes a significant loss of sugar in the gypsum. However, such processes could be replaced by an enzymatic hydrolysis process which is economical and environmental friendly.

4.2.2.2   Enzymatic Hydrolysis

Hydrolysis of lignocellulosic biomass by enzymes is an effective method. Enzymes are highly specific to the reactions that they catalyze, and do not form toxics as happens in dilute acid hydrolysis. [9, 80, 81]. Enzymes have the further advantage that they are naturally occurring compounds which are biodegradable. Enzyme based technology for bioethanol production has been used at a large scale which reduces the cost considerably. Thus, the large-scale application of ethanol production through enzymatic hydrolysis of lignocellulosic biomass is now beginning to appear economically advantageous. Enzymes known as cellulase catalyze the breakdown of cellulose into glucose for fermentation into ethanol. Cellulase hydrolyses the β-1–4 glucosidic bond in cellulose, leading to the formation of glucose and short cellodextrins. Cellulases are a group of three major enzymes, involved in the hydrolysis of cellulose namely, 1,4-β-Dglucancellobiohydrolyase (EC 3.2.1.91), 1,4-β-D glucanohydrolase (EC 3.2.1.3), and β-D glucosidase (EC 3.2.1.21). These enzymes are commonly referred to as exoglucanase, endoglucanase and cellobiase, respectively and work efficiently in synergistic manner

The degradation of crystalline cellulose is carried out mainly by exoglucanases which cleave β-1, 4-glycosidic bonds from chain ends. This enzyme has a tunnel-shaped closed active site which retains a single glucan chain and prevents it from re-adhering to the cellulose crystal [82, 83]. Endoglucanases cleave β-1, 4-glycosidic bonds internally only and appear to have cleft-shaped open active sites. Endoglucanases are typically active in the more soluble amorphous region of the cellulose crystal. There is a high degree of synergy seen between (exoglucanases) and endoglucanases, and it is this synergy that is required for the efficient hydrolysis of cellulose crystals. The products of exoglucanases and cellobiohydrolases are cellobiose and cellodextrans and are inhibitory to their activity. Thus, efficient cellulose hydrolysis requires the presence of β-glucosidases to cleave the final glycosidic bonds producing glucose. In this way, β-glucosidases plays a key role in the efficiency of an enzymatic lignocellulose-degrading system. For industrial biomass conversion targeting high feedstock loads, supplementing β-glucosidases to common microbial cellulolytic enzyme preparations can be imperative, because of high cellobiose level during the enzymatic conversion.

Several researchers reported the presence of cellulases in bacteria and fungi. Bacteria belonging to the genera of Clostridium, Cellulomonas, Bacillus, Thermomonospora, Ruminococcus, and Streptomyces are known to produce cellulase. Trichoderma ressei is known as the most commercial fungi for glucanase production. Species of genus Aspergillus, Schizophyllum and Penicillium are also reported for cellulose production [84–87]. Some newly isolated microorganisms showed high cellulase activity at extreme conditions (at high temperature pH and salt concentration). These enzymes have potential for industrial use in the hydrolysis of soluble and microcrystalline. A novel cellulase producing Paenibacillus campinasensis BL11 was isolated from black liquor Kraft pulping process. It is a themophilic, spore-forming bacterium which was found to grow at 60°C over a wide range of pH and produced multiple extracellular saccharide degrading enzymes including: a xylanase, two cellulases, a pectinase and a cyclodextrin glucano-transferase. More recently, a thermostable cellulase was extracted from a bacterium isolated Bacillus subtilis DR, isolated from a hot spring. This strain offers a potentially more valuable thermostable enzyme for the biorefining industry due to its extreme heat toler-ance [88]. Recently, a novel thermophilic, cellulolytic bacterium Brevibacillus sp. strain JXL was isolated from swine waste. It was found to use a broad spectrum of substrates such as crystalline cellu-lose, CMC, xylan, cellobiose, glucose and xylose [89]. Furthermore, a salt tolerant bacterium Bacillus agaradhaerens JAM-KU023 was isolated and showed an increased optimal thermostability [90]. The cultivation of thermophiles offers several advantages, it reduces the risk of contamination, reduces viscosity thus making mixing easier, and leads to a high degree of substrate solubility.

Many anaerobic microorganisms secrete biomass degrading mul-tienzyme complexes into their environment, called cellulosome. They are made of an array of multiple cellulose and hemicellulase enzymes, assembled by specific interaction protein scaffolding which allows the binding of the whole complex to crystalline cellu-lose via a nonspecific carbohydrate binding module and cohesions (allow binding of a wide variety of cellulolytic and hemicellulo-lytic enzymes). This spatial clustering of multiple lignocellulose degrading enzymes results in an increased synergy for the degra-dation of lignocellulosic biomass [91–93]. The cellulosome was first identified in 1983 from the anaerobic, thermophilic, spore-forming Clostridium thermocellum [68]. This enzyme complex has very high crystalline cellulose activity and can completely solubilize the cel-lulose. For example, a bifunctional endoglucanase/endoxylanase was isolated from Cellulomonas flavigena providing the potential to use it in industrial biofuel production processes. This enzyme was found to have cellulase and xylanase activity. Similarly, a multifunc-tional enzyme was found to be produced by Terendinibacter turnerae T7902. There are several other advantages to microorganisms that produce cellulosome specific characteristics, which give rise to effi-cient cellulose hydrolysis. The cellulosome eliminates the wasteful expenditure of energy on microorganisms continuously producing large amounts of free enzymes along with which, the products get diluted in the bulk solution. Researchers recognize the value of cel-lulosomes for their efficient hydrolysis of microcrystalline cellulose and have begun to focus research on creating designer cellulosomes for recombinant expression in industry. Murashima and colleagues created the first in vitro recombinant minicellulosomes [94].

It has been shown that recombinant cellulosomes can be transplanted to other industrially useful organisms, such as *S. cerevisiae* [95, 96] and *B. subtilis* [97]. The ability of cellulosomes to cluster activities may present unique capabilities, both in the synergistic breakdown of a substrate and in the targeted degradation of specific biomass components.

4.2.3   Detoxification

Many processes are used in the pre-treatment and hydrolysis of lignocellulosic biomass to produce a fermentable hydrolysate [9, 36, 98–100]. The overall goal of pre-treatment is to better expose cellulose for downstream hydrolysis, convert lignocellulosic biomass to a simple sugar and to remove lignin [9]. Physical and chemical pre-treatment and hydrolysis create a variety of toxic compounds that inhibit the fermentation performance. The generation of toxic inhibitory compounds is strongly dependent on the raw material, the hydrolysis method, and the hydrolysis conditions. Four groups of compounds that may act as inhibitors are phenolic compounds, aliphatic acids, and furan aldehydes.

A large number of phenolic compounds are formed from lignocellulosic biomass during dilute-acid hydrolysis [101, 102]. Common phenolic compounds found are vanillin, dihydroconiferyl alcohol, coniferyl aldehyde, vanillic acid, hydroquinone, catechol, acetoguaiacone, homovanillic acid, and 4-hydroxybenzoic acid [103]. Phenolic compounds interfere with the cell membrane, which will influence function and change the protein-to-lipid ratio which inhibit cell growth and ethanol productivity [104].

The common aliphatic acids in lignocellulose hydrolysates are acetic acid, formic acid, and levulinic acid. The most inhibitory of these three acids is formic acid followed by levulinic and acetic acid [105]. Aliphatic acids are formed from the degradation products of polysaccharides during the acid hydrolysis of lignocellulosic biomass. These aliphatic acids enter the cell through diffusion over the cell membrane and leads to a decrease in the intracellular pH. Maintaining constant intracellular pH by pumping out protons through the plasma membrane requires additional ATP, which negatively affects metabolic activity and may lead to cell death [106].

Synergistic effect occurs when these compounds are combined with several other compounds formed during lignin degradation.

Extractives (acidic resins, taninic, and terpene acids) and also acetic acid, derived from acetyl groups present in the hemicellulose, are released during the hydrolytic processes. In terms of toxicity, the extractives are considered less toxic to microbial growth than acetic acid. Other types of inhibitors are heavy metal ions (iron, chromium, nickel and copper) which come from reactor corrosion during the acidic hydrolysis pretreatment. Their toxicity acts at metabolic pathways level, by inhibiting enzyme activity [107].

Toxic compounds generated during the pre-treatment and hydrolysis process negatively affect the fermentation process. Several alternatives can be used to avoid the problems created by inhibitors. We can remove the inhibitors or reduce their formation during the process. The type and concentration of inhibitors in the sugar hydrolysates depend on the hydrolysis condition.

However, high sugar yield without inhibitor formation can be hard to achieve, especially if dilute acid hydrolysis is used. It is hard to accept a poor sugar and ethanol yield just because of avoidable inhibition problems. To reduce the negative effects of the inhibitors, several, physical chemical and biological detoxification methods were developed. The efficiency of detoxification methodology depends on the chemical composition of the hydrolysate, as well as on the microorganisms chosen for bioethanol production [30, 107, 108]. For this reason, detoxification methods cannot be directly compared since the mechanisms of inhibition and the degree of toxicity removal are completely different [98]. Significant efforts were taken to minimize the production of such highly toxic compounds. Vapour and vacuum evaporation are physical detoxification methods performed to reduce the concentration of volatile compounds (such as acetic acid, furfural and formaldehyde) present in the hydrolysate and to increase the concentration of sugar. However, these methods also increase the non-volatile toxic compounds content, such as extractives and lignin derivatives.

Chemical detoxification includes the precipitation of toxic compounds and the ionization of some inhibitors under certain pH values, the latter being able to change the degree of toxicity of the compounds [109]. Toxic compounds may also be adsorbed by activated charcoal [110, 111], and by ion exchange resins [103, 112]. Alkali treatment is the most common detoxification method and is considered one of the best technologies. This method consists of the addition of lime or other alkali compounds such as sodium or potassium hydroxide for the precipitation of toxic compounds [110–114]. Acetic acid, furfural, HMF, soluble lignin and phenolic compounds are mostly removed with the alkali treatment which increases the fermentability of hydrolysates. Treatment with the peroxidase and laccase enzymes, obtained from the ligninolytic fungus Trametes versicolor, has been shown to increase maximum ethanol productivity in a hemicellulose hydrolysate [115]. The filamentous soft-rot fungus Trichoderma reesei has been reported to degrade inhibitors in a hemicellulose hydrolysate obtained after steam pretreatment. Acetic acid, furfural, and benzoic acid derivatives were removed from the hydrolysate by treatment with Trichoderma reesei [116]. Exocellulase cleaves two to four units from the ends of the exposed chains produced by endocellulase, while β-glucosidase hydrolyses the exocellulase product into individual monosaccharide. In such a way, no degradation of glucose occurs by enzymetic hydrolysis.

4.2.4   Fermentation

This is the chemical transformation of an organic substance into a simpler compound by enzymes. Originally, the term fermentation was used to mean the enzymatic breakdown of carbohydrates in the absence of oxygen. In industrial practice, fermentation refers to any process by which raw materials are transformed by the controlled action of carefully selected strains of organisms into definite products. However, for the case of this chapter, it is a biological method of producing ethanol from lignocellulosic biomass. The fermentation reaction takes place by yeast or bacteria which feed on simple sugars. The glucose produced from the hydrolysis described above is fermented with yeast to produce ethanol.

4.2.4.1   Separate Hydrolysis and Fermentation (SHF)

This is a conventional two-step process where the lignocellulose is hydrolyzed using enzymes to reduce sugars in the first step and to ferment the sugars formed into ethanol in the second step using Saccharomyces or Zymomonas [117, 118]. In the SHF process, hydrolysis is performed separately from fermentation, which means that the optimal temperatures for both the enzymatic hydrolysis and fermentation can be applied. A drawback with SHF is that the generated cellobiose functions as a cellulase inhibitor [119]. It has also been proven that β - glucosidase can be inhibited by glucose [120]. Another drawback is that SHF is a two-step process which increases the overall cost of the process.

4.2.4.2   Simultaneous Saccharification and Fermentation (SSF)

In simultaneous saccharification and fermentation (SSF) processes, both cellulose hydrolysis and the fermentation of glucose are carried out in the presence of fermentative microorganisms in a single step. This technique reduces the number of steps in the process, and is a promising way to convert cellulose into ethanol [121]. In the SSF process, the cellulase enzymes hydrolyze cellulose to D-glucose, which ,in turn, is fermented into ethanol by yeast or bacteria [122]. The compatibility of both the saccharification and fermentation processes, with respect to various conditions, such as pH, temperature and substrate concentration, are important factors that govern the success of the SSF process. These enzymes are strongly inhibited by the simple sugars generated in hydrolysis by the enzymes themselves. The accumulation of ethanol in the fermentor does not inhibit cellulases as much as high concentrations of glucose [123]. The main advantages of using SSF for ethanol bioconversion are the enhanced rate of bioconversion of lignocellulosic biomass (cellulose and hemicellulose) due to the removal of sugars that inhibit cellulase activity, lower enzyme loading, higher product yield, reduced inhibition of the yeast fermentation in case of continuous recovery of the ethanol and the reduced requirement for aseptic conditions, resulting in improved economic conditions [124–127].

Because several inhibitory compounds are formed during the hydrolysis of the raw material, the hydrolytic process has to be optimized so that inhibitor formation can be minimized. SSF seems to offer a better option for commercial production of ethanol from lignocellulosic biomass. Penicillium funiculosum cellulase and Saccharomyces uvarum cells have reportedly been used for SSF [124].

The disadvantage of the SSF process is that the optimal temperatures for the cellulases and the fermenting microorganisms are not the same so the selected temperature is a compromise, which means that neither hydrolysis nor fermentation will be performed under optimal conditions.

4.2.4.3   Simultaneous Saccharification and Co-fermentation (SSCF)

The Simultaneous saccharification and co-fermentation processes combine the enzymatic hydrolysis of cellulose with the simultaneous fermentation of its main derived sugar (glucose) into ethanol. In this process, the stages are virtually the same as in the separate hydrolysis and fermentation system, except that both are performed in the same bioreactor. Thus, the presence of yeast together with the cellulolytic enzyme complex reduces the accumulation of the inhibiting sugars within the reactor, thereby increasing the yield and the saccharification rates. Another advantage of this approach is that a single bioreactor is used for the entire process, therefore reducing the investment costs. SSCF has been recognized as a feasible option for ethanol production from xylose-rich lignocellulosic materials. Recently SSCF was used by several researchers with genetically engineered yeast on different raw materials [128 -130].

4.2.4.4   Consolidated Bio-Processing (CBP)

Recently, efforts have been made to combine cellulase production, hydrolysis and fermentation into one single step. This concept is called Consolidated Bio-Processing (CBP) and the aim is to create a microorganism that is able to perform these three steps simultaneously [121]. In CBP, pretreated lignocellulosic biomass is neutralized and directly exposed to microorganisms that are capable of hydrolyzing cellulose and hemicelluloses into fermentable sugars and ferment sugars into ethanol. At present, there is no ideal CBP microorganism that can degrade lignocellulosic biomass proficiently and at the same time consume all the sugars released from biomass to yield ethanol. However, efforts are being made to engineer ethanol producing yeast (S. cerevisiae) capable of producing cellulases/hemicellulases and lignocellulose degrading bacteria (C. thermocellum) for efficient bioethanol production [121]. There are two different strategies to create a CBP microorganism. A naturally occurring cellulolytic microorganism can be modified by genetic engineering to gain important properties, such as the ability to give high ethanol yields, or alternatively a non-cellulolytic microorganism that gives high ethanol yields can be altered by genetic engineering to express heterologous cellulases [131]. The efforts refered to above via biotechnological engineering of yeast genome and protoplast fusion techniques have developed strains that show promising results [131]. The genome of C. phytofermentans (ATCC 700394) encodes for the highest number of enzymes for the degradation of lignocellulosic biomass among sequenced clostridial genomes [132]. This technology is superior to SSF technology in terms of cost effectiveness, producing higher yields, and having a shorter processing time [121, 127, 133]. The relatively lower tolerance of the ethanol is the main disadvantage of this process. A lower tolerance limit of about 3.5% has been reported as compared to 10% for ethanologenic yeasts. Acetic acid and lactic acid are also formed as by-products in this process in which a significant amount of carbon is utilized.

4.2.5 Distillation

After fermentation, a large amount of water is present in the bioreactor. To produce usable ethanol, the excess water contents from the liquid wash must be removed through a distillation process. Distillation is a commonly used method for purifying liquids and separating mixtures of liquids into their individual components. It is a method to separate two liquids utilizing their different boiling points. To separate a mixture of liquids which have different boiling points, they are heated to force the components into gaseous form and then condensed back into liquid form and collected. However, to achieve high purification, several distillations are required. This is because all materials have intermolecular interactions with each other, and the two materials will co-distill during distillation. Since ethanol evaporates faster than water, the ethanol rises through a tube, collects and condenses into another container and water is left behind. The final purity of the ethanol product is limited to 95–96% due to the formation of a low-boiling water-ethanol azeotrope. This mixture is called hydrous ethanol [134]. Many ethanol distillation approaches have been developed to improve system efficiencies [135].

4.3 Genetic Engineering for Bioethanol Production

Lignocellulosic biomass based ethanol production involves mixed sugar (pentose and hexose) fermentation. Hydolysate of lignocelluloseic biomass contains various toxic compounds such as low-molecular-weight organic acids, furan derivatives, phenolics, and inorganics released during the pretreatment or hydrolysis of biomass [136]. These toxic compounds negatively affect the fermentation process. There is a need for a suitable microorganism which can efficiently ferment both hexose and pentose without being affected by inhibiters. Saccharomyces cerevisiae is the most favored organism for ethanol production from hexoses. P. stipitis and Candida shehatae are capable of fermenting both hexose (glucose) and pentose (xylose) sugars to ethanol [137]. Strict anaerobic themophilic bacteria such as Clostridium sp. and Thermoanaerobacter sp. have been proposed [30, 37] to explore the benefits of fermentation at elevated temperatures. Some other thermo-tolerant microorganisms developed are K. marxianus, Candida lusitanieae and Z. mobilis [23].

For a commercially viable ethanol production method, an ideal microorganism should have broad substrate utilization, the ability to withstand high concentrations of ethanol, a high temperature tolerance, cellulolytic activity and a tolerance to inhibitors present in hydrolysate. Many research efforts have been devoted to the development of efficient microorganisms that are capable of fermenting glucose and xylose separately or simultaneously [136]. Industrial utilization of lignocelluloses for bioethanol production is hindered by the lack of ideal microorganisms which can efficiently ferment both pentose and hexose sugars [37].

Genetically modified or engineered microorganisms are thus used to achieve the complete utilization of the sugars in hydrolysate and better production benefits. In recent years, metabolic engineering concepts have been used for the production of fuel ethanol [138]. Genetic engineering has been employed to develop the various aspects of fermentation from higher yields to a better and wider substrate utilization to increased recovery rates. A number of genetically modified microorganisms such as P. stipitis BCC15191 [139], P. stipitis NRRLY-7124 [140, 141]), recombinant E. coli KO11 [142], C. shehatae NCL-3501 [143] and S. cerevisiae ATCC 26603 [140] have been developed.

Ethanol fermentation of pentose sugars is not as frequently performed by most of the microorganisms as that of hexose sugars. A limited number of bacteria, yeasts, and fungi can convert pentose sugar into ethanol with a satisfactory yield and productivity [144]. Genetically engineered strains of E.coli, S. cerevisiae, and Z. mobilis have been developed to ferment xylose [145]. Xylose fermentation has also been achieved by expressing xylose reductase and xylitol dehydrogenase together with the overexpression of the endogenous xylulokinase and xylose isomerase in P. stipitis and other bacterial or fungal species [146–148]. Ethanologenic recombinant bacteria such as Escherichia coli, Klebsiella oxytoca, and Zymomonas mobilis and the yeast Saccharomyces cerevisiae were successfully used in the fermentation of mixed sugars obtained from biomass containing glucose, xylose, arabinose, and galactose to produce ethanol [149]. The genetically engineered microbes Zymomonas mobilis AX101, Escherichia coli KO11, Klebsiella oxytoca P2, and

Saccharomyces cerevisiae are considered for commercial scale-up [150]. The recombinant yeast strain Saccharomyces cerevisiae MT8–1 was found to ferment xylose and cello-oligosaccharides by integrating genes from Pichia stipitis that express xylose reductase and xylitol dehydrogenase, and xylulokinase from Saccharomyces cerevisiae and a gene for displaying β-glucosidase from Aspergillus acleatus on the cell surface [151]. Recombinant Saccharomyces cerevisiae strains were used to utilize the hemicellulose components of lignocellulosic feedstocks. The hybrid strain MN8140XX showed a 1.3 and 1.9 fold improvement in ethanol production compared to its parent strains [152]. One other strain was constructed which expressed heterologous enzymes on the cell surface and produced ethanol from rice straw hydrolysate in a consolidating bioprocessing manner from hemicellulosic material containing xylan, xylo-oligosaccharides, and cellulo-oligosaccharides without requiring the addition of sugar hydrolyzing enzymes or detoxication [153]. Genetically engineered yeast strains were constructed by DNA shuffling with the recombination of the entire genome of P. stipitis with that of S. cerevisiae by using a modified genome shuffling method to improve fermentation [154]. The recombinant yeast strain obtained in this study was a promising candidate for industrial cellulosic ethanol production. The ethanol fermenting genes such as pyruvate decarboxylase and alcohol dehydrogenase II were cloned from Zymomonas mobilis and transformed into three different cellulolytic bacteria, namely Enterobacter cloacae JV, Proteus mirabilis JV and Erwinia chrysanthemi. After that, their cellulosic ethanol production capability was studied. It was found that the strain produced two times more ethanol than the wild type and also had an ethanol tolerance capability [155].

4.4   Future Perspective

Due to the increased demand for energy, the bioconversion of lingo-cellulosic biomass to bio-ethanol is a priority area of research. Ligno-cellulosic ethanol is produced from inexpensive and widely available raw materials that can reduce our dependence on fossil fuels. Bioethanol contributes to a clean environment because CO_2 emissions from bioethanol combustion are equal to the amount that plants absorb from the atmosphere during photosynthesis. Therefore, they can reduce the global warming problem. Bioethanol is a renewable and sustainable alternative to fossil fuel because raw material for its production is abundantly available and can be easily grown. The processing of agricultural residues will provide economic benefits to rural people. The bioconversion of lignocellulosic biomass into simple sugars for fermentation is a very complex process. Lignocelluloses are very complex in terms of their morphology and crystalinity is a major hurdle for bioconversion. The bioconversion of lignocellulosic biomass requires many enzymes that work synergistically for the production of simple sugars in order to degrade lignin. There is no known microorganism that has all of the enzymes required for the complete process. Many enzymes which can degrade lignocellulosic biomass and have end product inhibition, cannot utilize biomass completely. The co-cultivation of microorganisms can solve the problem of the bioconversion of complex lignocellulossic biomasses. Microorganisms from extreme environmental conditions may provide enzymes with improved properties suitable for industrial application. The combination of several approaches like metagenomics, biodiversity studies and metabolic engineering have the potential to improve the properties of enzymes. Hydrolysis of lignocelluloses produces a complex mixture of sugar and many other toxic compounds, which inhibit the activity of fermenting microorganisms. The approach of genetic engineering can be used for the development of fermentative microorganisms that are capable of fermenting pentose and hexose without being affected by inhibitors. The development of genetically engineered microorganisms and the designing of novel bioreactors with improved yields and productivity will reduce the cost of bioethanol production in future.

References

1. P. Raven, R. Evert, S. Eichhorn, Biology of plants, New York, NY, USA, 5th edition, p. 791, 1992.
2. J.L. Faulon, and P. G. Hatcher, Energy and Fuels, Vol. 8, p. 402–407, 1994.
3. A.J.Stamm, Wood and Cellulose Science, The Ronald Press Co, New York, p. 142–165, 1964.
4. V. Balan, L.D.C. Sousa, S.P.S. Chundawat, D. Marshall, L.N. Sharma, C.K. Chambliss, and B.E. Dale, Biotechnology Process, Vol. 25, p. 365–375, 2009.
5. R.L. Howard, E. Abotsi, E.L. Jansen van Rensburg, and S. Howard, African Journal of Biotechnology, Vol. 2 (12), p. 602, 2003.
6. B.C. Qi, C. Aldrich, L. Lorenzen, and G.W. Wolfaardt, Chemical Engineering Communications, Vol. 192(9), p. 1221–1242. 2005.

7. A. Roig, M.L. Cayuela, and M.A. Sanchez-Monedero, Waste Management, Vol. 26(9), p. 960–969, 2006.
8. G. Rodriguez, A. Lama, R. Rodriguez, A. Jimenez, R. Guillena, and J. Fernandez-Bolanos, Bioresource Technology, Vol. 99(13), p. 5261–5269. 2008.
9. Y. Sun, and J. Cheng, Bioresource Technology, Vol. 83, p. 1–11, 2002.
10. G, Mtui, and Y. Nakamura, Tanzania Biodegradation, Vol. 16(6), p. 493–499, 2005.
11. L.R. Lynd, M.S. Laser, D. Bransby, B.E. Dale, B. Davison, R. Hamilton, M. Himmel, M. Keller, J.D. McMillan, J. Sheehan, and C.E. Wyman, *Nature Biotechnology*, Vol. 26, p. 169–172, 2008.
12. R. Kumar, S. Singh, and O.V. Singh, *Journal of Indian Microbiology and Biotechnology*, Vol. 35, p. 377–391, 2008.
13. L.P. Wackett, *Current Opinion in Chemical Biology*, Vol. 12, p. 187–193, 2008.
14. A. Margeot, B. Hahn-Hagerdal, M. Edlund, R. Slade, and F. Monot, *Current Opinion in Biotechnology*, Vol. 20, p. 372–380, 2009.
15. M. Dashtban, H. Schraft, and W. Qin, *International Journal of Biological Science*, Vol. 5, p. 578–595, 2009.
16. C. Sanchez, *Biotechnology Advance*, Vol. 27, p. 185–194, 2009.
17. F. Xu, "Biomass-converting enzymes and their bioenergy applications", In *The Manual of Industrial Microbiology and Biotechnology*, 3rd ed. Washington, DC, USA, p. 495–508, 2010.
18. F.M. Girio, C. Fonseca, F. Carvalheiro, L.C. Duarte, and S. Marques, *Bioresource Technology*, Vol. 101, p. 4775–4800, 2010.
19. R.E.H. Sims, W. Mabee, J.N. Saddler, and M. Taylor, *Bioresource Technology*, Vol. 101, p. 1570–1580, 2010.
20. A.K. Chandel, and O.V. Singh, *Applied Microbiology and Biotechnology*, Vol. 89, p. 1289–1303, 2011.
21. Y.H.P. Zhang, *Process Biochemistry*, Vol. 46, p. 2091–2110, 2011.
22. C.A. Cardona, and O.J. Sanchez, Bioresource Technology, Vol. 98(12), p. 2415–2457, 2007.
23. A.B. Bjerre, A.B. Olesen, and T. Fernqvist, Biotechnology and Bioengineering, Vol. 49, p. 568–77, 1996.
24. A. Pandey, Handbook of plant-based biofuels, New York, CRC Press, p.1–3, 2009.
25. D.S. Scott, L. Paterson, J. Piskorz and D. Radlein, Journal of Analytical and Applied Pyrolysis, Vol. 57, p. 169–176, 2001.
26. S. Prasad, A. Singh, and H.C. Joshi, Resources, Conservation and Recycling, Vol. 50, p.1–39, 2007.
27. G.Y.S. Mtui, African Journal of Biotechnology, Vol. 8(8), p. 1398–415, 2009.
28. E. Takacs, L. Wojnarovits, C. Foldvary, P. Hargittai, J. Borsa, and I. Sajo, Radiation Physics and Chemistry, Vol. 57, p. 399–403, 2000.
29. M.A. Neves, T. Kimura, N. Shimizu, and M. Global Science Books, p.1–13, 2007
30. O.J. Sanchez, and C.A. Cardona, Bioresource Technology, Vol.99, p. 5270–95, 2008.
31. M. Balat, H. Balat, and C. Oz, Progress in Energy and Combustion Science, Vol.34, p.551–73, 2008.
32. R. Vanholme, K. Morreel, J. Ralph, and W. Boerjan, Current Opinion in Plant Biotechnology, Vol. 11 (3), p. 278–285, 2008.

33. M. T. Holtzapple, A. E. Humphrey, and J. D. Taylor, Biotechnology and Bioengineering, Vol. 33, p. 207–210, 1989.
34. K. L. Mackie, H. H. Brownell, K. L. West, and J. N. Saddler, Journal of Wood Chemistry and Technology, Vol. 5, p. 405–425, 1985.
35. J. D. McMillan, "Pretreatment of lignocellulosic biomass", In Enzymatic ConVersion of Biomass for Fuels Production, Washington, DC, p. 292–324, 1994.
36. N. Mosier, C. Wyman, B. Dale, R. Elander, Y.Y. Lee, M. Holtazapple, Bioresource Technology ,Vol. 96, p. 673–86, 2005.
37. F. Talebnia, D. Karakashev, and I. Angelidaki, Bioresource Technology, Vol. 101(13), p. 4744–53, 2010.
38. C.N. Hamelinck, G.V. Hooijdonk, and A.P.C. Faaij, Biomass and Bioenergy, Vol. 28, p. 384–410, 2005.
39. D. Ben-Ghedalia, and J. Miron, Biotechnology and Bioengineering, Vol. 23, p. 823–831, 1981.
40. W. C. Neely, Biotechnology and Bioengineering, Vol. 26, p. 59–65, 1984.
41. D. Ben-Ghedalia, and G. Shefet, Journal of Agricultural Science, Vol. 100, p. 393–400, 1983.
42. P. F. Vidal, and J. Molinier, Biomass, Vol. 16, p. 1–17, 1988.
43. J. Quesada, M. Rubio, and D. Gomez, Journal of Wood Chemical Technology, Vol. 19, p. 115–137, 1999.
44. C. I. Ishizawa, M. F. Davis, D. F. Schell, and D. K. Hohnson, *Journal of Agricultural Food Chemistry*, Vol. 55, p. 2575–2581, 2007.
45. C.A. Cardona, J.A. Quintero, and I.C. Paz, Bioresource Technology, Vol. 101(13), p. 4754–66, 2009.
46. M. V. Sivers, and G. Zacchi, *Bioresource Technology*, Vol. 51, p. 43–52, 1995.
47. A. Pandey, C.R. Soccol, P. Nigam, and V.T. Soccol, Bioresource Technology, Vol. 74, p. 69–80, 2000.
48. L. T. Fan, M. M. Gharpuray, and Y.H. Lee, Cellulose Hydrolysis, Biotechnology Monographs, Berlin, Vol. 3, p. 57, 1987.
49. J. I. Botello, M. A. Gilarranz, F. Rodriguez, and M. Oliet, Journal of Chemical Technology and Biotechnology, Vol. 74, p. 141–148, 1999.
50. X. Zhao, K. Cheng, and D. Liu, Applied Microbiology and Biotechnology, Vol. 82, p. 815–27, 2009.
51. H.V. Scheller, and P. Ulvskov, *Annual Review of Plant Biotechnology*, Vol. 61, p. 263–289, 2010.
52. J. Van den Brink, and R.P. de Vries, *Applied Microbial Biotechnology*, Vol. 91, p. 1477–1492, 2011.
53. J. Wu, Y.Z. Xiao, and H.Q. Yu, Bioresource Technology , Vol. 96, p.1357–1363, 2005.
54. D.J. Vocadlo, and G.J. Davies, Current Opinion in Chemical Biology, Vol. 12, p. 539–555, 2008.
55. Y.S. Liu, J.O. Baker, Y. Zeng, M.E. Himmel, T. Haas, and S.Y. Ding, Journal of Biological Chemistry, Vol. 286, p. 11195–11201, 2011.
56. D.B. Wilson, Annals of the New York Academy of Sciences, Vol. 1125, p. 289– 297, 2008.
57. J. Langston, N. Sheehy, and F. Xu, Biochimica et Biophysica Acta, Vol. 1764, p. 972–978, 2006.
58. B.B. Ustinov, A.V. Gusakov, A.I. Antonov, and A.P. Sinitsyn, Enzyme Microbial Technology, Vol. 43, p. 56–65, 2008.

59. D.B. Jordan, and K. Wagschal, Applied Microbiology and Biotechnology, Vol. 86, p. 1647–1658, 2010.
60. P. Biely, M. Mastihubova, M. Tenkanen, J. Eyzaguirre, X.L. Li, and M. Vrsanska, Journal of Biotechnology, Vol. 151, p. 137–142, 2011.
61. T. Koseki, S. Fushinobu, S.H. Ardiansyah, and M. Komai, Applied Microbiology and Biotechnology, Vol. 84, p. 803–810, 2009.
62. M. Duranova, S. Spanikova, H.A.Wosten, P. Biely, and R.P. de Vries, Archives of Microbiology, Vol. 191, p. 133–140, 2009.
63. B.C. Saha, Biotechnology Advances, Vol.18, p. 403–423, 2000.
64. V.L.Y. Yip, and S.G. Withers, Journal of Biocatalysis and Biotransformation, Vol. 24, p. 167–176, 2006.
65. S.L. Chong, E. Battaglia, P.M. Coutinho, B. Henrissat, M. Tenkanen, and R.P. de Vries, Applied Microbiology and Biotechnology, Vol. 90, p. 1323–1332, 2011.
66. K. Martin, B.M. McDougall, S. McIlroy, J.Z. Chen, and R.J. Seviour, FEMS Microbiology Reviews, Vol. 31, p. 168–192, 2007.
67. L. Moreira, and E. Filho, Applied Microbiology and Biotechnology, Vol. 79, p. 165–178, 2008.
68. M.J. Baumann, Plant Cell, Vol. 19, p. 1947–1963, 2007.
69. V. Lombard, T. Bernard, C. Rancurel, H. Brumer, P.M. Coutinho, and B. Henrissat, Biochemical Journal, Vol. 432, p. 437–444, 2010.
70. A. Payasi, R. Sanwal, and G.G. Sanwal, World Journal of Microbiology and Biotechnology, Vol. 25, p. 1–4, 2009.
71. M. Yadav , S.K. Singh , and S. Yadava, Prikl Biokhim Mikrobiol, Vol. 48(6), p. 646–52, 2012.
72. L. Banci , S. Ciofi-Baffoni , and M, Tien, Biological Science, Vol. 38 (10), p. 3205–3210, 1999.
73. P. M. Coll, J. M. Fernandez-Abalos, J. R. Villanueva, R. Santamaria, and P. Perez, Applied Environmental Microbiology, Vol. 59(8), p. 2607, 1993.
74. M.E. Brown, T. Barros, and M.C. Chang, ACS Chemical Biology, Vol. 7(12), p. 2074–81, 2012.
75. A. Hernandez-Ortega, P. Ferreira, and A.T. Martinez, Applied Microbiology and Biotechnology, Vol. 93(4), p. 1395–410, 2012.
76. P. J. Kersten, and D. Cullen, PNAS, Vol. 90 (15), p. 7411–7413, 1993.
77. M. Ahmad , J.N. Roberts , E.M. Hardiman , R. Singh , L.D. Eltis , and T. D. Bugg, Biochemistry, Vol. 50(23), p. 5096–107, 2011.
78. A. Rodriguez-Chong, J. A. Ramirez, G. Garrote, and M. Vazquez, Journal of Food Engineering , Vol. 61, p. 143, 2004.
79. S. Gamez, J.J. Gonzalez-Cabriales, J.A. Ramirez, G. Garrote, and M. Vazquez, Journal of Food Engineering, Vol. 74, p. 78–88, 2006.
80. S. Banerjee, S. Mudliar, R. Sen, B. Giri, D. Satpute, and T. Chakrabarti, Biofuels, Bioproducts and Biorefining ,Vol. 4, p. 77–93, 2010.
81. M.J. Taherzadeh, and K. Karimi, Bioresources,Vol. 2(4), p. 707–38, 2007.
82. D.B. Wilson, Microbiology, Vol. 14, p. 259–263, 2011.
83. M. Maki, K.T. Leung, and W. Qin, International Journal of Biological Science, Vol. 5, p. 500–516, 2009.
84. V.S. Bisaria, Bioprocessing of agro-residues to value added products, Chapman and Hall, UK, p.197–246, 1998.

85. S.J.B. Duff, and W.D. Murray, Bioresource Technology, Vol.55, p.1- 33, 1996.
86. J. Kaur, B.S. Chandha, and B.A. Kumar, Electronic Journal of Biotechnology, Vol. 10, p. 260–270, 2007.
87. D. Sternberg, Biotechnology and Bioengineering Symposium, p. 35–53, 1976.
88. W. Li, W.W. Zhang, M.M. Yang, and Y.L. Chen, Molecular Biotechnology, Vol. 2, p. 195–201, 2008.
89. Y. Liang, J. Yesuf, S. Schmitt, K. Bender, and J. Bozzola, Journal of Indian Microbiology and Biotechnology, Vol. 36(7), p. 961–70, 2009.
90. K. Hirasawa, K. Uchimura, M. Kashiwa, W.D. Grant, S. Ito, T. Kobayashi, and K. Horikoshi, Antonie Van Leeuwenhoek.;Vol. 89(2), p. 211–219, 2006.
91. S. Morais, Y. Barak, Y. Hadar, D.B. Wilson, Y. Shoham, R. Lamed, and E.A. Bayer, MBio, Vol. 15;2(6), p. 00233–11, 2011.
92. S.Y. Ding, Q. Xu, M. Crowley, Y. Zeng, M. Nimlos, R. Lamed, E.A. Bayer, and M.E. Himmel, Current Opinion in Biotechnology, Vol. 19, p. 218–227, 2008.
93. E.A. Bayer, R. Lamed, and M.E. Himmel, Current Opinion in Biotechnology, Vol. 18, p. 237–245, 2007.
94. K. Murashima, A. Kosugi, and R.H. Doi, Molecular Microbiology,Vol. 45(3), p. 617–626, 2002.
95. S.L. Tsai, and G. Goyal, Applied Environmental Microbiology, Vol. 76, p. 7514– 7520, 2010.
96. M. Lilly, H.P. Fierobe, W.H. van Zyl, and H. Volschenk, FEMS Yeast Research, Vol. 9, p. 1236–1249, 2009.
97. T.D. Anderson, S.A. Robson, X.W. Jiang, G.R. Malmirchegini, H.P. Fierobe, B.A. Lazazzera, and R.T. Clubb, Applied Environmental Microbiology, Vol. 77, p. 4849–4858, 2011.
98. E. Palmqvist, and B. Hahn-Hagerdal, Bioresource Technology, Vol. 74, p. 17–24, 2000.
99. R. Kumar, S. Singh, and O.V. Singh, Journal of Indian Microbiology and Biotechnology, Vol. 35, p. 377–391, 2008.
100. F. Niehaus, C. Bertoldo, M. Kahler, and G. Antranikian, Applied Microbiology and Biotechnology, Vol. 51, p. 711–729, 1999.
101. T.A. Clark, and K.L. Mackie, Journal of Chemical Technology and Biotechnology, Vol. 34, p. 101–110, 1984.
102. T. Popoff, and O. Theander, Acta , Vol. 30, p. 397–402, 1976.
103. S. Larsson, A. Reimann, N.O. Nilvebrant, and L.J. Jonsson, Applied Biochemistry and Biotechnology, Vol. 77, p. 91–103, 1999.
104. S. Larsson, A. Quintana-Sainz, A. Reimann, N.O. Nilvebrant, and L.J. Jonsson, Applied Biochemistry and Biotechnology, Vol. 84, p. 617–632, 2000.
105. J. Zaldivar, and L.O. Ingram, Biotechnology and Bioengineering, Vol. 66, p. 203– 210, 1999.
106. C. Verduyn, E. Postma, W.A. Scheffers, and J.P. Van Dijken, Journal of General Microbiology, Vol. 136, p. 305–319, 1990.
107. S.I. Mussatto, J.C. Santos, and I.C. Roberto, Journal of Chemical Technology and Biotechnoloyg, Vol. 79, p. 590–596, 2004.
108. S. Helle, D. Cameron, J. Lam, B. White, and S. Duff, Enzyme Microbiology and Technology, Vol. 33, p. 786–792, 2003.
109. S. I. Mussatto, Influencia do Tratamento do Hidrolisado Hemicelulosico de Palha de Arroz na Producao de Xilitol por Candida guilliermondii, Brasil, p. 1–9, 2002.

110. J. M. Dominguez, C. S. Gong, and G. T. Tsao, *Applied Biochemistry and Biotechnology*, Vol. 57–58, p. 49–56, 1996.
111. S. I. Mussatto, and I. C. Roberto, *Biotechnology Letters*, Vol. 23, p. 1681–1684, 2001.
112. N. O. Nilvebrant, A. Reimann, S. Larsson, and L. J. Jonsson, *Applied Biochemistry and Biotechnology*, Vol. 91–93, p. 35–49, 2001.
113. W. Lee, J. Lee, C. Shin, S. Park, H. Chang, and Y. Chang, *Applied Biochemistry and Biotechnology*, Vol. 78(1), p. 547–559, 1999.
114. L. Canilha, J. B. de Almeida e Silva, and A. I. N. Solenzal, *Process Biochemistry*, Vol. 39(12), p. 1909–1912, 2004.
115. L. J. Jonsson, E. Palmqvist, N. O. Nilvebrant, and B. Hahn-Hagerdal, *Applied Microbiology and Biotechnology*, Vol. 49, p. 691–697, 1998.
116. E. Palmqvist, B. Hahn-Hagerdal, Z. Szengyel, G. Zacchi, and K. Reczey, *Enzyme Microbial Technology*, Vol. 20, p. 286–293, 1997.
117. V. S. Bisaria, and T. K. Ghose, *Enzyme Microbial Technology*, Vol. 3, p. 90–104, 1981.
118. G. P. Philippidis, "Cellulose bioconversion technology". *Handbook on Bioethanol: Production and Utilization*, Ed. C. E. Wyman, p. 253–285, 1996.
119. M. Mandels, and E.T. Reese, "Inhibition of cellulases and J-glucosidases". *In: Advances in enzymatic hydrolysis of cellulose and related materials*, London, p. 115–157, 1963.
120. M. Holtzapple, M. Cognata, Y. Shu, and C. Hendrickson, *Biotechnology and Bioengineering*, Vol. 36, p. 275–287, 1990.
121. L.R. Lynd, W.H. Van Zyl, J.E. McBride, and M. Laser, *Current Opinion in Biotechnology*, Vol.16, p. 577–583, 2005.
122. S.H. Krishna, T.J. Reddy, and G.V. Chowdary, *Bioresource Technology*, Vol. 77, p. 193–196, 2001.
123. D. R. Shonnard, *CM4710 Biochem. Processes, Course*, Michigan Technology University, Michigan, p. 1–13, 2003.
124. V. V. Deshpande, H. Sivaraman, and M. Rao, *Biotechnology and Bioengineering*, Vol. 25, p. 1679–1684, 1981.
125. D. J. Schell, N. D. M. Hinman, C. E. Wyman, and P. J. Werdene, *Applied Biochemistry and Biotechnology*, Vol. 17, p. 279–291, 1988.
126. G. P. Philippidis, and T. K. Smith, *Applied Biochemistry and Biotechnology*, Vol. 51/52, p. 117–124, 1995.
127. J. D. Wright, C. E. Wyman, and K. Grohmann, *Applied Biochemistry and Biotechnology*, Vol. 18, p. 75–90, 1988.
128. K. Olofsson, A. Rudolf, and G. Liden, *Journal of Biotechnology*, Vol. 134, p. 112– 120, 2008.
129. K. Ohgren, O. Bengtsson, M.F. Gorwa-Grauslund, M. Galbe, B. Hahn-Hagerdal, and G. Zacchi, *Journal of Biotechnology* , Vol. 126, p. 488–498, 2006.
130. K. Olofsson, M. Wiman, and G. Liden, *Journal of Biotechnology*, Vol. 145, p. 168–175, 2010.
131. A.L. Demain, M. Newcomb, and J.H.D. Wu, *Microbiology and Molecular Biology Reviews*, Vol. 69, p. 124–154, 2005.
132. C. Weber, A. Farwick, F. Benisch, D. Brat, H. Dietz, T. Sub Til, and E. Boles, *Applied Microbiology and Biotechnology*, Vol. 87, p. 1303- 1315, 2010.

133. A.K. Chandel, E.C. Chan, R. Rudravaram, M.L. Narasu, L.V. Rao, and P. Ravindra, *Biotechnology Molecular Biology Reviews*, Vol. 2, p. 14–32, 2007.
134. R.W. Rousseau, and R.F. James, *Handbook of Separation Process Technology*, NJ, USA, p. 261–262,1987.
135. V. Griend and D. Lee, Ethanol Distillation Process. US Patent 7,297,236, November 20, p.1–12, 2007.
136. B. Hahn-Hagerdal, M. Galbe, M.F. Gorwa-Grauslund, G. Liden and G. Zacchi, *Trends in Biotechnology*, Vol. 24(12), p. 549–556, 2006.
137. S. Parekh, and M. Wayman, *Biotechnology Letters*, Vol. 8, p. 597–600, 1986.
138. T. W. Jeffries, and Y.S. Jin, *Applied Microbiology and Biotechnology* ,Vol. 63, p. 495–509, 2004.
139. B. Buaban, H. Inoue, S. Yano, S. Tanapongpipat, V. Ruanglek, and V. Champreda, *Journal of Bioscience and Bioengineering*, Vol. 110(1), p. 18 – 25, 2010.
140. M. Moniruzzaman, *World Journal of Microbiology and Biotechnology*, Vol. 11, p.646, 1995.
141. J.N. Nigam, *Journal of Biotechnology*, Vol. 87, p. 17–27, 2001.
142. C.M. Takahashi, K.G.C. Lima, D.F. Takahashi, and F. Alterthum, *World Journal of Microbiology and Biotechnology*,Vol. 16, p. 829–34, 2000.
143. M. Abbi, R.C. Kuhad, and A. Singh, *Journal of Industrial Microbiology*, Vol. 17, p. 20 - 3, 1996.
144. P. Sommer , T. Georgieva , and B.K. Ahring, *Biochemical Society Transactions*, Vol. 32(Pt 2), p. 283–9, 2004.
145. S. Kim, and M.T. Holtzapple, *Bioresource Technology*, Vol. 96, p. 1994–2006, 2005.
146. A. Eliasson, C. Christensson, C.F. Wahlbom, and B. Hahn-Hägerdal, *Applied Environmental Microbiology*, Vol. 66(8), p. 3381–3386, 2000.
147. M. Kuyper, H.R. Harhangi, A.K. Stave, A.A. Winkler, M.S. Jetten, W.T. De Laat, J.J. Den Ridder, H.J. Op Den Camp, J.P. Van Dijken, and J.T. Pronk, *FEMS Yeast Research*, Vol. 4, p. 69–78, 2003.
148. K. Karhumaa, B. Hahn-Hägerdal, and M.F. Gorwa-Grauslund, *Yeast*, Vol. 22, p. 359–368, 2005.
149. R.J. Bothast, N.N. Nichols, and B.S. Dien, *Biotechnology Progress*, Vol. 15, p. 867–875, 1999.
150. B. S. Dien, M. A. Cotta, and T. W. Jeffries, *Applied Microbiology and Biotechnology*, Vol. 63, p. 258–266, 2003.
151. S. Katahira , A. Mizuike , H. Fukuda, and A. Kondo, *Applied Microbiology and Biotechnology*, Vol. 72(6), p. 1136–43, 2006.
152. K. Hiroko, S. Hiroaki, Y. Ryosuke, H. Tomohisa, and K. Akihiko, *Applied Microbiology and Biotechnology*, Vol. 94, p. 1585–1592, 2012.
153. S. Takatoshi, H. Tomohisa, H. Yoshimi, Y. Ryosuke, and K. Akihiko, *Journal of Biotechnology* , Vol. 158, p. 203– 210, 2012.
154. Z. Wei, and G. Anli, *Biotechnology for Biofuels* , Vol. 5, p. 46, 2012.
155. P. Sobana Piriya, P. Thirumalai Vasan, V. S. Padma, U. Vidhyadevi, K. Archana, and S. John Vennison Biotechnology Research International, Article ID 817549, p.1–8, 2012.

5

Recent Progress on Microbial Metabolic Engineering for the Conversion of Lignocellulose Waste for Biofuel Production

Shubhangini Sharma[1], Reena[2], Anil Kumar[3] and Pallavi Mittal[4],*

[1]*CDL, Intas Biopharmaceutical, Ahmedabad, Gujrat, India.*
[2]*Institute of Genomics and Integrative Biology, Mall Road, New Delhi, India.*
[3]*National Institute of Immunology, New Delhi, India.*
[4]*ITS Paramedical College, GZB, UP, India*

Abstract

A key strategy for biofuel production is making use of the chemical energy stored in plant cell walls by releasing sugars from lignocellulose and converting them into fuel. Lignocelluloses are the most abundant renewable organic resources on earth that are readily available for conversion to ethanol and other value-added products. Substances formed during the pretreatment of lignocellulosic feedstock inhibit enzymatic hydrolysis as well as microbial fermentation steps. Therefore, production of these biofuels in huge yields requires the engineering of the microorganism's metabolism. Recent advancements in science make it possible to produce sustainable fuels from renewable energy sources by engineering microorganisms. There are microorganisms that can convert cellulosic biomass directly into ethanol and there are also attempts to genetically engineer good ethanol producing microorganisms by metabolic engineering to give them the ability to grow on cellulose. Metabolic engineering is a powerful tool to improve microbial fuel production, either through engineering the metabolic pathways within the native microorganism to encourage high fuel synthesis or though transferring the fuel production pathway into

Corresponding author: mittal_pallavi@yahoo.com

Dr.Vikash Babu, Dr. Ashish Thapliyal & Dr. Girijesh Kumar Patel (eds.) Biofuels Production, (119–146) 2014 © Scrivener Publishing LLC

a model organism for optimization. The future of biofuel is expected to lay in metabolic engineering and synthetic biology efforts entailing the engineering of plant biomass, ethanol, and cellulolytic enzyme-producing microorganisms.

Keywords: Biofuels, lignocellulose, microorganisms, metabolic engineering, *E. coli*.

5.1 Introduction

The increased use of fossil fuels has caused greenhouse gas emissions and created undesirable damage to the environment. Fossil fuels accounted for 88% of the primary energy consumption, with oil (35% share), coal (29%) and natural gas (24%) as the major fuels, while nuclear energy and hydroelectricity account for 5% and 6% of the total primary energy consumption, respectively [1]. The current instability of oil supplies and the continuous fluctuation of prices have further ignited widespread interest in alternative energy sources in order to reduce oil dependence and increase energy production by exploring solar, wind, hydraulic and other natural phenomena. In addition, biomass also possesses a potential target for fuel and power production, although its primary use is as a feedstock. These factors, which revolve around economic, environmental and geopolitical issues, are central to the current interest in renewable energy sources. Rising oil prices in the last few years and environmental concerns about climate change have lead to an increasing interest in biofuel production. Biofuels are renewable, can substitute fossil fuels, reduce greenhouse gas emissions and they can be produced, where they are needed and reduce the dependence on oil producing countries. In 1925, Henry Ford observed that fuel is present in all vegetative matter that can be fermented and predicted that Americans would someday grow their own fuel. Last year, global biofuel production reached 28 billion US gallons, and biofuel accounted for 2.7% of the world's transportation fuel. Bioethanol, a popular type of biofuel, is largely derived from sugary food crops such as corn and sugarcane [2]. Thus, biomass can efficiently replace petroleum-based fuels for the long term [3–9]. While the concept of biofuels was conceived in the 1970s when the world faced a large-scale oil crisis, recent advances in synthetic biology [10, 11], metabolic engineering [11–14], and systems biology [15, 16] have generated a renewed interest in the production of biofuels. At the present, a variety of possible fuel sources are being examined. Among these, many have proposed using hydrogen as an energy carrier in a future hydrogen economy. However, a sustainable, renewable supply of hydrogen to power this economy is required. One option would be to use biological means to produce hydrogen. One approach would be to recruit the power of photosynthesis to capture sunlight and split water, a process that is called biophotolysis [17, 18]. This approach, although attractive, suffers from major challenges that may require years of research to overcome. Photo-fermentation, the use of bacterial photosynthesis to capture light energy and use it to drive hydrogen evolution from otherwise inaccessible substrates, can be used to show nearly complete substrate conversion to hydrogen [19–21].

At this juncture, biotechnology is playing a significant role that embraces the bio-production of fuels and chemicals from renewable sources [22]. These technologies use living cells and enzymes to synthesize products that are easily (bio) degradable, require less energy and create less waste during their production or use than those produced from fossil resources.

This idea is widely spread throughout the entire renewable energy field, but it is specifically pronounced within the realm of biofuels development, where engineers, researchers, and industry leaders are confronted with the significant task of displacing over 4.3 billion barrels of crude oil and petrochemical imports annually while maintaining price parity with foreign fossil imports [23]. Considering these facts, many countries in Europe, North and South America and Asia are replacing fossil fuels by biomass-based fuels according to international regulations. One of the directives of the European Union (2009/28/CE) imposes a quota of 10% for biofuels on all traffic fuels until 2020 [22, 24].

Many biomass feedstocks such as food and non-food crops, cellulosic materials and waste residues from different sources can be used as energy sources for biofuel production. To make the cost of microbial fuel competitive with petroleum based products, the primary goal should be the production of these compounds directly from lignocellulosic biomass because of their abundance, non interference with the food chain, and production of bioethanol without using arable (to plough) lands [25, 26]. Lignocellulosic biomass is a complex raw material that can be processed in different ways to obtain various value-added compounds such as lactic acid, acetic acid, furfural, methanol, hydrogen and many other products which can be obtained from its sugars. But the complicated chemical structure of lignocellulosic biomass makes the processing difficult. Therefore, current processes include different pretreatments and enzymes to liberate sugar for further processing. An alternative processing technique called consolidated bioprocessing (CBP), is being developed, in which cellulase production, hydrolysis and fermentation are accomplished in a single step using genetically altered microorganisms [27]. The modified microbe possesses enzymes that hydrolyze the pretreated biomass and eliminates the need to add the enzymes, thus lowering the production cost.

5.2   Role of Genetic and Metabolic Engineering in Biofuel Production

The consideration of lignocellulose for the production of low cost biofuel gives a new scope for biofuel production. To date, no organism has been found that is capable of transforming lignocellulosic biomass into biofuel at high rates. To overcome such problems, microbiologists are using genetic techniques to change the pathways for the production of cellulose. This can be done in two ways, firstly by improving the yields from cellulase-producing microbes and second, by transferring microbial cellulase genes into standard strains of bacteria used in the industry (by producing a recombinant strain), to enable large-scale production of cellulase. Once the celluloses are broken down, the sugars can be fermented by yeast or other microorganisms. New strains have been identified that are suitable for fermentation in the industry, for example, heat-loving (thermophilic) bacteria that can ferment sugars into ethanol very efficiently at high temperatures, making the process more cost-effective.

Traditional microbiology was merged with molecular biology to yield an improved recombinant processes for the production of biofuels. The reshuffling of genes between two DNA molecules, forming recombinant DNA, occurs naturally in microorganisms. It is also possible to manipulate DNA artificially to combine genes from two different sources, even from vertebrates to bacteria. Artificial gene manipulation is known as genetic engineering, and the term biotechnology usually means the industrial use of genetically engineered microorganisms. The ultimate goal of recombinant fermentation research is the cost-effective production of the desired product by maximizing the volumetric productivity, i.e., to obtain the highest amount of bio-fuel in a given volume in the shortest period of time. Such bioprocessing for recombinant proteins using genetically modified organisms requires (1) a stable high-yielding recombinant clone, (2) a highly productive & optimized fermentation process and (3) cost-effective recovery and purification procedures.

Unfortunately, the proprietary nature of much research on genetically engineered algae makes it difficult to know which species and novel traits are in the research and development pipeline. Relevant details are just beginning to emerge from patent applications, funding agencies and published interviews with chief executive officers (CEOs). These organisms have been genetically engineered to expand their substrate range to include cellulose or the sugars freed from cellulose or hemicellulose degradation, as in the case of ethanologenic organisms such as *Escherichia coli* [28, 29], *Zymomonas mobilis* [30, 31], *Clostridium cellulolyticum, C. Reinhardtii* and *Saccharomyces cerevisiae* [32, 33].

Furthermore, improvements in the overall efficiency and yield of producing soluble sugars from lignocellulosic biomass are being pursued with different angles including selective breeding and genetic engineering of more labile energy crops [34, 35], improved pretreatment technologies [36], and the directed evolution of hydrolytic enzymes for improved activity and stability [37, 38]. It is anticipated that the largest gain in cost competitiveness in terms of producing fuels from biomass could be realized through the consolidation of several production steps into a streamlined process where hydrolytic enzymes are simultaneously produced in-situ by a solventogenic, fermentative microbe [39]. In contrast to starch ethanol production where robust biocatalysts are available (*Saccharomyces cerevisiae*), a consolidated bioprocess (CBP) requires a highly engineered microbial workhorse that has been developed for several different process-specific characteristics. These desirable traits include enzyme production/stability, balanced growth on hexoses and pentoses tolerance to pretreatment inhibitors, maximal product yield and production rates, solvent tolerance, and the ability to persevere through process fluctuations. Naturally occurring microbes possess many intrinsic qualities (i.e. cellulase production) that may advance them as candidates for use in an industrial scale consolidated processing scheme but deficits in other necessary phenotypes must be evolved or directly engineered into a host.

5.3   Problems with Different Biofuels and Areas of Improvement

Second generation biofuels rely on microbes to convert carbon feedstock into desired hydrocarbon fuels. Microorganisms have been identified that are capable of producing a range of fuel molecules and fuel precursors, yet the natural rates of microbial fuel synthesis are typically too low to support industrial-scale production. Bioconversion of lignocellulose by microbial fermentation is typically preceded by an acidic thermochemical pretreatment step designed to facilitate enzymatic hydrolysis of cellulose. Substances formed during the pretreatment of the lignocellulosic feedstock inhibit enzymatic hydrolysis as well as microbial fermentation steps. Due to the fact that the amount of biofuel produced from old techniques is less and not sufficient for industrial and long term use. This leads to an increse in production costs which is the main problem with biofuel production. Novel methods and developments are required to improve the future of biofuel. Metabolic engineering is a powerful tool to improve microbial fuel production, either through engineering the metabolic pathways within the native microorganism to encourage high fuel synthesis or though transferring the fuel production pathway into a model organism for optimization.

5.3.1   Ethanol

Ethanol is produced from pyruvate in two steps: pyruvate decarboxylase converts pyruvate to acetaldehyde, and alcohol dehydrogenase reduces the acetaldehyde to ethanol. Baker's yeast (*Saccharomyces cerevisiae*) has long been used in the brewery industry to produce ethanol from 6-C sugars but this organism is unable to ferment 5-C sugars. Many microorganisms, including bacteria and yeasts, can produce ethanol as the main fermentation product from carbohydrates. Since neither *S. cerevisiae* nor *Z. mobilis*, are currently used to carry out industrial ethanol fermentation, can use xylose or arabinose, microorganisms other than *S. cerevisiae* have come to the forefront in bioethanol production from lignocelluosic biomass. Indeed, many microorganisms are able to efficiently utilize pentose sugars but cannot naturally produce ethanol for a sufficient yield and productivity.

Synthetic biology and metabolic engineering have been extensively used in *S. cerevisiae*, *Zymomonas mobilis* and *Escherichia coli* to enhance ethanol fermentation [40–44]. Microorganisms that utilize pentoses, such as the bacteria *E. coli* and *Klebsiella oxytoca* and the yeast *Pichia stipitis*, have been successfully engineered for ethanol production [45]. Alternatively, pentose catabolic pathways have been expressed in ethanologenic microorganisms, such as the conventional yeast *S. cerevisiae* [46–48] or the ethanologenic bacterium *Z. mobilis*. This pathway is commonly exploited in yeast hosts, or in the Gram-negative gamma-proteobacteria, *Z. mobilis* or recombinant *E. coli* [49].

Distillation represents a higher percentage of the cost of production for corn ethanol than for cellulosic ethanol, because feedstock pretreatment and cellulase production still dominate projected costs for cellulosic ethanol.

5.3.2   Butanol

Recently there has been an increased interest to convert sugars from lignocellulosic biomass into butanol. Due to its physical properties, the four-carbon alcohol is a better replacement for gasoline than ethanol [50]. Other chemicals in the alcohol family include methanol (1-carbon), ethanol (2-carbon), and propanol (3-carbon). N-Butanol is a fermentation product of *Clostridium acetobutylicum* and *Clostridium bjerinkci* [51]. It is produced from acetyl-Co A through the dimerization of two acetyl-CoAs into acetoacetyl-CoA. Subsequent enzymes catalyze the four-electron reduction and dehydration of acetyl-CoA to butyric acid. In clostridia, acetone is often produced concomitantly with butanol by decarboxylation of acetoacetyl-CoA. The total pathway from glucose to butanol is, in principle, redox balanced. However, in some organisms, including *C. acetobutylicum* and *E. coli*, hydrogen is produced via formate-hydrogen lyase [52] and represents a competing outlet for reducing equivalents. In principle, hydrogenase activity or alternate formate dehydrogenases [53] could recover these reducing equivalents. During the first half of the 20th century, the production of butanol from biological sources was a commercial reality. As mentioned above, various clostridia have been utilized in butanol fermentation, although these gram-positive anaerobes coproduce butanol with a few byproducts, such as butyric acid, acetone, ethanol, therefore lowering its yield [54]. From a biotechnology perspective, the lack of efficient genetic tools to manipulate clostridia hinders metabolic engineering endeavors for the optimization of butanol synthesis and the reduction of by-product formation. Because of this, E. coli [55–58, 47, 48] and S. cerevisiae [59] were recently engineered for butanol synthesis from sugars. The *clostridial* pathway has been functionally expressed in recombinant E. coli hosts, although the productivity was less than what observed for clostridia [60]. Still the efforts are going on to make an improved strain which is more productive.

5.3.3 Higher lipids

Long-chain fatty acids, methylesters [61] carboxylated derivatives [62] are attractive as renewable substitutes for diesel fuel. Lipid production by algae has the potential to get around current biodiesel feedstock limitations [63, 64]. More recently, interest has developed in the generation of biofuels via lipogenic (or oleaginous) yeasts and other microbes [65]. This technology would allow the microbial conversion of cellulosic materials into lipids. An alternative metabolic engineering strategy is to engineer a secretion of the lipid products. Many details of fatty acid secretion even in model organisms like E. coli and S. cerevisiae are poorly understood [66].

5.3.4 Terpenoids

Isoprenoids are a broad class of metabolites synthesized from isoprenyl pyrophosphate (IPPP), the pyrophosphate ester of 3-methylbut-3-en-1-ol, or its isomer dimethylallyl pyrophosphate (DMAP), the pyrophosphate ester of 3-methylbut-2-en-1-ol. These molecules are synthesized either from glyceraldehyde-3- phosphate and pyruvate via the methylerythritol pathway or from acetyl-CoA via the evalonate pathway.

Strains and processes capable of converting sugars to terpenoids at yields similar to the ethanol process have not yet been reported in scientific literature. Many experiments have been conducted and in one of them, a representative titer for these squiterpene hydrocarbon amorphadiene has been reported as about 0.5 g/l [67]. Terpenoid production has usually been examined only under aerobic culture conditions, but for large-scale, high-yield production, anaerobic processes are desired. Obviation of the oxygen requirement for terpenoid overproduction is one key metabolic engineering objective for biofuels applications of this pathway.

5.4   General Process of Metabolic Engineering

Metabolic engineering approaches are adapted to address the challenges for optimization of cellular pathways that currently limit the construction of an integrated, efficient desired product. Although much progress has indeed been made in the past 20 years towards lignocellulosic biomass conversion by yeasts, the approach is to identify and optimize limiting steps. In the future, more emphasis will be on the host genome and regulatory structure to understand the full effect of the biological complexity on a pathway. The process is shown in fig. 5.1.

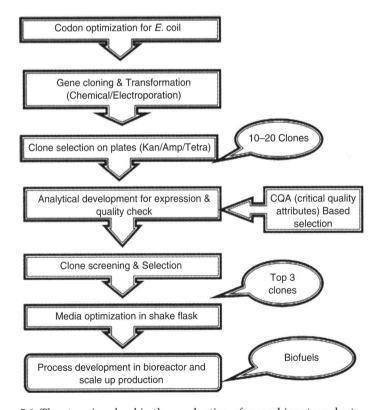

Figure 5.1 The steps involved in the production of recombinant product.

5.4.1 Host Organism Selection

Selection of a suitable host organism plays a key role in the productivity and yield of the biofuel production process. As the mechanisms associated with fuel production are host specific, finding a microorganism with desirable properties is the pre-requisite for a metabolic engineering effort. Generally, microbes are chosen based on a number of properties and can include industrial scalability, natural ability to produce a fuel of interest, tolerance to fuels, the ability to grow on unique feedstocks, and the ability to genetically modify them. Salient examples include the choice of yeast and *E. coli* for their proven industrial scalability and their ease in performing genetic manipulations and algae for its ability to naturally accumulate high-levels of fatty acids and use sunlight as a "feedstock". Examples of industrially relevant microbes with efforts from various companies include algae – Solazyme, Solix, Joule, Sapphire; yeast – Cargill, Gevo, Amyris; and bacteria– LS9. The following are the most commonly used microorganisms for biofuel production.

Different microbes such as bacteria, algae and yeast are selected depending on the different hydrocarbon producing pathways they have for the production of biofuels

5.4.1.1 E.coli and Yeast

Though numerous host strains could be employed for fatty acid production, it is desirable to take advantage of the existing metabolic capability in a genetically traceable host like *Escherichia coli* (*E.coli*) to achieve maximal productivity (g/l/h), titer (g/l), and yield (g substrate/g-fuel product) [68]. Because of the tractability of microbial hosts like *E. coli*, one can use a variety of genetic and metabolic engineering technologies to rapidly modify metabolic pathways to increase fatty acid production and to ultimately achieve industrially relevant production metrics that approach price-parity with petroleum fuels. Among bacteria, *E.coli* is a well-studied organism and it has been used to produce valuable chemicals and biofuels. Although it can be altered genetically to produce a wide variety of fuels such as ethanol, isopropanol, hydrogen and biodiesel [69], it has proven its ability to scale-up only in the case of 1, 3 propanediol production [70]. Though, the natural level of lipid stored in *E. coli* cell is low, it plays a very crucial role in its rapid growth rate, ability to grow in an anaerobic environment, efficient utilization of various biomasses as feedstocks for biofuel production and high fatty acid synthesis rate (0.2 g/l/h/g dry cell mass) [71, 72]. As described earlier, the heterotrophic bacteria (*E. coli*) can ferment glucose and other lignocellulose derived sugars as an energy substrate to produce different types of fuels. However, the recalcitrant nature of cellulosic biomass requires separation of polysaccharides from lignin and subsequent depolymerization of the polysaccharides by enzyme [73].

On the other hand, natural ethanologenic organisms such as *Saccharomyces cerevisiae* and *Zymomonas mobilis* have the ability to produce ethanol from sugars and result in a theoretical maximum yield of 98% [74]. However, these organisms lack the ability to ferment complex sugars like pentoses. To overcome this problem, research efforts have focused on introducing pentose-metabolizing pathways into *S. cerevisiae* and *Z. mobilis* or on engineering microorganisms to secrete cellulases and hemicellulases to depolymerize the complex sugars into simple sugars and their subsequent conversion into fuels [75].

5.4.1.2   *Light-utilizing Organisms: Photobacteria*

The use of photosynthetic organisms offers an alternate approach for the production of fuel compounds, in which they capture light energy and subsequently convert that into high-energy organic compounds using water as the final donor. Many photosynthetic organisms such as algae and bacteria have been used for the production of valuable metabolites including fuels, chemicals and organic acids. It is reported that employing photosynthetic bacteria for fuel and chemical production can overcome the energy intensive and costly biomass recovery [76, 77]. Among the photosynthetic bacteria, *Cyanobacterium* offers unique advantages due to the following reasons: their photosynthetic capability, rapid growth rate, efficient solar energy conversion, amenable to effective genetic manipulation, tolerance to high CO_2 content and ability to thrive in marginal environments [78].

5.4.2   Codon Optimization, Transformation and Selection

5.4.2.1   *Codon Optimization*

Codon optimization is a technique to maximize the protein expression in living organism by increasing the translational efficiency of genes of interest by transforming DNA sequences of nucleotides of one species into another species. When the mRNA of heterologous target genes is overexpressed in *E. coli*, differences in codon usage can impede translation due to the demand for one or more tRNAs that may be rare or lacking in the population. Insufficient tRNA pools can lead to translational stalling, premature translation termination, frame shifting and amino acid misincorporation. Heterologous protein expression is very low when a gene encodes clusters of numerous rare *E. coli* codons. The most severe effects on expression have been observed when multiple consecutive rare codons are near the N-terminus of a coding sequence. To overcome these problems, codon optimization is necessary. Codon optimization can be used to produce a synthetic version of a foreign gene for more efficient expression. Codon preferences reflect natural selection in organisms for translational optimization.

5.4.2.2   Transformation

Transformation is the uptake of foreign DNA by host cells. The plasmid can replicate its DNA using the bacterium (*Escherichia coli*) as a host organism. "Transformation" may also be used to describe the insertion of new genetic material. Plasmids are small circular DNA molecules found in bacteria. They independently replicate the bacterial chromosome and, depending on the plasmid, there may be from 1–100 copies per cell. If a given plasmid used for cloning is present in a hundred copies, then one can isolate a large amount of cloned DNA from a small number of cells by isolating the plasmid. In order for *E. coli* to take up DNA they must be made competent. Mainly two methods are used to make competent cells 1) Electoporation, 2) Calcium chloride.

5.4.2.3   Selection

Creating efficient biofuel producing cell lines for high yield production is a very important step. It requires the generation of a range of suitable constructs, transformation of these into host cells and the selection of appropriate clones. These clones must be of single cell origin, express the target protein in the right quality and quantity and grow rapidly in a simple medium. The most critical and variable step of the process is selecting optimal clones from a pool of randomly generated cell lines. Automation can be applied to limited-dilution plating, plate replication, assessing single cell clone origin, clone picking (antibiotic resistant based), expression screening using ELISA, protein quantization and a range of protein characterization methods.

5.4.3   Fermentation Procedure

The fermentation process basically consists of three parameters: microbes, fermentation media and fermentation.

The selected microbe, chosen for the production of a desired product, should have the following properties as well: i) It should not be pathogenic (except when used for the production of vaccine/toxin) and should not cause any allergic reaction. ii) It should not lose its desired property during sub culturing. iii) It should be genetically stable. iv) It should be amenable to mutation as and when desired.

The lignin-hemicellulose-pectin complex forms one of the most stringent seals around cellulose. The first step in the overall process of lignocellulosic fermentation is breaking this barrier (pretreatment). This is the most important and rate limiting step in the overall process. Further steps involve isolation and hydrolysis of cellulose and hemicellulose to generate fermentable sugars (saccharification) followed by fermentation and distillation. The pretreatment processes involve the use of acids, alkalis and/or organic solvents. The aim of this process is to separate lignin, cellulose, hemicellulose and pectin from lignocellulosic biomass. Post pretreatment, the recalcitrant lignocellulosic biomass becomes susceptible to acid and/or enzymatic hydrolysis as the cellulosic microfibrils are exposed and/or accessible to hydrolyzing agents. In the pretreatment process, small amounts of cellulose and most of hemicellulose is hydrolyzed to sugar monomers; mainly D-xylose and D-arabinose. The pretreated biomass is then subjected to filtration to separate liquids (hemicellulose hydrolysate) and solid (lignin+cellulose). The liquid is sent to a xylose (pentose) fermentation column for ethanol production.

Solids are subjected to hydrolysis (also called second stage hydrolysis). This process is mainly accomplished by enzymatic methods using cellulases. Mild acid hydrolysis using sulfuric and hydrochloric acids is an alternative procedure. The hydrolyzed sugars such as D-glucose, D-galactose, and D-mannose, can be readily fermented to ethanol using various strains of *Saccharomyces cerevisae*. The pentoses (D-xylose and D-arabinose) from hemicellulose hydrolysis are not easily utilized by saccharomyces strains; therefore, genetically modified strains of *Pichia stipitis, Zymomonas mobilis*, are used for their fermentation. *Candida shehatae* is capable of co-fermenting both pentoses and hexoses to ethanol and other value-added products at high yields.

5.4.4   Fermentation Media

The growth medium in which microbes grow and multiply is called fermentation medium. The selected microbe should be able to utilize and grow on cheap sources of carbon and nitrogen. Care should be taken to avoid the use of such microbes that require expensive nutrients like vitamins for their growth. An optimal process for fermentation uses a broth containing *saccharomyces cerevisiae* supplemented with 22% (w/v) sugar, 1% (w/v) of each of ammonium sulfate and potassium dihydrogen phosphate respectively, and fermented at pH 5.0 and 30°C [79]. Under such conditions a typical strain of *S. cerevisiae* is capable of producing 46.1 g ethanol/l broth [80]. Cane molasses conditioned with EDTA, ferrocyanide or zeolites, and fermented under similar conditions has been shown to enhance ethanol production [81]. Further, addition of minimal concentrations of hops acids to the fermentation broth has been shown to prevent bacterial growth and thus enhances ethanol yields [82]. Fermentation using immobilized yeast and broth supplemented with Mg, Zn, Cu or ca pantothenate has also been shown to increase fermentation efficiency by almost 20% [83].

5.4.5   Fermentation Process

In general, the fermentation process is divided into two parts i.e. Up Stream Processing (USP) and Down Stream Processing (DSP), as shown in fig. 5.3. USP is basically involved in the a) preparation of a seed culture (i.e. inoculums preparation), b) preparation of fermentation medium and its sterilization and c) the fermentation process for optimal production of a metabolite. DSP is concerned with the recovery of the product, its concentration, purification and the disposal of bio-waste generated during the fermentation process.

During the fermentation process, both cellulose hydrolysis and the fermentation of glucose are carried out in the presence of fermentative microorganisms in a single step and the process optimally operates at 37 to 38°C. This technique reduces the number of steps in the process, and is a promising way for converting cellulose to ethanol [84, 85]. During the fermentation process, lignocellulosic biomass is first pretreated with a dilute acid (1.1% sulfuric acid at 160°C for 10 min) to breakdown the complex of lignin and hemi-cellulose and pectin. The resulting broth is filtered to drain the liquid from the system. The drained liquid containing pentose sugars is neutralized with lime and processed via xylose fermen-tation process. Remaining solids containing cellulose and lignin are then hydrolyzed and fermented simultaneously using cellulase enzymes and yeast. The cellulase enzymes hydrolyze cellulose to D-glucose, which in turn is fermented into ethanol by yeast [86]. This combined process improves both the kinetics of fermentation as well as the economics of biomass processing via the reduction in the accumulation of hydrolysis product (glucose) that is inhibitory to cellulases. The disadvantages of using this process include (i) the operating temperature should be around 37 to 38°C and (ii) much of the sugar released by cellulose hydrolysis is used for the growth of yeast necessary to ensure good ethanol production.

5.5   Metabolic Engineering in Different Microorganisms

Producing biofuels directly from cellulose, known as consolidated bioprocessing, is believed to reduce costs substantially compared to a process in which cellulose degradation and fermentation into fuel are accomplished in separate steps.

While these microorganisms are capable of efficiently hydrolyz-ing cellulose, their biofuel productivities are significantly lower than those of existing industrial strains. In addition to improv-ing biofuel productivity [87], research efforts are also focused on increasing ethanol yields [88], eliminating competing pathways [89] and improving ethanol tolerance [90].

Synthetic biology has enabled researchers to biosynthesize these fuels with properties superior to ethanol and butanol, resulting in higher energy density, heating value and compatibility with the existing transportation infrastructure including use in combustion, compression and jet engines and the ability to transport them in existing pipeline infrastructure. There are a limited number of met-abolic pathways, which are being commercialized, that produce hydrocarbons relevant to the fuel chemistry and these include deri-vations of the amino acid pathway to produce isobutanol (Gevo), the mevalonate pathway to produce farnesene (Amyris), the polyketide pathway to produce a variety of fuels (Lygos) and the fatty acid pathway to produce biodiesels (Solazyme, Solix, Joule, Sapphire, & LS9). These pathways all naturally exist in many different microorganisms and can be genetically manipulated or transported into a "naïve" host to increase an organism's biosynthesis capacity for a specific fuel product.

In addition to the different hydrocarbon-producing pathways there are different microbes used for the production of fuels such as bacteria, algae and yeast.

5.5.1   E. Coli

Generally, microbes are chosen based on a number of properties and can include industrial scalability, the natural ability to produce a fuel of interest, tolerance to fuels, the ability to grow on unique feedstocks, and the ability to genetically modify them. Salient examples include the choice of yeast and *E. coli* for their proven industrial scalability and the ease of performing genetic manipulations. Here we focus on the fatty acid pathway for hydrocarbon fuel biosynthesis. Firstly, we give a general overview of hydrocarbon biosynthesis pathways, discuss the fatty acid pathway biochemistry and efforts to engineer the pathway.

5.5.1.1   Fatty Acid Biosynthesis for Biofuel Production

The fatty acid biosynthesis pathway yields a range of energy rich molecules suitable for use as biofuels and extensive research has been done to utilize this particular pathway, also reviewed elsewhere. Fatty acid biosynthesis is generally used by organisms to make their cell membranes and in microbes like *E.coli,* which is catalyzed by an enzyme system consisting of discrete proteins that play an important role in growing and fully reducing a linear hydrocarbon chain.

Researchers successfully showed that *E. coli* can be metabolically engineered to produce isobutanol by manipulating *E. coli's* amino acid biosynthesis pathway by diverting the 2-keto acid intermediates toward biofuel production. Using the same metabolic engineering strategy, we were able to achieve an isobutanol titer of 660 mg/l by the cellulolytic mesophile *C. cellulolyticum* by expressing *kivd yqhD alsS ilvCD*. To our knowledge, this is the first demonstration of isobutanol production directly from cellulose.

In bacteria, fatty acid biosynthesis, acetyl-CoA carboxylase (ACC) is the enzyme responsible for the first committed step, catalyzing malonyl-CoA synthesis from acetyl-CoA and carbon dioxide. Malonyl-CoA is then transacylated to a fatty acyl carrier protein (acyl-ACP) by malonyl-CoA:ACP transacylase. Subsequently, acetyl-ACP and malonyl-ACP are condensed to yield acetoacetyl-ACP in the presence of β-ketoacyl-ACP synthase with concomitant release of carbon dioxide. Then the β-ketoacyl substrate undergoes a cycle of reduction and dehydration that results in the formation of the four-carbon fatty acyl-ACP (butyryl-ACP). This cycle continues with two carbon additions to the fatty acid chain provided by malonyl-ACP condensations and results in a 14 to 18 carbon fatty acyl-ACP final product (Fig 5.2).

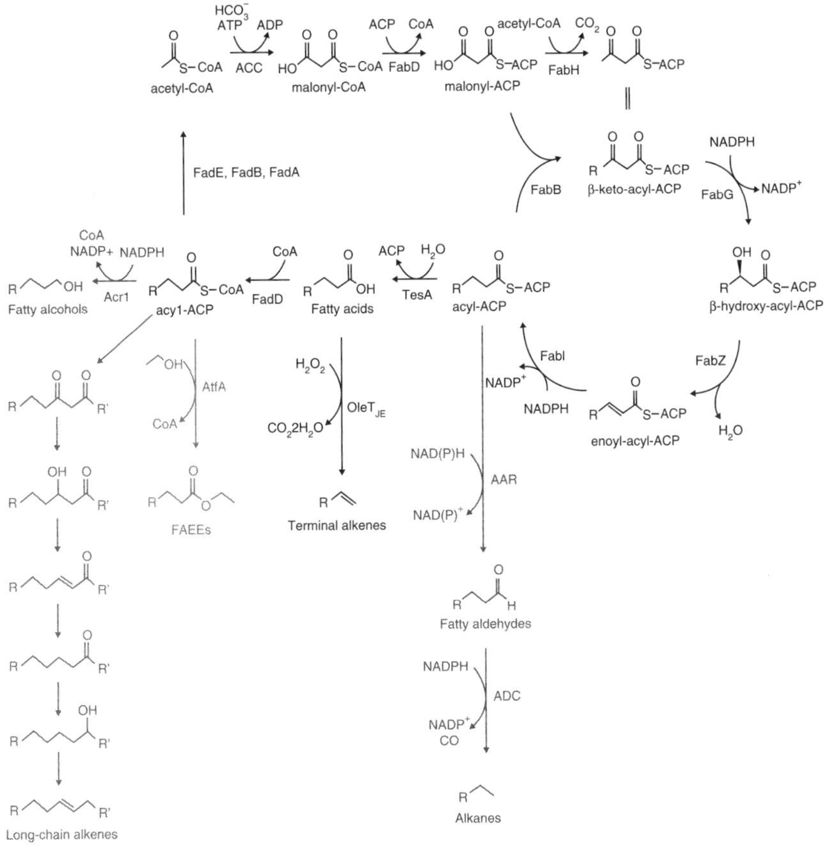

Figure 5.2 Pathways for the production of fatty acid-based biofuels. Native *E. coli* fatty acid pathway is colored black. The proposed pathway for long-chain alkene biosynthesis is colored red. Engineered pathways for the production of other derivatives are in different colors [92].

The fatty acid pathway in *E. coli* is tightly linked with biosynthesis of phospholipids and the pathway's final product is transferred to glycerol derivatives by glycerol-3-phosphate acyl transferase in order to build the cell membrane [91, 92]. Fatty acid biosynthesis is energetically expensive for the cell and thus tightly regulated. Transcriptional and translational regulation balances the presence of pathway enzymes, while flux regulation prevents further, unnecessary biosynthesis of fatty acids under nutrient limiting conditions.

5.5.2   Yeast

Yeast is the oldest microbe used for biofuel production. Even though many organisms are capable of natively converting these sugars, the most commonly selected organism is baker's yeast (*Saccharomyces cerevisiae*). The genetic tractability, widespread industrial use and endogenous ethanol production capacity of yeast motivates its use [93]; however, baker's yeast must be engineered to convert xylose and arabinose. Traditional pathway engineering approaches have enabled xylose and arabinose catabolism in yeast, but continued optimization of these strains requires novel metabolic engineering tools and strategies. Specifically, novel approaches should target and exploit additional cellular mechanisms influencing metabolic pathways, such as molecular transport, catabolite sensing and cellular tolerances.

To efficiently arrive as an optimized strain, metabolic engineering tools must be expanded to modify multiple interdependent steps, an approach we term 'panmetabolic engineering'. Metabolic engineering tools must also expand to incorporate recent breakthroughs in modifying catabolite sensing and in increasing cellular tolerances to significantly affect biofuel-producing organisms. Once developed, these tools will enable the addition of other substrates to the yeast carbon source portfolio.

Interestingly, the *S. cerevisiae* genome is encoded with a putative xylose metabolic pathway, although the expression levels of these genes are often too low to permit growth on xylose as the sole carbon source [94–96]. Recent work suggests that some strains of *Saccharomyces* may possess a latent oxidoreductase pathway with an active xylitol dehydrogenase [97] (fig 5.3).

Even so, yeast lack effective xylose and arabinose utilization pathways, and therefore require heterologous complementation

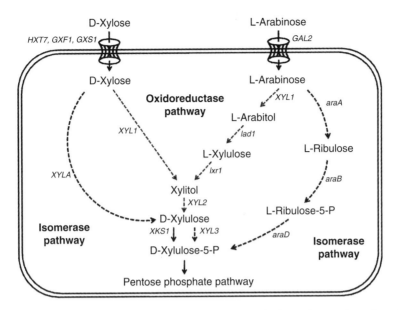

Figure. 5.3 Successful exogenous transport and metabolic pathways introduced in *S. cerevisiae*. Reported improvements of pentose utilization in yeast (as described in the text) are depicted in this schematic. (A) Bacterial xylose isomerase pathway; (B) fungal xylose oxidoreductase pathway; (C) Fungal arabinose oxidoreductase pathway; (D) Bacterial arabinose isomerase pathway. The genes used to accomplish the enzymatic step are italicized. Heterologous steps are indicated by dashed lines [106].

or significant genetic modification. The advent of recombinant DNA technology enabled the transfer of genes from native pentose catabolizing organisms into baker's yeast, thus facilitating novel pentose catabolism [98–101]. Two types of pentose pathways have been constructed in yeast: the oxidoreductase pathway and the isomerase pathway. Both xylose and arabinose can be metabolized through each of these pathways, although arabinose assimilation involves additional steps in both cases [102]. All four possible pathway variants have been previously constructed [102–105], and all feed into native yeast metabolism via D-xylulose or D-xylulose-5-phosphate (P). Once converted to xylulose 5-P, these sugars are further metabolized through the native pentose phosphate pathway (PPP). The following sections give a brief review of these pathways. Combined with molecular transporter engineering, they form the foundation of the panmetabolic engineering described in more detail later.

5.5.3   Clostridium Cellulolyticum

The metabolic engineering of a *Clostridium cellulolyticum* strain for isobutanol synthesis is described here. This strategy exploits the host's natural cellulolytic activity and the amino acid biosynthesis pathway and diverts its 2-keto acid intermediates towards alcohol synthesis [107].

Most studies employing the native cellulolytic strategy have been conducted with the thermophilic, cellulolytic *Clostridium thermocellum*. This strain is particularly attractive because it is able to thrive in high-temperature fermentations, which are conducive to high-level substrate conversion, low contamination risk, and high-level product recovery [108]. Although *C. thermocellum* has the potential to be a CBP organism, issues such as low transformation efficiency [109] and the lack of publications demonstrating successful overexpression of foreign proteins in *C. thermocellum* significantly impede the engineering progress of this organism to produce synthetic biofuels, such as isobutanol.

One way to produce biofuel is to first establish and optimize the desired metabolic pathways in a closely related, more amenable organism. Once the specifics, such as identifying which genes to over express, mutate, and/or delete, have been determined, the same strategy can then be adapted to *C. thermocellum*. *Clostridium cellulolyticum*, which was originally isolated from decayed grass [110], is a useful candidate for this initial metabolic engineering work, as it is also a member of *Clostridium* group III, based on 16S rRNA phylogenetic analysis [111]. Another advantage is its mesophilic nature that is associated with the heterologous expression of proteins in thermophiles that are circumvented. In addition, *C. cellulolyticum* has a sequenced genome (GenBank accession NC_011898.1) and where well-established DNA transfer techniques [112] and gene over-expression methods [113] for it exist. As a potential CBP organism in its own right, *C. cellulolyticum* can not only utilize cellulose similar to *C. thermocellum* but it can also utilize additional sugars released from hemicelluloses degradation, including xylose, arabinose, fructose, galactose, mannose, and ribose [114].

Previously, *C. cellulolyticum* has been genetically engineered for improved ethanol production [113]. Similarly, most of the research concerning the construction of a CBP organism has focused on ethanol production. Despite this, it has been asserted that higher alcohols (i.e., alcohols with more than two carbons), such as isobutanol, are better candidates for gasoline replacement because they have energy density, octane value, and reid vapor pressure that are more similar to those of gasoline [115]. Unlike ethanol, isobutanol can also be blended at any ratio with gasoline or used directly in current engines without modification [116]. *C. cellulolyticum* are metabolically engineered to produce isobutanol. By expressing enzymes that direct the conversion of pyruvate to isobutanol by using an engineered valine biosynthesis pathway, we were able to produce up to 660 mg/l of isobutanol by using *C. cellulolyticum* growing on crystalline cellulose. Specifically, we have confirmed the first production of isobutanol to approximately 660 mg/l from crystalline cellulose by using this microorganism [117].

In order to achieve isobutanol production directly from pyruvate, the genes encoding acetolactate synthase from *B.subtilis* , acetohydroxyacid isomeroreductase, dihydroxy acid dehydratase from *E.coli and* ketoacid decarboxylase,-alcohol dehydrogenases from *Lactococcus lactis* (Fig. 5.4) were cloned into a pAT187 derivative plasmid [118]. These specific genes were chosen because they were the same genes utilized for isobutanol production in *E. coli* [119] and *Synechococcus elongatus* [120]. The different combinations of the genes were cloned as single synthetic operon driven by the constitutive ferredoxin (Fd) promoter from *Clostridium pasteurianum*. The activities of the first three enzymes in the isobutanol pathway were examined by transforming plasmid expressing *alsS* or *alsS ilvCD* into *C. cellulolyticum*. While *C. cellulolyticum* was successfully transformed with the empty vector, no *C. cellulolyticum alsS* or *alsS ilvCD* transformants were obtained. The same results were observed after repeated transformation efforts. Due to the fact that *alsS* and *alsS ilvCD* transformants could not be obtained, the complete isobutanol pathway was then examined. *C. cellulolyticum* was transformed with a plasmid expressing *alsS ilvCD kivd adhA*. While transformation were obtained, sequence confirmation of the plasmid revealed that a single adenine insertion, which is not found in the wild type *alsS* sequence, was present 54 bp downstream of the start ATG. This single insertion, by shifting the reading frame, results in a downstream premature stop codon (TGA) and, subsequently, a truncated 37-amino-acid protein. This spontaneous mutation in *alsS* (*alsS) was found to have originated in the *E. coli* strain used for cloning.

The frameshift mutation in the *alsS* sequence was a cause for great concern because of the effect it could have on AlsS activity. Thus, to determine the activities of AlsS and the other enzymes

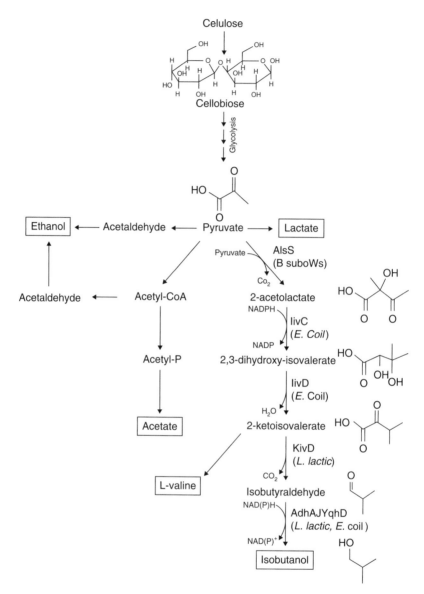

Figure. 5.4 The pathway for isobutanol production in *C. cellulolyticum* [117].

expressed from the synthetic operon, enzymatic assays were performed on lysates of the *C. cellulolyticum* strain expressing *alsS
ilvCD kivd adhA*. Surprisingly, for the AlsS assay, the *alsS ilvCD kivd
adhA* lysates were found to demonstrate an activity of 282 nmol/
min mg, which was significantly higher than the 11 nmol/min mg
demonstrated by the strain transformed with the vector.

5.6 Conclusion

Lignocelluloses biomass has a considerable potential to contribute as an environmentally sustainable alternative for the future production of biofuels. Because biofuels are clean, green and renewable, they could displace gasoline, diesel and jet fuel on a gallon-for-gallon basis by being directly dropped into today's engines and infrastructures. In the present era of technology various milestones in scientific development like genomic, transcriptomic and metabolomics techniques which undoubtedly carved a niche and provided deeper information of biofuel production from various microorganisms. These can convert lignocellulosic biomass into biofuels efficiently, and the exploration of the natural biodiversity is a promising strategy to identify and improve novel microbial species.

However, several challenges still exist in the production, harvesting, and conversion aspects of lignocellulose, and these must be resolved in order to reach economic viability. To overcome these challenges various numerous recombinat engineering tools will support the characterization of new mutants and new metabolic pathways for the production of the desired fuel-grade products, thus contributing to the strain optimization process. The improvement of metabolic models will provide a better description of the physiological behavior of the cells and a faster identification of targets for genetic modifications and further metabolic engineering. In summary, the field stands ready for the paradigm shift to panmetabolic engineering, which must include novel, global and cellular engineering tools.

5.6.1 Areas of Research to Improve Biofuel Production

Microbiologists are currently working on a number of areas to make biofuel production more efficient. These include:-

- Scaling-up the production of microbial cellulase that will break down celluloses into fermentable sugars.
- Engineering yeast to tolerate higher concentrations of alcohol to increase bioethanol production.
- Genetically modifying microorganisms to ferment sugars more efficiently to increase bioethanol yields.
- Optimizing microbial strains that will convert sugars into biobutanol as an alternative to bioethanol.
- Finding algae that produce high yields of oils or are otherwise well-adapted for biodiesel production.

References

1. BP. BP statistical review of world energy; 2009.
2. S, Gille., *et al.*, *The plant cell online*, 2011; DOI:10.1105/tpc.111.091728
3. O. J. Sanchez and C. A. Cardona, *Bioresource Technology*, Vol. 99, No. 13, pp. 5270–5295, 2008.
4. Alvarado-Morales *et al.*, *Chemical Engineering Research & Design*, Vol. 87, No. 9A, pp. 1171–1183, 2009.
5. B. Brehmer, R. M. Boom and J. Sanders, *Chemical Engineering Research & Design*, Vol. 87, No. 9A, pp. 1103–1119, 2009.
6. S. Gonzalez-Garcia, *et al.*, *Renewable & Sustainable Energy Reviews*, Vol. 13, No. 9, pp. 2613–2620, 2009.
7. R. R. Singhania, B. Parameswaran and A. Pandey, *Handbook of Plant-Based Biofuels*, CRC Press, 2009.
8. S. I. Mussatto *et al.*, *Biotechnology Advances*, Vol. 28, No. 6, pp. 1873–1899, 2010.
9. P. Sannigrahi, Y. Pu and A. Ragauskas, *Current Opinion in Environmental Sustainability*, Vol. 2, No. 5–6, pp. 383–393, 2010.
10. J. Pleiss, *Appl Microbiol Biotechnol*, Vol. 73(4), pp. 735–739, 2006.
11. S. K. Lee, *et al.*, *Current Opinion in Biotechnology*, Vol. 19(6), pp. 556–563, 2008.
12. J. E. Bailey, *Science*, Vol. 252(5013), pp. 1668–1675, 1991.
13. JD. Keasling and H. Chou, *Nature Biotechnology*, Vol. 26(3), pp. 298–299, 2008.
14. G. Stephanopoulos, *Metabolic Engineering*, Vol. 10(6), pp. 293–294, 2008.
15. A. Mukhopadhyay, *et al.*, *Current Opinion in Biotechnology*, Vol. 19(3), pp. 228– 234, 2008.
16. M. Rodriguez-Moya and R. Gonzalez, *Biofuels*, 2010, doi: 10.4155/BFS.10.5.
17. P. C. Hallenbeck, J. R. Benemann, *International Journal of Hydrogen Energy*, Vol. 27, pp. 1185–1193, 2002.
18. P. C. Hallenbeck, *et al.*, *Advances in Chemistry*, J. C. Taylor ed., vol. 6 pp. 125– 154, 2011.
19. T. Keskin, M. Abo-Hashesh, and P. C. Hallenbeck, *Bioresource Technology*, Vol. 102, pp. 8557–8568, 2011.
20. A. Adessi, R. De Philippis, *Springer*, New York, pp. 53–75, 2012.
21. H. Schepens, *EuropaBio*. Lyon 2003.
22. D. Rutz and R. Janssen, *Biofuel Technology Handbook*, WIP Renewable Energies München, Germany, 2008.
23. http://www.eia.gov/dnav/pet/pet_move_imp_dc_NUS-Z00_mbblpd_a.htm
24. A. M. R. B. Xavier, *et al.*, *Bioresource Technology*, Vol. 101, No. 8, pp., ISSN 0960–8524(print) | 1873–2976(electronic), 2012.
25. O. J. Sanchez and C. A. Cardona, *Bioresource Technology*, Vol. 99(13), pp. 5270– 5295, 2008.
26. Y. H. P. Zhang, *Journal of Industrial Microbiology & Biotechnology*, Vol. 35, No. 5, pp. 367–375, 2008.
27. G. Bokinsky, *et al.*, *Proceedings of the National Academy of Sciences*, 2011.
28. R. Srivastava, G. P. Kumar, and K. K. Srivastava, *Gene*, Vol. 164, pp. 185–186, 1995.

29. J. S. Yoo, *et al.*, *The Journal of Microbiology*, Vol. 42, pp. 205–210, 2004.
30. N. Brestic-Goachet et al., *Journal of General Microbiology*, Vol. 135, pp. 893–902, 1989.
31. M. Moniruzzaman *et al.*, *World Journal of Microbiology and Biotechnology*, Vol. 13, pp. 341–346, 1997.
32. D. C. la Grange, R. den Haan, and W. H. van Zyl, *Applied Microbiology and Biotechnology*, Vol. 87, pp. 1195–1208, 2010.
33. S. L. Tsai *et al.*, *Applied and Environmental Microbiology*, Vol. 75, pp. 6087–6093, 2009.
34. X Li , J. K. Weng, C. Chapple, *Plant J*, Vol. 54, pp. 569–581, 2008.
35. H. Hisano, R. Nanda kumar, and Z. Y. Wang, *In Vitro Cellular & Devlopmental Biology-Plant*, Vol. 45, pp. 306–313, 2009.
36. M. H. Studer *et al.*, *Biotechnology and Bioengineering*, Vol. 105, pp. 231–238, 2010.
37. E. Hardiman *et al.*, *Applied Biochemistry and Biotechnology*, Vol. 161, pp. 301–312, 2010.
38. Y. C. Li, D. C. Irwin and D. B. Wilson, *Applied and Environmental Microbiology*, Vol. 76, pp. 2582–2588, 2010.
39. L. R. Lynd *et al.*, *Nature Biotechnology*, Vol. 26, pp. 169–172, 2008.
40. A. Aristidou and M. Penttila, *Current Opinion in Biotechnology*, Vol. 11(2), pp. 187–198, 2000.
41. J. Zaldivar, J. Nielsen and L. Olsson, *Applied Microbiology and Biotechnology*, Vol. 56(1–2), pp. 17–34, 2001.
42. L. O. Ingram *et al.*, *Biotechnology and Bioengineering*, Vol. 58(2–3), pp. 204–214, 1998.
43. S. Ostergaard, L. Olsson and J. Nielsen, *Microbiology and Molecular Biology Reviews*, Vol. 64(1), pp. 34–50, 2000
44. M. Kuyper *et al.*, *FEMS Yeast Research*, Vol. 4(6), pp. 655–664, 2004.
45. L. O. Ingram *et al.*, *Biotechnology and Bioengineering*, Vol. 58(2–3), pp. 204–214, 1998.
46. E. Nevoigt, *Microbiology and Molecular Biology Reviews*, Vol. 72(3), pp. 379–412, 2008.
47. A. Matsushika *et al.*, *Applied Microbiology and Biotechnology*, Vol. 84(1), pp. 37–53, 2009.
48. M. Zhang *et al.*, *Science*, Vol. 267(5195), pp. 240–243, 1995.
49. J. C. Johnson, *Technology assessment of biomass ethanol: a multi-objective, life cycle approach under uncertainty.* In: Johnson, J.C. (Ed.), Chemical Engineering. MIT, Cambridge, MA, p. 280, 2006.
50. J. Carper, *The CRC Handbook of Chemistry and Physics.* Library Journal, Vol. 124(10), pp. 192-+, 1999.
51. T. C. Ezeji *et al.*, *Current Opinion in Biotechnology*, Vol. 18, pp. 220–227, 2007b.
52. X, Liu *et al*, *Biotechnology Progress.*, Vol. 22, pp. 1265–1275, 2006.
53. A. M. Sanchez *et al.*, *Journal of Biotechnology*, Vol. 117, pp. 395–405, 2005.
54. D. T. Jones and D. R. Woods, *Microbiological Reviews*, Vol. 50(4), pp. 484–524, 1986.
55. D. R. Nielsen *et al.*, *Metabolic Engineering*, Vol. 11(4–5), pp. 262–273, 2009.
56. J. M. Clomburg and R. Gonzalez, *Applied Microbiology and Biotechnology*, 2010.
57. S. Atsumi *et al.*, *Metabolic Engineering*, Vol. 10(6), pp. 305–311, 2008.
58. M. Inui *et al.*, *Applied Microbiology and Biotechnology*, Vol. 77(6), pp. 1305–1316, 2008.

59. E. J. Steen *et al.*, *Microbial Cell Factories*, Vol. 7. pp. 8, 2008.
60. S. Atsumi *et al.*, *Metabolic Engineering*, 2008.
61. Y. Zhang *et al.*, *Bioresource Technology*, Vol. 89, pp. 1–16, 2003.
62. P. Maki-Arvela *et al.*, *Energy Fuels*, Vol. 21, pp. 30–41, 2007.
63. Y. Chisti, *Biotechnology Advances*, Vol. 25, pp. 294–306, 2007.
64. B. Hankamer *et al.*, *Physiologia Plantarum*, Vol. 131, pp. 10–21, 2007.
65. T. A. Voelker, and H. M. Davies, *Journal of Bacteriology,*. Vol. 176, pp. 7320–7327, 1994.
66. E. M. Grima *et al.*, *Biotechnology Advances*, Vol. 20, pp. 491–515, 2003.
67. D. K. Ro *et al.*, *Nature*, Vol. 440, pp. 940–943, 2006.
68. J. Clomburg and R. Gonzalez, *Applied Microbiology and Biotechnology*, Vol. 86, pp. 419–434, 2010.
69. T. Liu *et al.*, *Metabolic Engineering*, Vol. 12, pp. 378–386, 2010a.
70. L. Katz, Dupont Develops Green 1,3-Propanediol (PDO), 2007.
71. M. R. Connor and S. Atsumi, *Journal of Biomedicine and Biotechnology*, 2010.
72. P. Handke *et al.*, *Metabolic Engineering*, Vol. 13, pp. 28–37, 2011.
73. X. Lu, *Biotechnology Advances*, Vol. 28, pp. 742–746, 2010.
74. L. P. Yomano, S. W. York, and L. O. Ingram, *Journal of Industrial Microbiology and Biotechnology*, Vol. 20, pp. 132–138, 1998.
75. M. Zhang *et al.*, *Science*, Vol. 267, pp. 240–243, 1995.
76. X. Liu *et al.*, Fatty acid production in genetically modified cyanobacteria. Proceedings of the National Academy of Sciences.
77. X. Tan *et al.*, *Metabolic Engineering*, Vol. 13, pp. 169–176, 2011.
78. J. Zhou and Y. Li, *Protein & Cell*, Vol. 1, pp. 207–210, 2010b.
79. M. M. Junior *et al.*, *Journal of the Institute of Brewing*, Vol. 115, pp. 191–197, 2009.
80. S. A. Maziar, *African Journal of Biotechnology*, Vol. 9, pp. 2906–2912, 2010.
81. M. Ergun, S. F. Mutlu, and O. Gurel, *Journal of Chemical Technology and Biotechnology*, Vol. 68, pp. 147–150, 1997.
82. J. P. Maye, Use of hop acids in fuel ethanol production. US patent. Patent US2006263484, 2006.
83. S. Nikolic *et al.*, *Food Technology and Biotechnology*, Vol. 47, pp. 83–89, 2009.
84. L. R. Lynd *et al.*, *Current Opinion in Biotechnology*, Vol. 16, pp. 577–583, 2005.
85. A. L. Demain, M. Newcomb, JHDWu, *Microbiology and Molecular Biology Reviews*, Vol. 69, pp. 124–154, 2005.
86. S. H. Krishna, T. J. Reddy, and G. V. Chowdary, *Bioresource Technology*, Vol. 77, pp. 193–196, 2001.
87. S. B. Roberts *et al.*, *BMC Systems Biology*, Vol. 4, pp. 31.
88. S. A. Tripathi *et al.*, *Applied and Environmental Microbiology*, 2010.
89. T. I. Williams *et al.*, *Applied Microbiology and Biotechnology*, Vol. 74, pp. 422–432, 2007.
90. C. Xu *et al.*, *Bioresource Technology*, Vol. 101, pp. 9560–9569, 2010.
91. K. Magnuson *et al.*, *Microbiology and Molecular Biology Reviews*, Vol. 57, pp. 522–542, 1993.
92. F. Zhang *et al.*, *Current Opinion in Biotechnology*, Vol. 22, pp. 775–783, 2011.
93. N. Ho, Z. Chen, and A. Brainard, *Applied Environmental Microbiology*, Vol. 64, pp. 1852–1859, 1998.

94. P. Richard, M. H. Toivari, and M. Penttilä, *FEBS Letters*, Vol. 457(1), pp. 135–138, 1999.
95. K. Traff, L. Jonsson, and B. Hahn-Hagerdal, *Yeast*, Vol. 19, pp. 1233–1241, 2002.
96. P. Richard, M. H. Toivari, and M. Penttilä, *FEMS Microbiology Letters*, Vol. 190(1), pp. 39–43, 2000.
97. J. W. Wenger, K. Schwartz, G. Sherlock, *PLoS Genetics*, Vol. 6(5), pp. e1000942.
98. A. Matsushika *et al.*, *Applied Microbiology and Biotechnology*, Vol. 84(1), pp. 37–53, 2009.
99. A. van Maris *et al.*, *Advances in Biochemical Engineering/Biotechnology*, pp. 179–204, 2007.
100. T. W. Jeffries, *Current Opinion in Biotechnology*, Vol. 17(3), pp. 320–326, 2006.
101. B. Hahn-Hagerdal *et al.*, *Advances in Biochemical Engineering/Biotechnology*, Vol. 108, pp. 147–177, 2007.
102. P. Richard *et al.*, *FEMS Yeast Research*, Vol. 3(2), pp. 185–189, 2003.
103. P. Kotter *et al.*, *Current Genetics*, Vol. 18(6), pp. 493–500, 1990.
104. M. Walfridsson *et al.*, *Applied and Environmental Microbiology*, Vol. 62(12), pp. 4648–4651, 1996.
105. J. Becker and E. Boles, *Applied and Environmental Microbiology*, Vol. 69(7), pp. 4144–4150, 2003.
106. Eric Young, Sun-Mi Lee and Hal Alper, *Biotechnology for Biofuels*, Vol. 3, pp. 24, 2010.
107. Wendy Higashide *et al.*, *Applied and Environmental Microbiology*, pp. 2727–2733, 2011.
108. L. Lynd, *In Lignocellulosic materials*, pp. 1–52, Vol. 38, 1989.
109. M. V. Tyurin, S. G. Desai and L. R. Lynd, *Applied and Environmental Microbiology*, Vol. 70, pp. 883–890, 2004.
110. Petitdemange *et al.*, *International Journal of Systematic Bacteriology*, Vol. 34, pp. 155–159, 1984.
111. M. D. Collins *et al.*, *International Journal of Systematic Bacteriology*, Vol. 44, pp. 812–826, 1994.
112. C. Tardif *et al.*, *Journal of Industrial Microbiology and Biotechnology*, Vol. 27, pp. 271–274, 2001.
113. E. Guedon, M. Desvaux, and H. Petitdemange, *Applied and Environmental Microbiology*, Vol. 68, pp. 53–58, 2002.
114. C. M. Gowen and S. S. Fong, *Chemistry and Biodiversity*, Vol. 7, pp. 1086–1097, 2010.
115. R. Cascone, *Chemical Engineering Progress*, Vol. 104, pp. S4–S9, 2008.
116. P. Du¨rre, *Biotechnology Journal*, Vol. 2, pp. 1525–1534, 2007.
117. Wendy Higashide *et al.*, *Applied and Environmental Microbiology*, pp. 2727–2733, 2011.
118. P. Trieu-Cuot *et al.*, *FEMS Microbiol. Lett.* Vol. 48, pp. 289–294, 1987.
119. S. Atsumi, T. Hanai, and J. C. Liao, *Nature*, Vol. 451, pp. 86–89, 2008.
120. S. Atsumi, W. Higashide, and J. C. Liao, *Nature Biotechnology*, Vol. 27, pp. 1177–1180, 2009.

6

Microbial Production of Biofuels

Panwar AS, Jugran J and Joshi GK*

*Department of Biotechnology, HNB Garhwal University,
Srinagar (Garhwal), Uttarakhand, India*

Abstract

The sources of conventional fuels are depleting at a very fast rate to meet the current energy demand. At the same time, their continuous use is posing a worldwide challenge by generating various environmental threats. As an alternative, biofuels are being seen as a renewable, sustainable, efficient, cost-effective and eco-friendly source of energy with potential to replace conventional petroleum based fuels. Microorganisms play an important role in the production of many of the biofuels by utilizing the versatility of their extraordinary metabolic capabilities. The potential of "fine-tuning" the existing microbial biofuel production processes as well as developing genetically modified species to be able to efficiently make use of otherwise useless materials and byproducts, makes microbial biofuels an appealing target for research. The present chapter describes the role of microorganisms in the conversion of biomass into five important types of biofuels.

Keywords: Biofuels, biomass, renewable, eco-friendly, genetically modified species

6.1   Introduction

Since its existence, mankind has been relying on renewable energy resources like wood, windmills, water wheels and animals such as horses and oxen. The development of new energy resources was a major driving force of the technological revolution. During the last two centuries fossil fuels have emerged as the major

---

Corresponding author: gkjoshi@rediffmail.com

Dr.Vikash Babu, Dr. Ashish Thapliyal & Dr. Girijesh Kumar Patel (eds.) Biofuels Production, (147–166) 2014 © Scrivener Publishing LLC

source of energy having utility in almost every industrial sector. Transportation worldwide depends heavily on the use of refined fossil fuels and without it the whole world will come to a standstill. The heavy use of fossil fuels has its own cost. Firstly, the natural resources for them are fast depleting and can meet the present energy requirements only for few more decades. Secondly, their increasing use has resulted in the emission and accumulation of various toxic chemicals that leads to many negative effects such as climate change, receding of glaciers, rise in sea level, loss of biodiversity, etc. This has led to a move towards alternative, renewable, sustainable, efficient and cost-effective energy sources with lesser emissions. Among other energy alternatives, biofuels are being seen as one of the strategically important sustainable fuel sources to significantly cater to world energy demand. Biofuels are renewable liquid or gaseous fuels made by and/or from living organisms or the wastes that they produce. There is a noticeable and growing trend worldwide in the use of biofuels for efficient bio-energy conversion. A range of biofuels are being used for this purpose. Both developing and industrialized countries are considering biofuels as a relevant technology for a variety of factors such as security reasons, environmental concerns, foreign exchange savings and socioeconomic issues related to the rural sector. Biofuels offer several advantages including their easy availability, ecofriendly combustion, biodegradability and sustainability. Considering the rising prices for crude oil and the increasing political instability in oil producing countries, the developments in the area of biofuels are of paramount importance. Compared to a near absence of oxygen in petroleum, biofuels have oxygen levels from 10% to 45% making the chemical properties of biofuels very different from petroleum [1]. In many of the countries, biofuels are increasingly used for heat and power production. Different measures have been adopted by different countries to introduce biofuels with varying production cost that depends on location, feedstock and several other factors. There is also a role of political agendas and environmental concerns in the utilization of biofuels. Plant biomass is an abundant and renewable source of energy that is rich in compounds that can be efficiently converted through thermochemical routes into biofuels of various types. Microorganisms through the process of fermentation can also transform the raw materials of various types into biofuels like alcohol, butanol etc. In the recent years, microorganisms particularly microalgae, have been reported to assimilate carbon in the form of various lipids that can be processed to be used as an excellent biofuel. One of the main problems associated with the production of the majority of biofuels is still the production cost, which can be minimized by developing more sophisticated technologies based on chemical conversions or the use of microorganisms.

6.2   Types of Biofuels Produced Through Microorganisms

Microorganisms can convert the carbonaceous matter of a raw material into biofuels such as ethanol, butanol etc. Plant waste is the ideal material to serve as the raw material for such a conversion. The general schematic presentation of biofuel production is given in figure 6.1. In the first step is that the complex raw material is reduced to simple sugars either by chemical means or by employing microbes secreting degradative enzymes. The simple sugars are then converted into biofuels through microbial fermentation. The following types of biofuels can be produced with the involvement of microorganisms which are summarized in table 6.1.

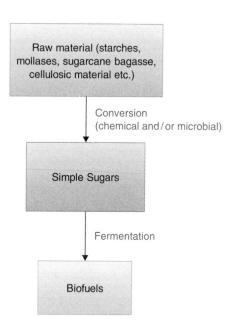

Figure 6.1 Outline of the fermentation route of biofuel production.

Table 6.1 Microorganisms and their substrates for biofuel production.

Bacterial strain	Substrate	Product
Clostridium thermocellum	Cellulose rich sources	Biogas [2]
Slackia heliotrinireducens	Cellulose, organic material	Biogas [2]
Carboxydothermus hydrogenoformans	Organic acids	Biogas [2]
Methanosarcina barkeri	Volatile organic acids, CO_2 and H_2	Biogas [2]
Methanoculleus marisnigri,	Volatile organic acids, CO_2 and H_2	Biogas [2]
Saccharomyces ceveresiae	Orange peel waste (sugars)	Bioethanol [3]
Staphylococcus warneri	Palm oil	Biodiesel [4]
Candida rugosa	Palm oil	Biodiesel [4]
Fusarium solani	Palm oil	Biodiesel [4]
Clostridium cellulolyticum	Cellulosic substance	Bioethanol [5]
Lactobacillus casei	Cellulosic substance	Bioethanol [5]
Zymomonas mobilis	Glucose and fructose	Bioethanol [6]
Plchia stipitis NRRL Y-7124	Wheat straw	Bioethanol [7]
Klebsiella oxytoca M5A1 (pLOI555)	Cellulosic substance	Bioethanol [8]
Clostridium acetobutylicum.	Straw, leaves, grass, spoiled grain and fruits etc	Biobutanol [9]
Enterobacter Aerogenes HU-101	Crude glycerol	Hydrogen [10]
Caldicellulosiruptor saccharolyticus	Carbohydrates rich sources	Hydrogen [11]

(Continued)

Table 6.1 (*Cont.*)

Bacterial strain	Substrate	Product
Thermotoga elfii	Carbohydrates rich sources	Hydrogen [11]
Caldicellulosiruptor owensensis	Carbohydrates rich sources	Hydrogen [12]
Thermoanaerobacter ethanolicus	Carbohydrates rich sources	Bioethanol [13]

6.2.1 Bioethanol

Alcoholic fuels are generally obtained from biological rather than petroleum sources. Since they originate from biological sources, they are sometimes referred to as bioalcohols. Ethanol or ethyl alcohol produced from biomass by hydrolysis followed by fermentation is called bioethanol. Biothanol can be used in combination with gasoline as fuel to drive internal combustion engines of automobiles. Bioethanol up to a concentration of 5 % can be added to the petrol with no engine modification required. The acceptable blend can have a bioethanol concentration as high as 85 % to be used with modified engines. Fuel containing a mixture of ethanol and petrol has a lower sulphur content than petrol alone. This results in lower emissions of sulphur oxide, a major component of acid rain, and a carcinogen.

Crops such as sugarcane, wheat and corn are the most essential types of natural bioresources that can be exploited for bioethanol production. Plant material contains a large variety of carbohydrates with a general formula of $(CH_2O)_n$ that can be transformed into ethanol by various chemical and biochemical methods. A major part of the ethanol produced as biofuel is derived from sugar or starch. Sugarcane, sugar beet and sorghum are the common crops used as feedstock for bioethanol production, whereas starchy feedstocks include maize, wheat and cassava. Plant cellulose may also be used as feedstock for sugar generation for subsequent bioethanol production. However, due to a complex structure, it is difficult to depolymerise cellulose into simple sugars. On the other hand, hemicelluloses are easily broken down into constituent sugars such as xylose and pentose but the subsequent fermentation process is difficult as it requires highly efficient microorganisms capable of fermenting 5- carbon sugars to ethanol. Although lignin itself can be used as a fuel, it is practically not fermentable.

Bioethanol production from plant material is done in three steps. In the first step a pretreatment is given to the plant material by employing the chemical techniques of stream explosion, ammonia fiber explosion, CO_2 explosion, acid or alkali hydrolysis. Generally these chemical treatments are operative at extreme conditions such as high temperatures, etc. To some extent, biochemical hydrolysis catalyzed by hydrolytic enzymes is a suitable alternative. In the next step the carbohydrates present in plant materials can be converted into simple sugars by the saccharificatoin process. Enzymes such as cellulase are employed for the saccharification of cellulose to simple sugars. The efficiency of cellulose hydrolysis depends on porosity, crystallinity and presence of lignin and hemicelluloses. Lignin and hemicelluloses exert a negative effect on cellulase by reducing the enzyme's accessibility for its substrate. Therefore, The hydrolysis process can be significantly improved by lignin and hemicelluloses removal, the reduction of cellulose crystallinity and the increase of porosity in the pretreatment processes [14]. Fermentation is the final step in which sugars are converted into alcohol by the action of microorganisms in the absence of oxygen. The bioethanol produced can be purified by distillation. Yeast is the microorganism normally employed for alcohol fermentation. It secretes an invertase enzyme, which converts the sucrose of the raw material into glucose and fructose. These new compounds are then converted to ethanol by 'zymase', another enzyme present in yeast. The convertibility of any particular type of biomass to sugars decides its value as feedstock for fermentation.

6.2.1.1 Microorganisms Involved in Bioethanol Production

Ethanol is largely consumed by the pharmaceutical, chemical and food industries for different purposes. Approximately, 80% of the production of ethanol in industries is carried out by fermentation. Depending on the raw material utilized, different species of microorganisms can be used for bioethanol production. Yeasts are the commonly preferred microorganisms for the industrial scale production of bioethanol; *Saccharomyces ceveresia is* commercialized worldwide for this. Some of the thermophiles that have been reported to be associated with ethanol production are *Clostridium* hermosaccharolyticum and *Thermoanaerobacter ethanolicus*. They preferably use pentose and hexose as a substrate. The highest ethanol yield (1.9 moles per mole of glucose consumed) has been reported for the thermophilic bacteria *Thermoanaerobacter ethanolicus* [15]. For bioethanol production, thermophiles offer advantages of fast growth rates and the utilization of a broad range of substrates. Along with this, many thermophiles produce fewer types of undesired end products compared to mesophiles. Fungi such as *Trichoderma reesei* and *Aspergillus niger* produce the enzyme cellulase which can excellently convert cellulose present in sugarcane bagasse into fermentable sugars for ethanol production. In the tropical area, bioethanol has been produced by using *Zymomonas mobilis*, but its spectrum of carbohydrate fermentation is not too broad which makes it a little unimpressive. Recently, genetic engineering is popular to express foreign genes associated with the catabolism of different types of organic matter. *Zymomonas mobilis* and E.coli have been successfully transformed into efficient ethanol producers [16, 17]. The company DuPont has recently started to use a genetically engineered Z. mobilis for cellulosic ethanol production [18].

6.2.1.2   Substrates for Bioethanol Production

So far, the majority of world bioethanol production is based on sugar crops such as sugarcane and sugar beet. Grains such as corn are also utilized in many areas as raw material for microbial ethanol production. Scientists are now exploring lignocellulosic biomass as a substrate because of its plentiful abundance. Lignocellulose is a mixture of lignin, hemicellulose and cellulose. The lignocellulose is the part of the plant that remains undigested by humans and most animals. It is a non-foodstuff e.g. stalks, sawdust and wood chips. Lignocellulosic biomass can be hydrolysed into sugar. Different hydrolysis technologies for different lignocellulosic feedstocks produce a mixture of sugars (such as glucose, xylose, arabinose), some of which are very difficult to ferment efficiently into bioethanol using conventional microorganisms. To achieve a high production of bioethanol, fermentation of the difficult sugars resulting from the hydrolysis of the lignocellulosic biomass is necessary. For this, microorganisms with the ability to ferment difficult sugars are required. Genetic engineering has resulted in the development of new strains of microorganisms that are capable of efficiently fermenting difficult sugars such as xylose.

There is a huge amount of non-edible plant waste that can be recycled for bioethanol production. Sugarcane bagasse is an abundantly and cheaply available byproduct from sugar industries. It is rich in cellulose which can be hydrolysed by the cellulase enzyme for bioethanol production. Similarly, other cellulosic biomass such as wood, paper pulp or agricultural waste can be used for bioethanol production. Orange peel, Agave leaves waste, Sago waste and potato flour is also being used for bioethanol production.

6.2.2   Butanol

Biobutanol is often misdescribed as a "new" fuel. This has been continuously produced since the beginning of 20th century as a solvent and a basic chemical. However, new uses for biobutanol have been emerging recently, e.g. as a diesel and kerosene replacement, as silage preserver, biocide and C4 compound for chemical industry. It is a four carbon alcohol that contains more carbon and hydrogen than ethanol and therefore, is easier to blend with gasoline and other hydrocarbon products. It is less corrosive than ethanol and therefore, can easily be shipped and distributed to the filling stations. An 85% butanol/gasoline blend can be used in unmodified petrol engines. In a diesel compression engine it can be blended up to at least 30 % (70 % diesel). Butanol is much less evaporative than gasoline or alcohol. This makes it safer to use as it generates fewer volatile organic compound emissions when consumed in internal combustion engines. It yields only carbon dioxide making it more environmentally friendly than biofuel [14]. Butanol exists in four isomeric forms, viz. n-butanol (normal butanol), 2-butanol (secondary butanol), i-butanol (iso-butanol) and t-butanol (ter-butanol). Their fuel efficiency in terms of energy release and blending properties when compared to gasoline is similar. However, their manufacturing processes are quite different. Mainly, n-butanol can be produced through fermentation on large scale. The industrial production of n-butanol through the fermentation of starch and sugar is, in fact, a century old practice. This is also known as "Acetone-butanol-ethanol" fermentation or ABE fermentation which involves anaerobic conversion of carbohydrates by strains of *Clostridium acetobutylicum* into acetone, butanol and ethanol (Fig. 6.2). Bacteria secrete numerous enzymes that facilitate the breakdown of polymeric carbohydrates into monomers. The recovery and purification of n-butanol from the

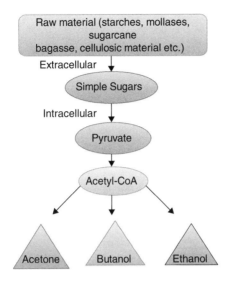

Figure 6.2 Schematic representation of ABE production pathway.

fermentation broth is costly and difficult as a low concentration of it is accumulated in the fermentation broth due to product inhibition. However, efforts are underway to increase the concentration by developing advanced technology mainly through genetically modified microbes. The t- butanol is a petrochemical product and no fermentation process is known to produce it. A small amount of iso-butanol can be produced by yeast as a by-product during wine making. However, it is undesirable in this process due to its toxic nature. 2-butanol production partially involves fermentation where initially a bacterial fermentation converts glucose of the raw material into an intermediate product. This intermediate product is chemically modified and converted into 2-butanol. Compared to n-butanol and iso-butanol, 2-butanol has a boiling point lower than that of water, which makes its recovery from fermentation cost effective.

6.2.2.1   Microorganisms Involved in Biobutanol Production

Butanol is produced by the bacterias *Clostridium acetobutylicum* and *Clostridium beijerinckii* under anaerobic conditions. During the fermentation process acetone, butanol and ethanol are produced at a ratio of 3:6:1. The main problem with these microbes is the low product yield as the accumulation of butanol in the fermentation broth inhibits their growth. Recently, genetically modified strains of *C. acetobutylicum* have been developed that can sustain high concentrations of butanol thereby resulting in a high butanol yield.

6.2.2.2 Substrate for Biobutanol Production

The use of crop products for butanol fermentation is not very economical primarily because of their high demand in food industries leading to fast price escalation of the raw material. Therefore, agricultural wastes such as straw, leaves, grass, spoiled grains and fruits etc are commonly used as raw material for butanol production through fermentation. Other sources for biobutanol production can be plant biomass and algae.

6.2.3 Biodiesel

Vegetable oil (m)ethyl esters, commonly referred to as "biodiesel" is a prominent candidate as an alternative fuel. Biodiesel production has considerably increased in the past few years and so have the amount of residues generated during production. Europe is still the biggest biodiesel producer, whereas Brazil had the highest increase in production rates in the last few years when compared with the United States and Europe. Vegetable oil itself can be used as a fuel for diesel engines but its high viscosity is a limitation and therefore, demands engine modification. Vegetable oils can be converted into biodiesel by the transesterification process in the presence of a catalyst. Potassium and sodium alkoxides are such catalysts used for preparing methyl, ethyl, propyl and butyl esters of vegetable oil. Transesterification lowers the viscosity of the oil. This is a potentially less expensive way of transforming the large, branched molecular structure of the bio-oils into smaller, straight-chain molecules of the type required in regular diesel combustion engines [1]. When used in vehicle engines, the biodiesel provides slightly lower power and torque with a high consumption rate compared to petroleum diesel. It however, offers the advantages of low sulphur content, flash point, aromatic content and biodegradability. A biodiesel-diesel blend of up to 5 % (v/v) can be used in a normal diesel engine. Conventional diesel engines can also be converted into those suitable for pure biodiesel at a reasonable price. This however, suffers from the disadvantage of the frequent requirement of engine oil changes.

6.2.3.1 Microorganism Involved in Biodiesel Production

Biodiesel is an environmentally friendly alternative liquid fuel. Biodiesel is made from renewable resources and has environmental benefits that make it more attractive. Microorganisms have no direct role in the synthesis of biofuels. In fact, the microbiology of biodiesel is a concern because of bacterial oxidation during storage as well as the unavoidable water content leading to corrosion problems. However, the conversion of vegetable oil to methyl- or other short chain alcohol esters can be catalyzed in a single transesterifcation reaction using microbial derived lipases in inorganic solvents. Lipases of high specific activity derived from a range of microorganisms, e.g. *Staphylococcus warneri, Candida rugosa, Fusarium solani etc.*, have been successfully tested at the laboratory scale for such conversion. The process at an industrial scale is still under development. Strategies like immobilization of the enzyme and cytoplasmic overexpression are developed to catalyze methanolysis in a solvent free reaction system. An entirely new concept of bacterial microdiesel has been proposed in recent years. There are experimental proofs of the esterified lipid production through bacteria by employing the advanced concept of metabolic engineering. Considerable further development is needed in this direction [19].

6.2.3.2 Substrates for Biodiesel Production

Biodiesel can be used in any diesel engine without modification in an environmentally friendly manner. Due to its less polluting and renewable nature, there is an increasing interest worldwide towards the development of biodiesel from different types of plant oils. Most of the biodiesel currently used is made from soybean oil. Soybean oil, however, being a food product, makes the cost effective production of biodiesel very challenging. Other sources of biodiesel are oils from mustard, jojoba, jatropha, coconut, sunflower etc. Low-cost fats and oils such as restaurant waste and animal fats have also been successfully used for biodiesel production but the yield is low because of the presence of free fatty acids that are hard to convert into biodiesel by any catalyst.

Some algae store a good amount of lipids in them and this property makes them a good source of biodiesel. Besides this, algae can be grown using waste materials such as sewage with no use of fertile land making it an attractive alternative to plant oil for biodiesel production. Prokaryotic microalgae, i.e. cyanobacteria (Chloroxybacteria), and eukaryotic microalgae, e.g. green algae (Chlorophyta), red algae (Rhodophyta) and diatoms (Bacillariophyta) can be used for microbial lipid production. *Neochloris oleabundans* (fresh water microalga) and *Nannochloropsis* sp. (marine microalga), due to their high oil content, are suitable for biodiesel production, [20].

6.2.4   Biohydrogen

Biohydrogen is a potential biofuel that can be obtained from organic waste materials. The use of hydrogen as an energy source could improve global climate change, energy efficiency, and air quality. It could be used as an alternative for fossil fuels because it is easily convertible to electric energy in fuel cells or burnt and converted to mechanical energy without excessive production of CO_2. Hydrogen is being looked to as a futuristic fuel to reduce dependence on petroleum, to decrease pollution and to minimize greenhouse gas emissions. Biomass can be converted into hydrogen by thermochemical conversion processes such as pyrolysis or steam gasification. Biologically, it can be produced by algae or cyanobacteria through bio-photolysis of water. Other biological routes of hydrogen production involve photo-fermentation and dark-fermentation. In photo-fermentation, hydrogen is produced from organic substances by photosynthetic bacteria. On the other hand, anaerobic organisms can produce H_2 from organic substances in 'dark-fermentation'. A combination of both dark- and photo-fermentation is now considered the most efficient method for hydrogen production through fermentation. It utilizes a series of steps, firstly the waste materials are broken down into organic acids in dark-fermentation, and these organic acids are then used as a substrate in the photo-fermentation process.

Hydrogen production by fermentation of biomass is a continuous process that requires a non-sterile substrate with a readily available mixed microflora. The ease with which a raw material can be processed to yield a fermentable feedstock decides the overall rate of conversion of biomass to hydrogen. A variety of reactor systems have been developed for hydrogen production from the bacterial fermentation of sugar. Hexose concentration in the medium, temperature, pH and reactor hydraulic retention time (HRT) are the important factors that decide the H_2 yield. The hydrogen fermentation can be operated in the liquid phase with immobilized cells or by enabling the formation of self-flocculated granular cells. Despite various reports at the laboratory scale, microbiological hydrogen is not yet developed into an economically viable technology at the industrial scale.

6.2.4.1 Microorganisms Involved in Biohydrogen Production

Three types of bacteria can be utilized for hydrogen production: cyanobacteria, anaerobic bacteria and fermentative bacteria. For photobiological hydrogen production, cyanobacteria are considered as the ideal source since they have simple nutritional requirements. They can grow in air (N_2 and CO_2), water and mineral salts with light as the only energy source and carry out efficient photoconversion of water to hydrogen. For immediate hydrogen production, the cultivation of cynobacteria in a nitrate-free media under air and CO_2 followed by incubation in light under argon and a CO_2 atmosphere is being practiced. Anaerobic bacteria such as *Clostridium* spp. have been employed for H_2 production. Among fermentative bacteria, the mixed acid fermentation of *E.coli* and butylene glycol fermentation by *Aerobacter* also produce hydrogen gas. A high yield of hydrogen (3.3–4.0 mole of H_2 per mole carbohydrate utilized) has been reported from the hyperthermophiles *Caldicellulosiruptor saccharolyticus* and *Thermotoga elfii* [21]. The purple non-sulphur (PNS) bacteria (e.g. genus *Rhodobacter*) also hold significant promise for the production of hydrogen by anoxygenic photosynthesis and photo-fermentation.

6.2.4.2 Substrate for Biohydrogen Production

A variety of biomass has been used for hydrogen production through thermochemical and fermentative processes. Steam gasification of bio-nut shell and black liquor yields hydrogen. Crop straw, tea waste and olive husk undergo pyrolysis for hydrogen production. Pulp and paper waste and manure slurry is used for hydrogen production through fermentation.

6.2.5 Biogas

Anaerobic digestion of an organic fraction of biomass can be broken into a methane and carbon dioxide mixture called "biogas". It is a renewable energy source, like solar and wind energy. Biogas is an environmentally friendly biofuel that can be produced from regionally available raw materials like dung or sewage in digesters. The digestion may take a period of ten days to a few weeks. Biogas is basically a mixture of methane (CH_4) and carbon dioxide

Table 6.2 Composition of biogas.

Compound	Percentage (%)
Methane (CH_4)	50–75
Carbon dioxide (CO_2)	25–50
Nitrogen (N_2)	0–10
Hydrogen (H_2)	0–1
Hydrogen sulphide (H_2S)	0–3

(CO_2) with a small amount of hydrogen sulphide (H_2S), moisture and siloxanes (Table 6.2).

The composition of biogas depends on the anaerobic digestion conditions. The chief component of biogas is methane, which amounts to 50% of the total volume. The advancement in waste treatment technologies has enhanced this limit up to 55–75% of methane. In reactors with free liquids, the methane concentration can reach as high as 80–90% using *in-situ* gas purification techniques [22]. In some cases, biogas contains siloxanes. These siloxanes are formed from the anaerobic decomposition of materials commonly found in soaps and detergents. For the production of biogas, a digester system, commonly referred to as an anaerobic digester, is employed. In the digester, the decomposition of biomass and its complex organic compounds into simpler compounds and gaseous products takes place. There are mainly three types of digesters: (1) vertical tank systems, (2) horizontal tank or plug-flow systems, and (3) multiple tank systems. Several other types of bio-digesters e.g. floating drum, fixed dome, plastic tubes have also been developed.

Biogas production through the anaerobic decomposition of organic matter is a complex process in itself. This involves a step-wise series of reactions requiring the cooperative action of several organisms. The overall process can be divided into four steps (Fig 6.3). In the first step, known as the hydrolysis process, carbohydrates are broken down into simple sugars, proteins into peptides and free amino acids, and fats into fatty acids by the extracellular enzymes, such as cellulases, amylases, proteases and lipases released by the microorganisms. In the second acidogenic step, the products from the hydrolysis are converted into organic acids like carbonic acid

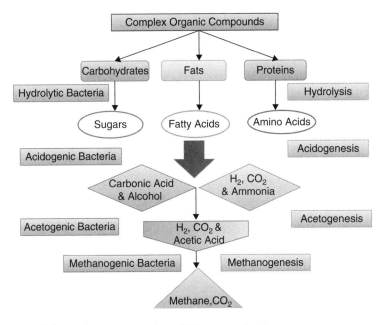

Figure 6.3 Schematic representation of biogas production.

and low alcohols by acid-producing bacteria. Since these bacteria are facultative anaerobic and are able to grow in an acidic environment, subsequently, the products of the acidogenic process are converted into acetic acid (CH_3COOH), carbon dioxide (CO_2) and hydrogen (H_2). Microorganisms need oxygen and carbon for acetic acid production and this creates an anaerobic environment suitable for methane-producing microorganisms. Finally, methane producing bacteria decompose the product of acidogenesis and produce methane. If the process is well balanced, these phases are synchronised. Temperature is an important factor that decides the rate of biogas production. Biogas can be produced by employing psychrophiles, mesophiles and thermophiles over a range of temperatures. As the temperature increases the rate of methane production also increases. But increased temperature is associated with an elevation in the concentration of free ammonia, which leads to the inhibition of methane production. Other factors such as the nature and the concentration of the substrate, feed rate, pH value (of input mixture in the digester should be between 6 to 7), bacterial population and chemical inducers etc. also influence the biogas production to a considerable extent.

6.2.5.1   Microorganism Involved in Biogas Production

Biogas is methane produced by the process of anaerobic digestion of organic material by anaerobes. A variety of microbes partici-
pate in the microbial food chain and gradually degrade the complex
molecules to produces the mixture of CH_4 and CO_2. The first step of
biomass hydrolysis involves a number of microbes like *Clostridium
thermocellum, Bifidobacterium longrum, Clostridium celluloliticum,
Bacteroides thetaiotaomicron, Enterococcus faecalis, Bacteroides capil-
losus* etc. In the second step of acidogenesis, *Slackia heliotriniredu-
cens, Cloacamonas acidaminovorans, Clostridium kluyveri, Clostridium
acetibutilicum, Clostridium perfingens, Clostridium saccharolyticum,
Caldanaerobactersubterraneus, Finegoldia magna, Enterococcusfaecium,
Lactobacillus helveticus* and *Streptococcus pneumoniae* play the important roles [2]. The microflora comprising mainly of *Arboxydothermus
hydrogenoformans, Morella thermoacetica* and *Pelotomaculum thermo-
propionicum* is involved in the third step of acetic acid and CO_2
production. Finally, methanogenesis is carried out by acetotrophs
comprising *Methanosarcina barkeri, Methanosarcina acetivorans*
and hydrogenotrophs which include *Methanoculleus marisnigri,
Methanoregula boonei, Methanosphaerula palustris, Methanospirillum
hungatei, Methanoplanus petrolearius* and *Methanocorpusculum labre-
anum*. These microorganisms are very sensitive to environmental
variations since they are strictly anaerobic.

6.2.5.2   Substrates for Biogas Production

Biogas is produced by the anaerobic digestion or fermentation of
biodegradable materials such as sewage, municipal waste, green
waste, plant material, and crops (Fig. 6.4 and 6.5). Farmers can produce biogas from cattle dung by using anaerobic digesters. Manure
from other farm animals like horses, chickens, pigs can also be
used. Besides this, waste oil, fat from slaughter waste, organic-rich
industrial waste water, organic household or municipal solid waste,
garden waste and rotten foodstuff, all have the potential of biogas
production. Organic waste from hospitals which may include paper
and cotton, waste from agriculture or food production, municipal sewage sludge, etc. are also consumable substrates for biogas
production. Some crops such as maize (whole plant including the
corn), grass, clover, young poplar and willow are especially grown
for biogas production. Homogeneous substrate quality can be
maintained throughout the year by storing the green plant material

Methane content in biogas (%)

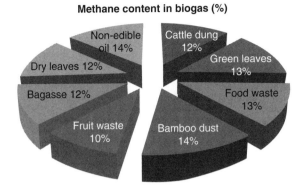

Figure 6.4 Methane content in biogas of different wastes [23, 24].

Biogas production liters/kg

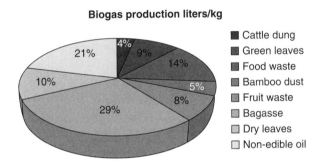

Figure 6.5 Biogas production potential of different wastes [23, 24].

as silage. Fig. 6.4 and 6.5 depict the methane content and biogas production potential of different types of waste materials.

6.3   Future Prospects and Conclusion

From an economic point of view, it is important for the industry that the feedstocks for biofuel production are abundant year-around. In this sense, some countries may not have sufficient biomass throughout the four seasons especially due to the competition of biomass for other uses. The diversification of biomass feedstocks in the industry and the regionalization of production plants may provide a solution to it. In the future, the availability of edible materials for biofuels at a global scale will become increasingly squeezed as the demand increases for food to provide greater food security and to feed the growing global population and satisfy the growing consumption of the protein-rich diet. Hence, increases in the quantity of biomass used for biofuel production of any significant scale would have to come principally from non-edible parts of plants, or from agricultural waste. There is potential for using municipal organic waste, agricultural wastes and residues from forestry and agriculture for production of liquid biofuels or biogas (second-generation biofuels). However, significant investment in waste treatment by anaerobic digestion will be required for biogas, and second-generation biofuels are not expected to play a major role in bioenergy until after 2020. There is reason to be cautious about the contribution that second-generation biofuels from agricultural waste and forestry will make, largely because these materials have important ecosystem functions that limit how much can be extracted. Some agricultural wastes are also very important sources of fodder.

Currently produced first generation biofuels appear to provide little or no greenhouse gas reduction once all the impacts of biomass cultivation and fuel production are taken into account. Despite the development of criteria for the protection of biodiversity and ecosystem services and certification schemes, there are serious concerns about the sustainability of biofuel production and its impacts on the natural environment.

The manipulation of microbes to allow the cost competitive commercial production of fuels and chemicals, such as ethanol, butanol, isoprenoids and others, on sugar and lignocellulose has advanced significantly in the last years. Indeed, several processes based on engineered microorganisms, especially yeast and bacteria species, have been developed and implemented. [25]. The task of obtaining microbial strains able to operate under industrial process conditions is still very challenging. The use of engineered strains commonly employed in industry, especially yeasts and E. coli, can solve this problem. However, metabolic engineering strategies have to be used to drive metabolite synthesis through homologous and heterologous pathways. Although in industry the most commonly used strains are engineered producers of fuels and chemicals, a careful screening of the biodiversity is generally advisable, since microorganisms can naturally produce a vast range of compounds. Once enzymes and metabolic pathways in such microorganisms are identified and isolated they can be used for the development of recombinant strains for the production of biofuels. It is expected that strains for industrial implementation of these bioprocesses should become available in the next years.

References

1. A. Demirbas, *Energy Conversation and Management*, Vol. 49 (8),p. 2106–2116, 2008.
2. R. Wirth, E. Kovacs, G. Maroti, Z. Bagi, G. Rakhely and K. L. Kovacs, *Biotechnology for Biofuels*, Vol. 5, p. 41, 2012.
3. M.R. Wilkins, L. Suryawati, N.O. Maness, and D. Chrz, *World Journal of Microbiology and Biotechnology*, Vol. 23, p. 1161–1168, 2007.
4. P. Winayanuwattikun, *African Journal of Biotechnology*, Vol. 10 (9), p. 1666–1673, 2011.
5. V.S. Kumar, P. Gunasekaran, *Journal of scientific and industrial research*,Vol.64, p. 845–853, 2005.
6. B.S. Dien, M.A. Cotta, and T.W. Jeffries, *Applied Journal of Biotechnology*,Vol. 63, p. 258–266, 2003.
7. J.N Nigam, *Journal ofBiotechnology*,Vol. 87, p. 1727, 2001.
8. B.S Dien, M.A. Cotta, and T.W. Jeffries, *Applied Journal of Microbial. Biotechnology*, Vol. 63, p. 258–266,2003.
9. N. Qureshi, and I.S Maddox, *Journal of Fermentation and Bioeng*ineering, Vol. 80 (2), p. 185–189, 1995
10. T. Ito, Y. Nakashimada, K. Senba, T. Matsui and N. Nishio, *Journal of Bioscience and Bioengineering*Vol.100, p. 260–265, 2005.
11. E.W. Van Niel, M.A. Budde, G.G. de Haas, F.J. Van der Wal, P.A. Claassen, and A.J. Stams, *International Journal of Hydrogen Energy*, Vol. 27, p. 1391–8, 2002.
12. A.A. Zeidan, Ed. W.J. Van Niel., *International Journal of Hydrogen Energy*, Vol. 35, p. 1128–37, 2010.
13. J. Wiegel, L.G. Ljungdahl, *Archives of Microbiology*, Vol. 128, p. 343–8, 1981.
14. P.S. Nigam and A. Singh, *Progress in Energy and Combustion Science*, Vol. 37(1), p. 52–68, 2011.
15. P. Sommer, T. Georgieva, B.K. Ahring,*Biochemical Society Trans*actions, Vol. 32, p. 283–9, 2004.
16. H. Yanas, M. Masuda, T. Tamaki and K. Tonomura, *Journal of Fermentation and Bioengineering*, Vol. 70, p. 1–16, 1990.
17. P. Su, S. F. Delaney and P. L. Rogers, *Journal of Biotechnology*, Vol. 9, p. 139–152, 1998.
18. DuPont Danisko Cellulosic Ethanol LLC, Fermentation, In: *Technology*, (April 4, 2011), Available from http://www.ddce.com/technology/fermentation. html), 2011.
19. Dominik, V.Z. Vladimir, W.H. Schwarz, *Applied Microbiology and Biotechnology*, Vol. 77, p. 23–35, 2007.
20. Luisa Gouveia · Ana Cristina Oliveira, *Journal of Indian Microbiology and Biotechnology*, Vol. 36, p. 269–274, 2009.
21. E.W. Van Niel, M.A. Budde, G.G. de Haas, F.J. van der Wal, P.A. Claassen, and A.J. Stams, *International Journal of Hydrogen Energy*, Vol. 27, p. 1391–8, 2002.

22. B. Richards, F.G Herndon, W.J Jewell, R.J Cummings, and T. E. White, *Biomass and Bioenergy*, Vol. 6 (4), p. 275–274, 1994.
23. B. Mursec, P. Vindis, M. Jansekovic, *Journal of achievements in Materials & Manufacturing Engineering*, vol. 37, p. 652–659, 2009.
24. A. Olukemi, Bolarinwa, and O Esther Ugoji, *Journal of Science Research & Development*, vol 12, p. 34–45, 2010.
25. J. Wiegel, L.G. Ljungdahl, *Archives of Microbiology*, Vol. 128, p. 343–8, 1981.

7

Microalgae in Biofuel Production-Current Status and Future Prospects

Navneet Singh Chaudhary

Department of Biotechnology, Sir Padampat Singhania University (SPSU),
Udaipur-313001, Rajasthan (India)

Abstract

Energy plays a pivotal role in the socio-economic development of any country leading to improved standards of living and quality of life. The current energy demand is fulfilled mostly by non-renewable fossil fuels. In the future, it may be extremely difficult to meet the increasing global energy demand from the emerging economies of the world because of the depletion and non-renewable nature of fossil fuels. Biofuel production from renewable sources can reduce fossil fuel dependence and assist in maintaining a healthy environment and economic sustainability. Multiple approaches are currently being adopted for the use of microorganisms such as microalgae/cyanobacteria in the production of various biofuels such as biodiesel, biohydrogen, biogas and bioethanol/biobutanol. Microalgae are very rich in biomass and fatty oil contents for biofuel production and are seen as potential sources of various biofuels. Many species of microalgae/cyanobacteria have been researched for selection, processing, extraction, separation and some are engineered genetically and metabolically for the production of adequate microalgal biofuels to satisfy the fast growing energy demand within the restraints of land and water resources. Microalgal biofuel production is potentially sustainable with some biotechnological, ecological and economic challenges.

Keywords: Biofuel, microalgae, biohydrogen, biodiesel, biogas (bio-syngas), bioethanol, cyanobacteria, genetic and metabolic engineering

Corresponding author: navneetiitr@gmail.com

Dr.Vikash Babu, Dr. Ashish Thapliyal & Dr. Girijesh Kumar Patel (eds.) Biofuels Production, (167–210) 2014 © Scrivener Publishing LLC

7.1 Introduction

Energy plays a pivotal role in the socio-economic development of any country leading to an improved standard of living and quality of life. Biofuel research is currently an area of immense interest due to the increasing global energy demand by emerging economies and the recent increases in global oil prices. According to the International Energy Agency, World Energy Outlook, the demand for energy will probably increase by 40% between now and 2030. Demand for liquid biofuels tripled between 2000 and 2007. Biofuels contribute to 10% of the global energy supply, whereas liquid biofuels account for a mere 1.5% of transport fuels. There are four major types of commercial biofuels: (i) biohydrogen, obtained from lignocellulosic biomass via microbial photolysis and fermentation (ii) biogas, obtained from agricultural and animal waste, or lignocellulosic biomass via anaerobic digestion (iii) biobutanol and bioethanol, obtained from sugar, starch crops and lignocellulosic biomass via saccharification and fermentation (iv) biodiesel, obtained from oil crops and lignocellulosic materials. Multiple approaches are currently being researched for the use of microorganisms in the production of various biofuels such as alcohol, hydrogen, biodiesel, and biogas from multiple starting materials. Both edible and non-edible oils can serve as raw materials for the production of biodiesel. When edible oils are used for the production of biodiesel it does not turn out to be cost effective because it also raises the price of food preparation when vegetable oil is mainly used. Research on the production of ethanol from plant materials started by German scientists as early as 1898, and continued in the United States during World War I. These processes involved the use of acidification to produce glucose from wood and subsequent fermentation by anaerobic microorganisms. But in countries like India where production of edible oil is not much higher, emphasis is given to non-edible oils like Jatropha, Pongamia, Mahua etc. These plants could be cultivated at a massive scale on agricultural, degraded, wasteland, so that the raw material may be available to produce biodiesel at the farm scale. The availability of land is considered the major limitation for biofuel production from non-edible oil. It creates a competitive situation for acreage due to which the prices may rise for food based on agricultural feedstock. Besides this, biofuel from oil crops, waste cooking oil and animal fat may not be reasonably sufficient to meet the existing demand of transportation fuels. The above limitations can be significantly avoided by exploring the generation of biodiesel from microalgae. Microalgae are a promising and potent source of triglycerides and in some cases, the oil productivity of microalgae greatly exceeds the oil productivity of the best oil producing crops [1]. Oil content in microalgae varies from 20–50% and sometimes may reach up to 80% by weight of dry biomass compared to 5% of the best agricultural oil crops. Like plants, microalgae also use sunlight to produce oils but they do so more efficiently than crop plants. Unlike oil crops, microalgae grow more rapidly and less land area is required for their cultivation. It commonly doubles its biomass within 24 hrs as compared to plants where several years are required for them to grow to a particular age when they start producing oil. Microalgae can be grown in a number of environments that are unsuitable for growing other crops, such as fresh, brackish or salt water or non-arable lands that are unsuitable for conventional agricultural. For biodiesel production, the use of microalgae will not compromise the production of food, fodder or other products derived from crop plants. Oil, extracted from microalgae, is converted into alkyl esters (biodiesel) in the presence of a catalyst and glycerol is released as the main byproduct, which has several other industrial applications. Finally, the biomass resulting after oil extraction can be processed into ethanol or methane to be used also as biofuel, livestock feed or as organic fertilizer. Biodiesel from microalgae has a tremendous potential to maintain the sustainability of energy reserves and to mitigate climatic changes due to the production of greenhouse gases. Research during the mid-twentieth century convincingly demonstrated the ability of various fungi and bacteria to degrade cellulose and other plant polymers. Such research was mainly of academic interest due to the presence of an abundant, secure, and inexpensive supply of fossil fuels. Biofuel production from renewable sources can reduce fossil fuel dependence and assist in maintaining a healthy environment and economic sustainability. The biomass of currently produced biofuel is human food stock, which is believed to cause the shortage of food and worldwide dissatisfaction, especially in developing nations. Therefore, microalgae can provide an alternative biofuel feedstock, thanks to their rapid growth rate, greenhouse gas fixation ability, and high production capacity of lipids, since microalgae do not compete with human and animal food crops. Moreover, they can be grown on non-arable land and in saline water.

In this chapter, the discussion is focused on the potential of microalgae in the production of liquid biofuels and co-products. Many advantageous features such as high lipid content, high growth rate, the ability to rapidly improve strains, co-product formation, and the avoidance of farming arable land, make microalgae an exciting addition to the sustainable fuel portfolio. Thus, there is a great requirement for establishing microalgae as an environmentally and economically viable platform, with an emphasis on combining biofuel production with the production of co-products, which is as an essential strategy for the economic viability and broad adoption of this potential biofuel source.

7.2   Microalgae in Biofuel Production

Biofuels are generally referred to as solid, liquid or gaseous fuels derived from organic matter. Much of the discussion over biofuel production has focused on higher plants and the problems associated with their use, such as the loss of ecosystems, global warming, the receding of glaciers, rising sea levels, the loss of biodiversity or increases in food prices [2, 3]. Several microorganism-based options have the potential to produce large amounts of renewable energy without disruptions. The current requirement of human societies for energy has driven critical fossil-fuel dependence. Therefore, interest in microalga-derived biofuels has been on the rise, which is triggered by crude oil price peaking, energy security, greenhouse gas emissions, and competition for food-oriented agricultural commodities. Microalgae have a strong potential for biofuels because their oil content may exceed 70% compared to 5% of the best agricultural oil crops. Microalgae are single-cell microscopic organisms naturally found in fresh water and marine environment. Microalgae are considered to be some of the oldest living organisms on our planet. The partnership between microalgae and biofuels is promising because of the intrinsic autotrophic capacity of microalgae. Microalgae require only water, atmospheric CO_2 and sunlight to synthesize biomass, which is available at essentially, zero cost. There are more than 300,000 species of microalgae, which is much more diverse than plants [4]. They are thallophytes lacking roots, stems, and leaves that have chlorophyll as their primary photosynthetic pigment and lack a sterile covering of cells around the reproductive cells [5]. The mechanism of photosynthesis in microalgae is similar to that of vascular plants and they are generally more efficient converters of solar energy because of the simplicity in their cellular structure and function. In addition, as microalgae cells grow in aqueous suspension, they have more efficient access to water, CO_2 and other nutrients [6]. Under normal conditions, autotrophic microalgae use sunlight and fix inorganic carbon from the atmosphere for assimilation in the form of carbohydrates and lipids, which can be exploited for biofuel production [7].

Generally, microalgae are classified in accordance with their colors. The current systems of classification of microalgae are based on the i) kinds of pigments, ii) chemical nature of storage products, and iii) cell wall constituents [6, 7]. It is clear that several species of microalgae can have oil contents of up to 80% of their dry body weight. As mentioned earlier, some microalgae can double their biomass within 24 hours and the shortest doubling time during their growth is around 3.5 hours, which makes microalgae an ideal renewable source for biofuel production.

7.3   Comparison of Cyanobacteria with Microalgae in Biofuel Production

Cyanobacteria also known as blue-green algae and their superior photosynthesis capabilities can convert up to 10% of the sun's energy into biomass, compared to the 1% recorded by conventional energy crops such as corn or sugarcane, or the 5% achieved by algae. Photosynthetic cyanobacteria like microalgae can potentially be employed for the production of biofuels in an economically effective and environmentally sustainable manner [8]. Cyanobacteria are oxygenic photosynthetic bacteria that have significant roles in global biological carbon fixation, oxygen production and the nitrogen cycle. Cyanobacteria can be developed as an excellent microbial cell factory that can harvest solar energy and convert atmospheric CO_2 into useful products. It is claimed that Cyanobacteria were found around 3.5 billion years ago and most probably played a key role in the formation of atmospheric oxygen, eventually evolving into chloroplasts of algae and green plants [9]. Cyanobacteria exhibit diversity in their metabolism, structure, morphology and habitat. Moreover, cyanobacteria and microalgae have simple growth requirements and use light, CO_2 and other inorganic nutrients efficiently. Cyanobacteria and microalgae are the only organisms known so far that are capable of both hydrogen production and oxygenic photosynthesis. The biological production of H_2 by microorganisms using solar radiation is of great public interest because it promises a renewable energy carrier from nature's most plentiful resources, solar energy and water. They have been investigated to produce different feed stocks for energy generation like hydrogen by direct synthesis in cyanobacteria, hydrocarbons and isoprenoids for gasoline production, lipids for biodiesel and jet fuel production and carbohydrates for ethanol production. In addition, the complete algal biomass can also be processed for bio-syngas production, hydrothermal gasification for hydrogen or methane production, methane production by anaerobic digestion, and co-combustion for electricity production. Since, cyanobacterial and microalgal systems may contribute to sustainable bioenergy production, different biotechnical, environmental and economic challenges have to be overcome before energy products from these systems enter the public domain. There are several aspects of cyanobacterial and microalgal biofuel production that have combined to capture the interest of researchers and entrepreneurs around the world [9]. The advantages are the following:

- Both are capable of performing oxygenic photosynthesis using water as an electron donor.
- Mass cultivation for commercial production of both cyanobacteria and microalgae can be performed efficiently with their high densities and high productivity capacities compared to typical terrestrial food and oil crops.
- Both are non-food based feedstock resources and use otherwise non-productive, non-arable land.
- Both utilize a wide variety of water sources such as fresh water, brackish water, seawater and wastewater.
- Both produce biofuels and valuable co-products.

7.4 Applications of Cyanobacteria and Microalgae in Biofuel Production

The macroalgae industry is primarily focused on food products for human consumption, which account for 83–90% of the global value of seaweed. Some important properties of microalgae which are typical of higher plants such as efficient oxygenic photosynthesis, simple nutritional requirements, fast growth rates and the ability to accumulate or secrete metabolites will provide the main rationale for microalgal biotechnology in the near future. Furthermore, the huge pool of existing species of microalgae (estimated to be 100,000 to 300,000) constitutes a unique reservoir of biodiversity that supports potential commercial exploitation of many value added products, e.g. vitamins, pigments and polyunsaturated fatty acids [10, 11]. These compounds may favorably contribute to making biofuel manufacturing from microalgae more competitive, based on a biorefinery approach. Research and development efforts focused on the development of algal biofuels have grown significantly in recent years because of the economic viability of microalgae operations in large photobioreactors, the handling of biomass and metabolites to sufficiently high levels and extractable forms [12, 13]. Research in the last six decades has demonstrated that cyanobacteria and microalgae produce a diverse array of chemical intermediates and hydrocarbons, precursors to biofuels. Hence, microalgae/cyanobacteria-to-fuel promise to be a potential substitute for products currently derived from fossil fuels. Microalgae and cyanobacterial biomass can be directly used as food sources or various feedstocks. Various important biomolecules such as antioxidants, coloring agents, pharmaceuticals and bioactive compounds can be obtained. Biomass can be converted to biogas (biomethane] on anaerobic digestion. A cyanobacterial photosynthetic system is able to diverge the electrons emerging from two primary reactions, directly into the production of H_2. Calvin cycle leads to the production of carbohydrates, proteins, lipids and fatty acids. Carbohydrates can be converted into bio-ethanol by fermentation. Lipids can be converted into biodiesel. Fatty acids in fermentation form acetate, butyrate and propionate, which when stabilized, form CH_4, H_2 and e^- [14].

7.4.1 Advantages of Microalgae over Higher Plants

Currently, the two most abundant and feasible biofuels for large-scale production are ethanol from corn or sugarcane and biodiesel from oil crops such as soy or palm oil [15]. These food crops are attractive and widely used as raw materials for biofuel production because of well established farming practices, simple and cheap processes for the release of starches, sugars and oils. However, using food crops for biofuel production is only a partial solution and may not satisfy the full demand for renewable fuels. For example, in order to achieve the 2020 federal mandate for renewable fuel in the United States with corn ethanol, approximately 100% of the domestic corn crop currently available would be required [16]. In recent years, an increase in the demand on food crops for fuel applications has resulted in concern about food scarcity, higher prices of food commodities and the pollution of agricultural land [7]. In the context of the challenges mentioned, microalgae are an attractive renewable source for biofuel production with many advantages over terrestrial biomass from food or cellulosic materials. The following are the advantages of marine algae (macroalgae and microalgae) as an attractive renewable source for biofuel production over biomass from traditional terrestrial plants [8, 17, 18]. These advantages are summarized below.

- Conventional terrestrial plants are hardly efficient in capturing solar energy as compared with microalgae may attain up to 10%.
- Their typical doubling time (~24 h) is significantly faster than that of any oil crop and essentially independent from weather provided that a minimum of sunshine hours exists.
- They require even less water than land crops for irrigation.
- Plain seawater supplemented with commercial nitrate and phosphate fertilizers, and a few extra micronutrients available in effluents of aquaculture are the only nutrients required by microalgae because atmospheric CO_2 is available for free and their halophilic character brings about ecological advantages against fast-growing microorganisms such as bacteria.
- Marine algae have relatively high photon conversion efficiency and can therefore rapidly synthesize biomass through assimilating abundant natural resources such as sunlight, CO_2 and inorganic nutrients.
- Since they are single-celled organisms that duplicate by division, high-throughput technologies can be used to rapidly evolve strains. This can reduce processes that take years in crop plants, down to a few months in algae.
- Production yields of algae per unit area are significantly higher than those for terrestrial biomass.
- They can be grown on land that would not be used for traditional agricultural, and are very efficient at removing nutrients from water. Thus, the production of algae biofuels minimizes land use compared with terrestrial plants.

- Microalgae have a greater potential for of CO_2 fixation compared to terrestrial biomass at a higher rate.
- Microalgae lack hemicellulose and lignin and thus can be depolymerized relatively easily as compared to lignocellulosic biomass in most terrestrial plants.
- Algae production strains also have the potential to be bioengineered, allowing the improvement of specific traits and the production of valuable co-products, which may allow algal biofuels to compete economically with fossil fuel.
- Microalgae do not require arable land and can be grown in a variety of marine environments including fresh water, salt water and municipal waste water.
- Microalgae's ability to grow in salt water or waste water is critical for sustainable biofuel production to avoid competition with food crops that require fresh water and cultivable land.
- The microalgal species being investigated as potential biofuel crops originate from groups whose ancestral relationships are significantly broader than the most diverse land plants
- Microalgae require lower rates of water renewal than terrestrial crops need as irrigation water, despite being grown in aqueous media.
- They can be cultivated in brackish water and they do not need the application of pesticides.

7.4.2 SWOT Analysis of Microalgae in Biofuel Production

The desirable features for biofuel production are limited by several bottlenecks toward large-scale, sustainable and economically feasible operation. An appropriate strengths, weaknesses, opportunities and threats (SWOT) analysis is briefly described here [18]:

Strengths:

- Microalgae use sunlight and CO_2 (a carbon source) that is readily available from atmosphere and flue gases.
- All microalgae have the capacity to produce energy-rich oils, and a number of microalgal species have

been found to naturally accumulate high oil levels in total dry biomass.

- Marine microalgae withstand highly salted media which is a competitive advantage over most non-halophilic microbial predators such as bacteria.
- Microalgal species grow in a wide range of aquatic environments, from freshwater through saturated saline
- With potentially millions of species, algal diversity gives researchers many options for identifying production strains and also provides sources for genetic information that can be used to improve these production strains.
- Microalgae are metabolically modifiable with physiological state in order to control bioactive compound synthesis.

Weaknesses:

- Biochemical optimization is difficult with microalgae as most metabolic pathways are not fully known and understood.
- Difficult recovery of microalgae from their growth media due to their small size and planktonic distribution.
- Cell wall breaking for endocellular metabolites of interest is hard and leads to unspecific mixtures.
- Rational control of endocellular reactions is hard because of incipient network of metabolic regulation in microalgae.

Opportunities:

- CO_2 is a major component of flue gases, and accounts for global warming via greenhouse effect.
- Inductions of microalgal secondary metabolism by several forms of stress stop growth and trigger lipid synthesis.
- The development of metabolic and genetic engineering tools to enhance the production of metabolites due to genome studies of microalgae.

Threats:

- Biomass and lipid yields by microalgae are low because of the excessive dilution and poor solubility of atmospheric CO_2 in water.
- CO_2 uptake by microalgae leads to unwanted pH increase which result in H_2CO_3 formation and ionization of water.
- Currently available media for microalgae have not been designed toward specific metabolite synthesis, so trial-and-error with generic media is standard practice.
- Sunlight is difficult to drive into high-density microalgal cultures as required to keep capital costs of occupied land and reactors tractable.
- Economic assessment should encompass the whole biorefinery approach therefore willingness to pay for goods and to accept environmental impacts are part of the decision process.
- Major challenges include strain isolation, nutrient sourcing and utilization, production management, harvesting, co-product development, fuel extraction, refining and residual biomass utilization.
- Disposal and incineration of microalgal biomass left after recovery of metabolites is a vector of pollution.

7.5   Selection of Microalgae for Biofuel Production

Microalgae hold clear advantages over higher plants in terms of oil productivity. Besides synthesizing storage lipids in the form of triacylglycerols, they can be induced to accumulate substantial amounts thereof via several forms of stress [19]. A multi standard based strategy is, however, to be considered for the successful selection of a specific wild microalgal strain [19, 20]. These standards are as follows:

- Growth rate of microalgal strains.
- Lipid quantity and quality, especially the fatty acid residue profile of acylglycerols.

- Response to processing conditions such as temperature, nutrient input, light requirement and competition with other microalgal and bacterial species.
- Nutrient requirements and rate of uptake of CO_2, nitrogen and phosphorus to a lesser extent (especially when brackish waters and agricultural effluents are to be sought).
- Ease of biomass harvesting, oil extraction and further processing.
- Possibility of obtaining high added-value chemicals in parallel for foods, cosmetics or pharmaceuticals.
- Finally, the selection of strain should be carried out interactively with growth medium and reactor design.

The strengths and opportunities account for the selection of microalgae for biodiesel production is as described in the previous section of SWOT analysis. However, the technological bottlenecks that materialize in weaknesses and threats need a rational approach with intrinsic and extrinsic parameters playing a role in microalgal metabolism. The specific microalga strains to be selected for biodiesel production should therefore address the following issues in a balanced manner [18, 19, 21].

- Optimum versus enhanced growth, so that structured models and objective optimization is to be sought instead of empirical approaches.
- Alkane versus glyceride synthesis, so that both the lipid quantity and quality can be considered.
- Conduction versus bubbling of CO_2, to maximize sequestration besides nitrogen and phosphorus uptake.
- Heterotrophic versus autotrophic metabolism, to take advantage of nutrients in wastes and effluents.
- Transmission versus incidence of light, to maximize photosynthetic efficiency and energy savings.
- Integration versus separation of processing, to decrease otherwise high costs of downstream handling.
- Excretion versus accumulation of lipids, to facilitate the recovery of product/s.
- Environmental versus economic sustainability, so that all relevant inputs are considered in biodiesel pricing.

7.6   Cultivation of Microalgae for Production of Biofuel and Co-Products

Harvesting solar energy via photosynthesis is one of the nature's remarkable achievements. Microalgae depend critically on a sufficient supply of carbon and light to carry out photosynthesis and to convert inorganic substances into simple sugars using captured energy. However, they use more than one type of metabolism, i.e. heterotrophic, mixotrophic, photoheterotrophic and photoautothrophic. They may also undergo metabolic shifts in response to growth conditions [19]. Typical examples are *Chlorella vulgaris*, *Haematococcus pluvialis* and *Arthrospira platensis* (*Spirulina*), all of which can grow under photoautotrophic, heterotrophic and mixotrophic conditions [20] or *Selenastrum capricornutum* and *Scenedesmus acutus*, which operate photoautotrophically, heterotrophically or photoheterotrophically [22]. Under phototrophic cultivation, there is a large variation in lipid content that ranges from 5 to 68% depending on the microalga species, with the highest lipid productivity reported to be 179 mg L^{-1} d^{-1} for *Chlorella* spp. [19]. The prime factors that determine the growth rate of microalgae are light, ideal temperature, growth medium, aeration, pH, CO_2 requirements and light and dark periods. Some of the important nutrients required for the growth of microalgae are NaCl, $NaNO_3$, $MgSO_4$, $CaCl_2$, KH_2PO_4, citric acid and some trace metals.

Extensive studies have been carried out for the cultivation of different microalgae using a variety of cultivation systems ranging from closely-controlled laboratory (enclosed photobioreactors) methods to less predictable methods in outdoor tanks (open ponds). Enclosed photobioreactors are more suitable for cultures that are easily contaminated, whereas open systems are preferable for microalgae able to survive in extreme growth conditions and environments, such as high pH (e.g. *Spirulina*) or salinity (e.g. *Dunaliella* spp.), or which grow very rapidly (e.g. *Chlorella* spp.).The most commonly used systems include shallow big ponds, tanks, circular ponds and raceway ponds [23–25]. One of the major advantages of open ponds is that they are easier to construct and operate than most closed systems [26]. However, major limitations in open ponds include poor light utilization by the cells, evaporative losses, diffusion of CO_2 into the atmosphere and the requirement of large amounts of water, land and low biomass productivity [27]. Land-based pond systems for the cultivation of micro/macroalgae have many advantages over farming on open marine water, such as, the ease of nutrient application and the avoidance of bad weather, disease and predation. Moreover, land-based systems can be integrated with the aquaculture of other species such as fish to provide waste materials as a cheap supply of nutrients for the micro/macroalgae. Open pond systems are indeed relatively non-expensive, and the basic requirements for microalgal phototrophic growth are reduced to atmospheric CO_2 and only a few readily available micronutrients (besides sunlight). In addition to these advantages, the application of open pond systems for providing biomass to a biofuels marketplace would require a reduction in pond construction costs and technology for a considerable scale-up from current practices [28].

Alternatives to open ponds are closed ponds (enclosed bioreactors) where the control over the environment is much better than that for the open ponds. The enclosed bioreactor systems are more cost intensive than the open ponds, and considerably less than photobioreactors. Enclosed photobioreactors have the ability to scrub power plant flue gases and/or remove nutrients from wastewater, but require operation under sterile conditions, thus maintaining stricter hygiene measures added to the final cost of biodiesel. Enclosed photobioreactors allow more species to be grown, grown species to stay dominant, and extend the growing season. Usually closed ponds are used in *Spirulina* cultivation [29]. On the other hand, they offer the opportunity to optimize the light path with distinct configurations built to improve light supply and biomass productivity. They can distribute the sunlight over a larger surface area, which can be up to 10 times higher than the footprint area of the reactor and evaporation can be avoided. This allows for the cultivation of microalgae also in arid areas, where classical terrestrial agriculture is not possible. Closed photobioreactors have different types such as vertical reactors, flat-plate reactors, annular reactors, arrangements of plastic bags, and various forms of tubular reactors, all of them stirred mechanically or by air-lifting [30]. Limiting factors are the high reactor costs and the need for auxiliary energy requirements. However, ongoing research in the reactor field is promising and will lead to cheaper and more energy-effective designs [27].

The cultivation of microalgae in sewage and wastewater treatment plants is expected to bring double benefit to the environment since this method can be used to extract nutrients from wastewater and convert it to fat for biodiesel production, therefore reducing pollution from the atmosphere [31]. Another economical way of cultivating microalgae is seawater (salt water). Seawater is a salt solution of nearly constant composition dissolved in variable amounts of water, which contains the required nutrients for microalgal growth. There are over 70 elements dissolved in seawater with six of them making up >99% of all the dissolved salts; all occur as ions electrically charged atoms or groups of atoms of sodium (Na^+), chlorine (Cl^-), magnesium (Mg^{2+}), potassium (K^+), sulfate (SO_4^{2-}) and calcium (Ca^{2+}) [32].

7.7   Harvesting and Drying of Microalgae

There are significant challenges for engineers to either design photobioreactors (PBRs) that are cheap enough for large-scale processing, or for engineers and biologists to combine efforts to develop species that grow efficiently in low-cost open systems [26]. There is a relatively low biomass concentration obtainable by microalgal cultivation systems due to the limit of light penetration (typically in the range of 1–5 g/L) and the small size of microalgal cells (typically in the range of 2–20 μm in diameter). Different technologies including chemical flocculation, sedimentation, filtration and centrifugation have been investigated for microalgal biomass harvesting. Microalgal cultures require processing steps such as harvesting, dewatering (drying) and the extraction of fuel precursors. The selection of downstream processes depends on the type of culture, feedstock and on the desired product. Obtaining fuels from high water content and high nitrogen (N) and phosphorous (P) content is the major limitation in the downstream processing of microalgae. Besides these, other economical and practical issues such as energy costs, plant site, transportation, water quality and recycling issues are of significant concern and need more attention and to be addressed properly in order to make a feasible microalgal-to-fuel strategy [14, 33].

The term harvesting refers to the concentration of algal suspensions until a thick paste/dry mass is obtained, depending on the need for the desired product. The main methods involve filtration, centrifugation, sedimentation and flotation. Filtration, a conceptually simple process, is carried out commonly on membranes of various kinds with the aid of a suction pump. The greatest advantage of filtration is that it is able to collect cells of very low density. However, various issues such as the clogging of filter, efficient cell mass recovery and washing requirements, have been the biggest hindrances until now [14, 34]. Several methods such as the reverse-flow vacuum, a direct vacuum with a stirring blade above the filter to prevent particles from settling and other changes in filtration design, are making this process economically feasible [35]. Centrifugation is a method of settling the algal cells to the bottom of a tube by applying centrifugal force. The biggest advantage for centrifugation technologies is the high throughput processing of large quantities of water and cultures. At a commercial and industrial scale, centrifugation techniques on a long-term basis are economically feasible [36]. Flocculation is a technique in which flocculants (chemical additives) are added to increase the size of the cell aggregates. Alum, lime, cellulose, salts, polyacrylamide polymers, surfactants, chitosan, etc. are some chemical additives that have been studied. Manipulating suspension pH and bio-flocculation (co-culturing with another organism) are the other options for the chemical additives. Flocculation is always followed by either sedimentation or flotation. Naturally, flocculation leads to sedimentation in many older cultures, otherwise, forced flocculation is required to promote sedimentation. To induce flotation, air is bubbled through the cell suspension causing cell clusters to float to the surface and the top layer is removed as scum [14, 36–38]. Biomass drying before further lipid/bioproduct extraction and/or thermochemical processing is another step that needs to be taken into consideration. Dewatering and drying are used to achieve higher dry mass concentrations. Drum dryers and other oven-type dryers are used to provide the heat required for drying [39, 40]. However, the costs climb steeply with the increase in time and temperature. Air and sun drying are also possible in low humidity environments. These are probably the cheapest drying methods that have been used for the processing of microalgal biomass. However, this method takes a long time to dry, requires large drying surface and risks the loss of some bioreactive products. Low-pressure shelf drying is another low-cost drying technology that has been investigated [14, 39].

7.8   Processing, Extraction and Separation of Microalgae

Microalgae differ from traditional biomass feedstocks in several respects, such as cell wall chemistry, the presence of large amounts of water and their smaller cell size. These differences highlight the importance of the specific extraction techniques. Various methods like mechanical, chemical and enzymatic are applicable for the extraction of biomass/biofuel. Biodiesel production requires the release of lipids from their intracellular location, which should be done in the most energy-efficient and economical way possible to avoid using large amounts of organic solvents. This should maximize the pool of liquid biofuel without significant recovery of other byproducts, e.g. DNA and chlorophyll [4]. In light of the above, cell disruption should first be applied because most microalgae possess a strong cell wall. The overall extraction yield depends largely on the extent and quality of cell disruption. Several methods can be followed and one's choice depends chiefly on the microalgal wall and the target metabolite/s. They are based on mechanical action (e.g. cell homogenizers, bead mills, ultrasound, autoclaving and spray drying) or non-mechanical action (e.g. freezing, organic solvent extraction, osmotic shock, and acid/base- or enzyme-mediated reactions) [20]. Cell structure presents a formidable barrier for access to biomolecules and the biomass must be mechanically disrupted prior to any further processing. The most common of these are (i) freezing and thawing [41, 42], (ii) grinding cells while frozen in liquid nitrogen [43], (iii) lyophilization followed by grinding, (iv) pressing (with expeller), (v) ultrasonication, (vi) bead beating and (vii) homogenizers. Chemical methods include (i) hexane solvent method [44], (ii) soxhlet extraction (hexane/petroleum ether) [45], (iii) two solvent systems [46], (iv) supercritical fluid extraction (methanol or CO_2) [47], (v) accelerated solvent extraction at high pressure (vi) subcritical water extraction [48, 49], (vii) milking (two phase system of aqueous and organic phases) [50] and (viii) transesterification [51]. Enzymatic extraction uses enzymes such as cellulose and xylanase to degrade cell walls, making fractionation much easier. However, costs of this extraction process are as of now making this process not viable. Osmotic shock is a sudden change in osmotic pressure, which can cause cells in a solution to rupture. Osmotic shock leads to release of cellular components. After cell disruption, lipids are to be extracted from cell debris. This process should be lipid-specific in order to maximize the recovery of neutral lipids containing mono-, di- and triacylglycerol moieties and to minimize the co-extraction of non-lipid materials. A typical solid/liquid extraction using organic solvents is mostly employed directly on the biomass as it is fast and efficient enough to preclude significant degradation. Several solvents can be used, e.g. hexane, ethanol

(96% v/v in water) or a mixture thereof. Meanwhile, a number of alternative extraction methods have gained attention, such as ultrasound and microwave-assisted ones, as well as supercritical carbon dioxide extraction [19, 52].

7.9 Biofuels and Co-Products from Microalgae

Microalgae/cyanobacteria are a diverse group of photosynthetic microorganisms that can grow rapidly due to their simple structures. They have been investigated for the production of different biofuels including biohydrogen, biodiesel, bioethanol and biomethane. To make biofuel production economically viable it is needed to use remaining algal biomass for co-products of commercial interests. It is possible to produce adequate microbial biofuels to satisfy the fast growing demand within the restraints of land and water resources.

7.9.1 Biodiesel

Biodiesel is a non-petroleum-based diesel fuel consisting of alkyl esters (mainly methyl, but also ethyl, and propyl) of long chain fatty acids. Biodiesel can be produced from various animal and plant sources (oleaginous crops, such as rapeseed, soybean, sunflower and from palm) by mono-alcoholic trans-esterification process, in which triglycerides reacts with a mono-alcohol (most commonly methanol or ethanol) with the catalysis of enzymes. However, the use of microalgae and cyanobacteria can be a suitable alternative because algae are the most efficient biological producer of oil on the planet, a versatile biomass source and may soon be one of the Earth's most important renewable fuel crops [8, 53]. Research on biodiesel from algae has been funded in US national laboratories through the aquatic species program, launched in 1978 and sponsored by the department of energy. The production of biodiesel from microalgae has multiple advantages and has been termed the third-generation biofuel. Unlike other oil crops, microalgae grow extremely rapidly and many are exceedingly rich in oil. Microalgae commonly double their biomass within 24 h, and biomass doubling times during exponential growth are commonly as short as 3.5 h. The oil content in microalgae can exceed 80% by weight of dry biomass and oil levels of 20–50% are quite common

[1, 54, 55]. Biodiesel from photosynthetic algae that grow on CO2 has great potential as a biofuel. Most importantly, autotrophic algae do not compete with starting plant materials for biofuel production, due to their photosynthetic nature. On the contrary, algae fix and thus reduce the amount of CO_2 in the atmosphere, a gas that contributes to the process of global warming. These organisms are being seriously considered as a substitute for plant oils to make biodiesel. Producing biodiesel from algae provides the highest net energy because converting oil into biodiesel is much less energy-intensive than methods for conversion to other fuels. This characteristic has made biodiesel the favorite end-product from algae. Producing biodiesel from algae requires selecting high-oil content strains, and devising cost effective methods of harvesting, oil extraction and conversion of oil to biodiesel. It is envisioned that algae could be grown to generate biodiesel in dedicated artificial ponds. Algae grow as a thin surface layer in ponds, so harvesting miles and miles of growth to get large amounts of biodiesel is needed. Huge ponds are required to grow microalgae in quantities that make the process commercially feasible. The growing of microalgae in natural lakes or ocean shores has been proposed [55]. In addition, research is currently being conducted to use heterotrophic algae for biodiesel production using sugars as substrates. Heterotrophic algae have the advantage of achieving much higher growth densities compared to phototrophic algae. In addition, dark growth of heterotrophic algae possesses no engineering challenge when compared to phototrophic algae [55, 56]. However, the invasiveness of algae could present an environmental hazard, since the grown algae will destroy and overtake the ecosystem.

The economics of biodiesel production could be improved by advances in production technology. Specific outstanding technological issues are efficient methods for recovering the algal biomass from the dilute broths produced in photobioreactors. A different and complimentary approach to increase the productivity of microalgae is via genetic and metabolic engineering. This approach is likely to have the greatest impact on improving the economics of the production of microalgal diesel [14, 53]. A promising alternative to the conventional process that may reduce processing costs is *in situ* trans-esterification. This process facilitates the conversion of fatty acids to their alkyl esters right inside the biomass, thereby eliminating the solvent extraction step and alleviating the need for biomass drying in harvesting. Such a form of integrated alcoholysis leads to higher biodiesel yields, up to 20% better than the conventional process and wastes are reduced as well [57].

7.9.2   Biohydrogen

Molecular hydrogen is one of the most promising biofuels because it does not evolve CO_2 (Greenhouse Gas) in combustion, but it liberates large amounts of energy per unit weight in combustion. Advances in hydrogen fuel cell technology, coupled with the realization that the combustion of H_2 releases plain water, make that feedstock particularly attractive. Hydrogen gas is seen as a future energy carrier because it is renewable. Biological hydrogen production has several advantages over hydrogen production by photoelectrochemical or thermochemical processes. Hydrogen can be produced biologically by a variety of means, including the steam-reformation of bio-oils, dark and photo fermentation of organic materials and photolysis of water catalyzed by special microalgal species.

Currently, there are three main processes in the center of current biohydrogen production research. The most direct approach involves using photosynthetic microorganisms (*cyanobacteria* and microalgae) for biohydrogen production. Photosynthetic microorganisms have the ability to split water into electrons and oxygen from one molecule of water using sunlight. The produced electrons are used for energy production (through electron transport chain), biomass production and sugar production using the Calvin cycle. However, they could also be converted to hydrogen by hydrogenase enzymes. The appeal of this system is that it uses water as a substrate and sunlight as an energy source. Therefore, in principle, this approach is extremely promising for low-cost hydrogen production. The second approach is about the use of nitrogenase enzymes in anoxygenic photoheterotrophic microorganisms (purple nonsulfur bacteria) for hydrogen production. The function of nitrogenase is to fix atmospheric N_2 gas to ammonia, thus enabling nitrogen-fixing microorganisms to grow in the absence of organic or inorganic nitrogen sources in growth media. Nitrogenase enzymes are also capable of producing hydrogen from electrons and protons in the absence of oxygen and in the presence of light. When grown in the light and in absence of oxygen, purple non-sulfur bacteria can obtain adenosine triphosphate (ATP) and electrons through cyclic anoxygenic photosynthesis, and carbon from organic substrates. Electrons extracted from organic substrates could be used for hydrogen production using nitrogenase enzymes. This photoheterotrophic versatility of purple non-sulfur bacteria makes it theoretically possible to divert 100% of the electrons produced during carbon metabolism to hydrogen production, since electrons required for anabolic, biosynthetic reactions could be obtained via photosynthesis. Research on this approach has been conducted by the Caroline Harwood group at the University of Washington using *Rhodopseudomonas palustris* as a model purple non-sulfur bacterium and via additional genetic manipulations, a strain of *R. palustris* capable of producing 7.5 ml of hydrogen/liter of culture has been obtained, and initial engineering designs have been proposed [18, 55, 58, 59]. The third approach is the production of hydrogen by fermentative bacteria using organic substrates, e.g. sugar, lingocellulosic biomass, industrial, residential, and farming waste for anaerobic fermentation. Several groups of microorganisms are known to produce hydrogen as an end product of fermentation, e.g. *E. coli, Enterobacter aerogenes*, and *Clostridium butyricum*. These "dark fermentation" reactions do not require light energy, so they are capable of constantly producing hydrogen from organic compounds through the day and night. However, production of hydrogen is only one of several electron sinks employed by fermentative microorganisms, since other fermentation end products are produced beside hydrogen [55, 60, 61].

Biological hydrogen production by photosynthetic microorganisms for example, requires the use of a simple solar reactor such as a transparent closed box, with low energy requirements whereas electrochemical hydrogen production via solar energy based water splitting, on the other hand, requires the use of solar batteries with high energy requirements. However, its technological viability is strongly dependent on the development of cost-effective, sustainable H_2 production systems at a large scale that are able to replace the classical processes of steam reforming of natural gas, petroleum refining and coal gasification [14, 19]. Hydrogen release is indeed a feature of many phototrophic organisms, including several hundred species from different groups of microalgae, cyanobacteria and anaerobic photosynthetic bacteria. Currently, *Chlamydomonas reinhardtii* remains the best photosynthetic eukaryotic hydrogen producer, with *Nostoc* and *Synechocystis* cyanobacteria also holding a promising status as candidates for H_2

production [62, 63]. Hydrogen release by microalgae was induced after anaerobic incubation in the dark; a hydrogenase (containing Fe as prosthetic group) is expressed during such incubation, and catalyzes light-mediated production of H_2 with a high specific activity. This enzyme is encoded in the nucleus, but the mature protein is localized and functions in the chloroplast stroma. Light absorption by the photosynthetic apparatus is essential for generation of hydrogen because it brings about oxidation of water that releases electrons and protons, and facilitates endergonic transport of said electrons to ferredoxin (Fd). This ferredoxin thus serves as physiological electron donor to the Fe-hydrogenase, so it links that enzyme to the electron transport chain in the chloroplasts of microalgae [19, 64, 65]. Cyanobacteria can be used for the production of molecular hydrogen (H_2), a possible future energy carrier. Cyanobacteria are able to diverge the electrons emerging from the two primary reactions of oxygenic photosynthesis directly into the production of H_2, making them attractive for the production of renewable H_2 from solar energy and water. In cyanobacteria, two natural pathways for H_2 production can be used. The first pathway is nitrogen fixation by nitrogenases that produces H_2 and the second pathway uses the activity of bidirectional hydrogenases for H_2 production. Nitrogenases require ATP whereas bidirectional hydrogenases do not require ATP for H_2 production, hence making them more efficient and favorable for H_2 production with a much higher turnover [9, 14, 66–68].

7.9.3   Bioethanol

Cyanobacteria and algae are capable of secreting glucose and sucrose. These simple sugars by anaerobic fermentation under dark conditions produce ethanol. If ethanol can be extracted directly from the culture media, the process may be use drastically less capital and energy than competitive biofuel processes. The process would essentially eliminate the need to separate the biomass from water and extract and process the oils. However, it could technically be argued that ethanol is not the best compound to be used for biofuel. For example, the water solubility of ethanol makes it less suited for pipeline transport, and makes it easier to be watered down. In addition, the energy content of ethanol is approximately two-thirds that of an equal volume of a standard petroleum mix, as opposed to 86% for longer chain alcohols [54]. Until now, only a handful of research information and data are available in the literature on the production of bioethanol from microalgae due to several reasons; (1) a lot of attention has been diverted to biodiesel production from microalgae since certain strains are capable to accumulate large quantity of lipid naturally inside their cells, (2) through nitrogen-deficient cultivation method (to save energy and cost), lipid content inside the microalgae cells is boosted up significantly by blocking carbohydrate synthesis pathway, while carbohydrate is the main substrate to produce bioethanol and (3) biodiesel has a higher calorific value than bioethanol, 37.3 MJ/kg and 26.7 MJ/kg, respectively. Nonetheless, microalgae are found to be superior feedstocks to produce bioethanol in comparison with other first and second generation bioethanol feedstocks. First generation bioethanol is derived from food feedstocks such as sugarcane and sugar beet, in which over exploitation of these feedstocks create "food versus fuel" issues and raise several environmental problems including deforestation and ineffective land utilization. Second generation bioethanol is produced from lignocellulosic biomasses such as wood, rice straw and corn stover. Initially, this lignocellulosic biomass must be subjected to pre-treatment to break down the complex structure of lignin and to decrease the fraction of crystalline cellulose by converting to amorphous cellulose [69, 70].

Bioethanol could be very important to foster energy independence and reduce greenhouse gas emissions. A very strong debate on the gradual substitution of petroleum by the use of renewable alternatives such as biofuels dominates the political and economic agenda worldwide [71]. Alternative bioethanol production methods from cyanobacteria and microalgae need to be developed so that the costs associated with the land, labor and time of traditionally fermented crops can be circumvented. In 1996, Ueda *et al.* [72] patented a two-stage process for microalgae fermentation. In the first stage, microalgae undergo fermentation in an anaerobic environment to produce ethanol. The CO_2 produced in the fermentation process can be recycled in algae cultivation as a nutrient. The second stage involves the utilization of the remaining algal biomass for the production of methane by an anaerobic digestion process, which can further be converted to produce electricity. In 2006, Bush and Hall pointed out that the patented process of Ueda *et al.*, 1996 was not commercially scalable due to the limitations of single cell free floating algae [72, 73]. They patented a modified fermentation process of yeasts, *Saccharomyces cerevisiae* and *Saccharomyces uvarum*, and were added to algae fermentation broth for ethanol production. In 2010, Harun *et al.* [74] studied the suitability of microalgae (*Chlorococum* sp.) as a substrate, using yeast for bioethanol production by fermentation. They achieved a productivity level of around 38% weight which supports the suitability of microalgae as a promising substrate for bioethanol production [14].

7.9.4   Biogas as Biofuel

When biomass is processed under high temperatures in the absence of oxygen, products are produced in three phases: the vapor phase, the liquid phase, and the solid phase. Biogas, a mixture of methane and carbon dioxide, is produced from the methanogenic decomposition of organic waste under anaerobic conditions. Biogas production could be achieved by a defined culture of a fermentor and/ or syntroph in association with an acetate degrading and hydrogenotrophic (hydrogen-consuming) methanogen. In addition, undefined cultures (e.g. microorganisms in cow dung or waste water sludge) could be used as an inoculum for biogas production [55, 75). Organic materials like biomass can be used to produce biogas via anaerobic digestion and fermentation. Organic biopolymers (i.e. carbohydrates, lipids and proteins) are hydrolyzed and broken down into monomers, which are then converted into a methane-rich gas via fermentation. Carbon dioxide is the second main component found in biogas (approximately 25–50%) and, like other interfering impurities, has to be removed before the methane is used [14, 53]. Methane in the form of compressed natural gas is used as a vehicle fuel and is claimed to be more environmentally friendly than fossil fuels such as gasoline/petrol and diesel. However, on a local level, biogas could be and is currently used and exploited. For example, biogas-producing facilities e.g. wastewater treatment plants and landfills can use biogas produced during operation for running the plant, thus becoming energy neutral. The use of biogas on a local, residential scale could be exploited in the countryside of developing countries. India had great success using biogas produced in pits associated with rural homes, with no utilities connected, for the generation of biogas for cooking and electricity [76]. Cow dung was used as an inoculum in this effort. Such approaches are currently being considered in Egypt for the treatment of rice straw and other low-nutrient agricultural waste that could not be fed to feedstock and is currently being burned.

7.9.5   Co-products

All primary components of algal biomass, i.e. carbohydrates, fats (oils), proteins and a variety of inorganic and complex organic molecules must be converted into different products either through chemical, enzymatic or microbial conversion means using appropriate technologies to make biofuels economically viable. The use of technologies will be determined primarily by the nature of the end products and economics of the system which may vary from region to region according to the cost of the raw material. A large number of different commercial products have been derived from microalgae. These include products for human and animal nutrition, poly-unsaturated fatty acids, anti-oxidants, coloring substances, fertilizers and soil conditioners and a variety of specialty products such as bio-flocculants, biodegradable polymers, cosmetics, pharmaceuticals, polysaccharides and stable isotopes for research purposes [14, 77].

7.9.5.1   Nutrition and Fertilizers

The consumption of microalgal and cyanobacterial biomass as a human health food supplement is currently restricted to only a few species, e.g., *Spirulina (Arthospira)*, *Chlorella*, *Dunalliella*, and to a lesser extent, *Nostoc* and *Aphanizomenon* . However, the market is expected to grow in the future. Microalgae and cyanobacteria are also used as feed in the aquaculture of mollusks, crustaceans (shrimp) and fish. The most frequently used species are *Chaetoceros, Chlorella, Dunaliella, Isochrysis, Nannochloropsis, Nitzschia, Pavlova, Phaeodactylum, Scenedesmus, Skeletonema, Spirulina, Tetraselmis* and *Thalassiosira*. Both the protein content and the level of unsaturated fatty acids determine the nutritional value of microalgal aquaculture feeds. Microalgal and cyanobacterial biomass have also been used with potential benefits such as better immune response, fertility, appearance, weight gain as a feed additive for cows, horses, pigs, poultry, and even dogs and cats. In poultry rations, biomass up to a level of 5–10% (wt) can be safely used as a partial replacement for conventional proteins. The main species used in animal feed are *Spirulina, Chlorella* and *Scenesdesmus* [14, 56, 78). Microalgal biomass is used to improve the water-binding capacity, mineral composition of depleted soils and as a plant fertilizer. Moreover the effluent generated during anaerobic digestion for biomethane production can also be used as a fertilizer [48].

7.9.5.2   Biomolecules

Phycobiliproteins, phycoerythrin, phycocyanin and allophycocyanin produced by the cyanobacteria are used as food dyes, pigments in cosmetics, and as fluorescent reagents in clinical or research laboratories. Microalgae-produced coloring agents are used as natural dyes for food, cosmetics and research, or as pigments in animal feed. A number of anti-oxidants, sold for the health food market, have also been produced by microalgae. The most prominent is β-carotene from *Dunaliella salina*, which is sold either as an extract or as a whole cell powder. Moreover, bioflocculants biopolymers and biodegradable plastics, cosmetics, pharmaceuticals and bioactive compounds, polysaccharides and stable isotopes for research are other important co-products obtained from microalgae and cyanobacteria species [14, 41, 42, 56, 78–83]. Microalgae and cyanobacteria can also be cultured for their high content in polyunsaturated fatty acids (PUFAs): PUFAs, which may be added to human food and animal feed for their health promoting properties. The most commonly considered PUFAs are arachidonic acid (AA), docohexaenoic acid (DHA), γ-linolenic acid (GLA) and eicosapentaenoic acid (EPA). arachidonic acid has been shown to be synthesized by *Porphyridium*, DHA by *Crypthecodinium* and *Schizochytrium*, GLA by *Arthrospira* and EPA by *Nannochloropsis*, *Phaeodactylum* and *Nitzschia* [56, 78, 81].

7.10   Challenges and Hurdles in Biofuel Production

Worldwide industries have focused on economically feasible processes. Many factors such as the price of available raw materials, land costs, water resources, transportation costs and others influence the commercial price of the product. As a result, a successful strategy at one location might not be successful at another location or vice versa. The PetroSun Company situated in Arizona, USA, use saltwater ponds for cultivation whereas Aquaflow Binomics is trying to become the first company to produce biofuel from wild algae. Solazyme Inc. situated in San Francisco, USA grows algae in the dark where they are fed sugar for growth. To make the biofuel economical, companies focus on the remaining algal biomass for co-products. Nearby industries and their raw material requirements, food sources, social acceptability and other such points can help in deciding which biofuels and co-products will be good choices. Neptune industries situated in Boca Raton, USA has a patented Aqua-Sphere system wherein fish waste is used to create additional revenue streams through the growth of algae for biofuel and methane. GreenFuel Technologies in Cambridge, Massachusetts in the US have developed a system through which they can capture up to 80% of the CO_2 emitted from a powerplant. The major research in companies is focused on manipulating the cyanobacteria or microalgae by genetic engineering or other approaches to increase the productivity and make the recovery of desired products easy and less expensive. Aurora Biofuels use the genetically modified algae to efficiently create biodiesel using a patented technology, developed at University of California, Berkeley, that claims to create biofuel with yields 125 times higher and at costs 50% less than other production methods [14, 19]. The production of biofuels from microalgae has proven technologically feasible and the use of microalga biomass, rich in lipids, may significantly reduce the use of arable land when compared to crops. However, several issues relating to the quality and quantity of such land have to be addressed on a case-by-case basis. Unfortunately, microalgal biodiesel has not yet reached a clear-cut economic feasibility. The biggest challenge is the relatively high costs of production of microalgal biomass and the extraction/separation of lipids for biodiesel. Microalgal systems could contribute to a sustainable bioenergy production however different biotechnological, environmental and economic challenges have to be overcome before energy products from these systems can enter the market.

7.10.1 Biotechnological Challenges

The main biotechnical challenges are cultivation, large scale production, harvesting, recovery, extraction and genetic engineering of microalgae and cyanobacteria.

7.10.1.1 Cultivation and Large-Scale Production Challenges

- The majority of commercial production occurs in unsophisticated, artificial open ponds with low productivity and low yields [1].
- Sustained open pond production is successful only for a limited number of cultures like *Spirulina* and

Dunaliella with extreme conditions such as very high salinity or high pH.

- Future advances in cultivation might require closed systems since not all microalgal species of interest grow in highly selective conditions. However, their high costs have largely precluded their commercial application until recently.
- The extreme light intensity may result in photo-inhibition or overheating. Thus, light distribution and its utilization inside a photo-bioreactor is one of the major biotechnical challenges in bioreactor design.

7.10.1.2   Harvesting, Recovery and Extraction Challenges

- The final costs, energy demand and processing time for recovery and harvesting of much diluted suspension culture of microalgae and cyanobacteria are very high because these cultures require filtration, centrifugation, sedimentation and flocculation to recover high density and quantity of biomass [84].
- The present harvesting techniques are not applicable for large-scale and low-cost harvesting to produce low-value energy products.
- A technique with low-energy demand is the settling of algae by induced flocculation. However, flocculation of algal biomass is still poorly understood, which makes it difficult to control this harvesting process [14].
- Extracting lipids from microalgae is a biotechnical challenge due to the sturdy cell wall making oil hard to get out.
- An alternative to this is the use of super-critical fluids, but the process requires special machinery adding to the expense.

7.10.1.3   Genetic Engineering Challenges for Obtaining Modified Strains

- Among the 10,000 algal species that are believed to exist, only a few thousand are kept in collections for strain modification due to lack of much research information.

- Only few hundred species of microalgae are investigated for chemical content and just a handful are cultivated in industrial quantities.
- Metabolic engineering using genetic engineering techniques now seems to be necessary in order to enhance productivity and to optimize them for cultivation and harvesting.
- The large-scale cultivation of genetically modified strains of microalgae carries the risk of escape and contamination of the surrounding environment and of cross breeding with native strains.
- Genetically modified strains could be transported in the air over long distances and survive a variety of harsh conditions in a dormant stage.
- The cultivation of genetically modified strains can have unintended consequences to public health and environment.
- The development of a number of transgenic algal strains boasting recombinant protein expression, engineered photosynthesis and enhanced metabolism, encourage the prospects of engineered microalgae.

7.10.2   Ecological Challenges

A major advantage of cyanobacteria and microalgae is their ability to capture additional environmental benefits such as the recycling of CO_2 and wastewater treatment. However, to realize these benefits, some hurdles need to be overcome. These challenges are addressed below.

7.10.2.1   Recycling of CO_2

- A cheap source of CO_2 to fuel the photosynthetic process is needed because atmospheric CO_2 concentrations limit the growth of photosynthetic organisms.
- If the purpose of algae cultivation is to sequester the industrial CO_2 outputs of fossil-fueled power plants, it has to be taken into account that during night time and during cloudy days the algae slow down their reproduction rate and thus take up less CO_2.

- There is considerable need for the installation of gas storage facilities to cope with the influx of CO_2 during night.
- Before the commercial-scale preparation of microalgae systems becomes feasible, the challenge of limited land availability for large scale CO_2 capture from industrial or power plants by microalgae have to be overcome by sophisticated area-efficient techniques to recycle CO_2 by microalgae [85].
- It is worth noting that sequestering industrial CO_2 outputs through algae cultivation is temporal storage since it is emitted during the conversion of the algae and its use as energy.

7.10.2.2   Requirements of Nutrients and Land

- Microalgae have nutrient requirements much higher than higher plants, especially high contents of N and P.
- Cultivation on the large scale may involve huge quantities of N and P for which environmental and economic impact may not be sustainable.
- Microalgae ponds have been utilized for the treatment of sewage and wastewater since they provide dissolved oxygen for the bacterial composition of organic wastes.
- The major limitations in recycling nutrients from wastewater are relatively low loadings that can be applied per unit area-time, limited nitrogen and phosphorous removal, increasing land requirements and the high cost of removing algal cells from pond effluent.
- Recycling nutrients via anaerobic digestion could be an answer to the nutrient challenge but this process can mineralize algal waste containing organic N and P, resulting in a flux of ammonium and phosphate that can be used for the cyanobacteria and microalgae.
- Genetic engineering of photosynthetic algal trains is required so that they are capable of fixing nitrogen to minimize the demand of N fertilizer.
- Microalgae produce much higher yields than traditional energy crops and thus need much less land.
- It is unclear how much land is available and suitable to produce high yields and utilize waste CO_2 and nutrients.

7.10.3   Economic Challenges

The development of cyanobacteria and microalgae for mass energy production is in its infancy. Because of that, it seems critical to base cost assumptions on state-of-the-art techniques used for the small-scale production of high-value products. Growing and processing algae consumes energy, both in infrastructure and operation. Depending on the cultivation and the process of harvesting on yields, the energetic inputs of microalgae production could exceed the energetic output. However, ongoing research in the reactor design is promising and will lead to cheaper and more energy-efficient designs. The economics of biofuel production from microalgae can be improved by capturing additional revenues from the co-production of food, feed and high-value products, wastewater treatment and net fertilizer value, in the case of nitrogen fixing algae [14, 27]. The following points need to be considered to improve the use of microalgal systems for designing economical biofuel production strategies.

- The capital costs for starting algal biofuel projects may include expenses for land, infrastructure establishment, bioreactors, labor and various overhead expenses.
- Significant funding in research would be required to obtain maximum levels of productivity for a successful commercial-scale production.
- The production costs may include expenses for cultivation including the expenses for nutrients, harvesting and dewatering, extraction and separation, maintenance, components replacement, transportation and overhead expenses.
- Worldwide, a number of companies and government organizations have developed different methodologies and designs and prepared cost estimates for commercial-scale production.
- Algae to biofuel plants may be effectively developed on the land adjacent to power stations (to convert CO_2 from exhaust into fuel), in wastewater treatment plants, or in seawater (to save land and fresh water).
- Global warming will accelerate unless we take action to reduce the net addition of CO_2 to the atmosphere.

- Fossil-fuel use will decline only when society comes up with renewable energy sources such as alternative fuels, in very large quantity.
- One of the best options in the long term is bioenergy, in which the sun's energy is captured as biomass and converted to useful energy forms.
- Successful bioenergy faces two serious challenges. The first is producing enough biomass-derived fuel to replace a significant fraction of the ~13 TW of energy generated today from fossil fuels. The second challenge is producing bioenergy without incurring serious damage to the environment and to the food-supply system.
- Of the many bioenergy options on the table today, most fail on both counts. However, microalgal and cyanobacteria based bioenergy options have the potential to produce renewable energy on a large scale, without disrupting the environment or human activities.

7.11 Genetic and Metabolic Engineering of Microalgae for Biofuel–Bioenergy Production

To date, only a few algal species have been successfully genetically manipulated. These modifications have come in the form of induced random mutagenesis to identify new genes that play important roles in processes of interest. Heterologous gene expression has also been used as a means to modify biological function and the nuclear genomes of a number of algae have been transformed, with a variety of reporter genes, as well as drug-resistant genes. Although there are natural microorganisms able to utilize algal sugar extracts, economically feasible biofuel production from microalgae requires a more efficient conversion of mixed sugars in algae hydrolysates by robust strains. Both hydrogen and sugar based biofuel production could potentially impact the use of petroleum fuels. Microalgae have the potential to produce a number of secondary metabolites and nutrients that are beneficial for humans or animals and have characteristics much closer to existing petroleum fuels. The most promising of these are the terpenes, which offer a potential new fuel source outside of fatty acids that are compatible with our existing fuel framework. Other valuable and potential co-products include long-chain polyunsaturated fatty acids (LCPUFAs) such as omega-3 and omega-6 and a wide variety of useful carotenoids, such as lutein, zeaxanthin, lycopene, bixin, β-carotene and astaxanthin. Microalgae can synthesize many other unique molecules with commercial potential, such as toxins, vitamins, antibiotics, sterols, lectins, mycosporine-like amino acids, halogenated compounds and polyketides. Additional minor commercial products from microalgae are phycobiliproteins, used as food and research dyes (*Arthrospira* and *Porphyridium*), extracts for cosmetics (*Nannochloropsis* and *Dunaliella*) and stable isotope biomolecules used for research. Therefore, genetic and metabolic engineering is a potential technique to obtain the optimized concentration of these many value added products. With rising concerns of energy sustainability and climate change, genetic and metabolic engineering strategies must be applied to advent the development of biofuels. Photosynthetic microorganisms offer a promising solution to these challenges, while at the same time, addressing growing environmental concerns through CO_2 mitigation.

7.11.1 Potential Avenues of Genetic and Metabolic Engineering

Although the application of genetic engineering to increase biofuel and energy production in microalgae is in its infancy, significant advances in the development of genetic tools have recently been achieved with microalgal model systems and are being used to manipulate central carbon metabolism in these organisms. It is likely that many of these advances can be extended to industrially relevant organisms. The following section is focused on potential avenues of genetic and metabolic engineering that may be undertaken in order to improve cyanobacteria/microalgae as a biofuel platform for the production of bioenergy.

- Sequencing the genome of cyanobacteria will examine their potential as one of the next great sources of biofuel.
- Manipulation of metabolite pathways can redirect cellular functions towards the synthesis of preferred products.
- Metabolic engineering allows direct control over the organism's cellular machinery through mutagenesis or through the introduction of transgenes [86].

- Many research works are focused on altering the micro-algal/cyanobacterial cell wall properties, transforming novel genes for hydrogen or other products, increasing the lipid synthesis, finding novel precursors and many more interesting and useful areas. Researchers from Arizona State believe that they have found a way to make biofuels cheaper and easier to produce by genetically programming microbes to self-destruct after photosynthesis, thus making the recovery of biofuel precursors easier and potentially less costly. The genes were taken from the bacteriophage [87–90].

- In recent years, there have been attempts to improve hydrogen production, mainly by the targeted genetic engineering of cyanobacterial strains by targeting activity and overexpression of H_2 evolving enzymes such as nitrogenase, hydrogenases, introducing a synthetic polypeptide based on proton channel into thylakoid membranes to dissipate proton gradients across thylakoid membranes, increasing quantum efficiency of both PS I and PS II and directing the electron flow towards the H_2 producing enzymes and away from any other competing pathway [9, 14].

- Asada and co-workers attempted to overexpress hydrogenase from *Clostridium pasteurianum* in a cyanobacterium, *Synechococcus* PCC7942, by developing a genetic engineering system for cyanobacteria. They demonstrated that *clostridial* hydrogenase protein, when electro-induced into cyanobacterial cells is active in producing hydrogen by receiving electrons produced by photosystems [91].

- Photosynthetic cyanobacteria can be redesigned for highly efficient ethanol production by the combination of gene transformation, strain/process development and metabolic modeling/profiling analysis. Dexter and Fu, 2009 have transformed pyruvate decarboxylase (*pdc*) and alcohol dehydrogenase II (*adh*) genes from *Zymomonas mobilis* into *Synechocystis* sp. PCC 6803. This strain can phototrophically convert CO_2 to ethanol [92].

- Unlike algae, cyanobacteria have well established methods for genetic engineering, as evidenced by the genetic engineering of cyanobacteria for the production of first generation biofuels including ethanol and butanol. Furthermore, cyanobacteria will secrete free fatty acids (FFA), a biodiesel precursor, into extracellular media, simplifying downstream product isolation. These attributes motivate the investigation of cyanobacteria as a potential source for biodiesel feedstock.

- Microalgae, natural photosynthetic oil producers, are the focus of most biodiesel research efforts. Cyanobacteria do not naturally produce oil like algae. Overexpression of acetyl-CoA carboxylase (ACCase) has been tried for increasing the lipid biosynthesis [86].

- Metabolic engineering has been applied to improve the fermentation of sugars that can be recovered from seaweed. Galactose is a major sugar compound in the hydrolysate of red marine algae, making up to 23% of the hydrolysate from red seaweed Ceylon moss. Therefore, research efforts have been made to enhance galactose fermentation to ethanol by engineering *S. cerevisiae*. The wild type yeast *S. cerevisiae* is capable of galactose fermentation but there are two major issues that limit the ethanol yield and productivity. First, ethanol production rate and yield from galactose are considerably lower than from glucose [93–97].

- Genetic transformation appears to hold a major potential in attempts to improve H_2 production, especially in the case of *C. reinhardtii* since its genome has been fully elucidated and it is one of the best H_2 producers known to date. Several mutants have been obtained, and the mutations were performed at various levels, e.g. hydrogenase, sulfate permease and ribulose-1,5-bisphosphate carboxylase (RuBisCO) enzyme, as well as at the PSI and PSII photosystems [98]

- Technical and physiological parameters of microalga cultivation have also been optimized to increase hydrogen production efficiency. For example, Scoma and Torzillo reported on the interplay between light intensity, chlorophyll concentration and culture mixing upon H_2 production in *C. reinhardtii* [98, 99].

- The waste of light energy can be minimized by reducing the cross-sectional area of the light harvesting antenna of Photosystem II (PSII) via genetic engineering which prevents oversaturation of the photochemically active reaction centers. This strategy has the further advantage of reducing photodamage, while allowing light, which would otherwise be dissipated, to penetrate deeper into the culture thereby increasing the overall H_2 yield. *In vitro* testing has been shown to double the biomass production efficiency and outdoor trials are currently underway to assess the efficiency of such strains under real-world conditions. One illustrative example is *C. reinhardtii* strain Stm6, which is able to produce 5-fold more H_2 than its wild type [19, 99–101].

7.11.2 Developments and Efforts Made by Many Countries

Advances in the genetic manipulation of crucial metabolic networks will form an attractive platform for production of numerous high-value compounds. The development of a number of transgenic strains boosting recombinant protein expression, engineered photosynthesis and enhanced metabolism encourage the prospects of modified microalgae and other seaweeds for biofuel generation. Compared to terrestrial biomass, marine macroalgae has received less attention so far, however, with urgent challenges in the pursuit of cost-effective renewable fuels as well as concerns over terrestrial biofuel crops, increasing effort and investment have been put into developing microalgal biofuels. Following are the some of the developments and efforts made by many countries across the globe for biofuel production from microalgae and seaweeds [86, 96]:

- In Europe, the UK and Irish joint project Sustainable Fuels from Marine Biomass (the BioMara Project; [102] aims to 'demonstrate the feasibility and viability of producing third generation biofuels from marine biomasses. The sea microalgae cultivation and harvest process has been successfully established in Scotland, and the project is investigating ethanol production and/or methane generation from seaweed. Statoil and Bio-Architecture Lab (BAL) also aim to commercialize the production of ethanol and co-products from sea-weed (microalgae) in Norway and in Europe.

- In the United States, the DOE's Advanced Research Projects Agency-Energy (ARPA-E) is supporting the development of a process to convert seaweed biomass into isobutanol, which is undertaken by DuPont and BAL under The Technology Investment Agreement. The program focuses on improving seaweed aquaculture, converting seaweed biomass to fermentable sugars, isobutanol production from sugars, and economic and environmental optimization of the production process [103].

- In Asia, a project lead by the Mitsubishi Research Institute in Japan plans to start the demonstration of ethanol production with waste seaweed in 2012, to develop cultivation technologies by 2016, and to set up a production process by 2020. The South Korea National Energy Ministry has started a 10-year project with the aim of producing nearly 400 million gallons a year of ethanol by 2020. The Philippine government also has invested more than $5 million US dollars to build an ethanol plant with seaweed bioethanol technology from South Korea.

- There are several biotechnology companies that are in the process of commercializing bioethanol production from seaweed biomass, including Butamax, Seaweed Energy Solutions, Green Gold Algae, and Seaweed Sciences, Inc. Because seaweed and microalgae cultivation is well established in some Asian countries and has been grown at a commercial scale for food products for decades, collaborative efforts that can bring together their advanced knowledge in converting seaweed biomass to biofuels and their established large-scale cultivation and harvesting technologies will be beneficial in promoting seaweed biofuel development. For example, Novozymes is collaborating with India's Sea6 Energy to develop a seaweed bioethanol process, with the former focusing on the development of bioconversion process and the latter sharing its knowledge of offshore seaweed cultivation.

7.12   Conclusion and Future Prospectus

Microalgae including marine algae and cyanobacteria are a diverse group of prokaryotic and eukaryotic photosynthetic microorganisms that can grow rapidly due to their simple structure. They have been investigated for the production of different biofuels including biodiesel, bioethanol, bio-syngas, and bio-hydrogen. Microalgal biofuel production is potentially sustainable. It is possible to produce adequate microalgal biofuels to satisfy the fast growing energy demand within the restraints of land and water resources. To effectively address current bottlenecks associated with microalgal biodiesel, a leap in both fundamental knowledge and technological applicability is required, and as a result, microalga-mediated manufacture of biodiesel will be a fully competitive process. For optimization of biomass and biodiesel production from microalgae, the preferences should be given to following approaches:

- Optimization of lipid productivity should use mechanistic, rather than empirical models.
- Direct synthesis of alkanes should be prompted instead of glycerides should be promoted.
- CO_2 should be conducted through porous hollow fibres into the bulk of the medium, rather than bubbled.
- Heterotrophic metabolism might be more applicable in the presence of large amounts of organic waste.
- Maximization of the light adsorption by the cells.
- Integrated separation of lipids should be replaced by downstream processing.

Technological developments, including advances in photobioreactor design, microalgal biomass harvesting, drying, and other downstream processing technologies are important areas that may lead to enhanced cost-effectiveness and therefore, effective commercial implementation of biofuel from microalgae is an achievable strategy. With regard to biocatalyst engineering, efforts should focus on:

- The collection of novel genetic resources from microalgae, including genome sequencing to widen the availability of suitable hosts and gene libraries.
- An improvement in the traditional and the design of new methods of nuclear transformation.

- The study of marker and reporter genes, suitable for controlled overexpression of lipid metabolites.
- The elucidation of metabolic pathways of cellular synthesis and accumulation/excretion of lipids.
- The design of novel photobioreactor configurations specifically targeted at high rates of mass transfer and light transmission.
- The design and optimization of separation protocols, encompassing harvesting, extraction and the purification of lipids.

Bioprospecting is of importance to identify algal species that have desired traits (e.g. high lipid content, growth rates, growth densities and/or the presence of valuable co-products), while growing on low-cost media. Despite the potential of this strategy, the most likely scenario is that bioprospecting will not identify species that are cost competitive with petroleum, and subsequent genetic engineering and breeding will be required to bring these strains to economic viability. The range of potential for engineering algae is just beginning to be realized, from improving lipid biogenesis and improving crop protection, to producing valuable enzymes or protein co-products. No sustainable technology is without its challenges but blind promotion of those technologies without honest consideration of the long-term implications may lead to the acceptance of strategies whose long-term consequences outweigh their short-term benefits. Microalgae can produce a large variety of novel bioproducts with wide applications in medicine, food, and the cosmetic industries. Combining microalgal farming and the production of biofuels using biorefinery strategies is expected to significantly enhance the overall cost-effectiveness of biofuel from a microalgae approach. Major breakthroughs are indeed necessary regarding the design and development of technologies that are able to reduce processing costs, while increasing product yields. Integrated studies in the form of well-funded R&D programs could eventually aid in selecting microalga strains specifically adapted to regional conditions, genetic improvement and process optimization. In particular, the biorefinery issue will be central, as it allows upgrades of spent biomass via the production of alternative bulk or fine chemicals, which will contribute positively to the overall economic feasibility of microalgal biotechnology.

Responses to environmental issues including global warming cannot be delayed much longer otherwise mankind may be at risk.

Biofuels produced from microalgae are one piece of the puzzle in what concerns energy security, since they may eventually replace traditional fossil fuels. Combined with measures on energy efficiency and savings and with educated changes in consumer behavior, they could be helpful to meet the energy demands of the near future in a sustainable manner.

References

1. Y. Chisti, *Biotechnol. Adv*, Vol. 25, pp. 306–394, 2007.
2. A. Pandey, *Handbook of Plant-Based Biofuels*, CRC Press, Francis & Taylor's, Boca Raton, USA, p. 297, 2008.
3. E. Gnansounou, C. Larroche, and A. Pandey, Special issue of *J. Sci. Ind. Res*, 67, pp. 837–1040, 2008.
4. S.A. Scott, M.P. Davey, J.S. Dennis, I. Horst, C.J Howe, D. Lea-Smith, and A.G. Smith, *Cur. Opin. Biotechnol*, Vol. 21, pp. 277–286, 2010.
5. L. Brennan and P. Owende, *Renew. Sust. Energy Rev*, Vol.. 14, pp. 557–577, 2010.
6. G. Dragone, B. Fernandes, A.A. Vicente, and J.A. Teixeira, "Third generation biofuels from microalgae in Current Research", Technology and Education, Topics in Applied Microbiology and Microbial Biotechnology, A. Mendez-Vilas (ed.), Formatex, pp.1355–1366, 2010.
7. P.J. Rojan, G.S. Anisha, K. Madhavan, and P. Ashok, *Bioresour. Technol.*, Vol. 102, pp. 186–193, 2011.
8. Y. Li, M Horsman, N. Wu, C.Q. Lan, and N. Dubois-Calero, *Biotechnol. Prog*, Vol. 24, pp. 815–820, 2008.
9. P. Tamagnini, E. Leitao, P. Oliveira, D. Ferriera, F. Pinto, D.J. Harris, T. Heidom, and P. Lindblad, *FEMS Microbiol. Rev*, Vol. 31, pp. 692–720, 2007.
10. T.R. Lorenz and G.R. Cysewski, *Trends Biotechnol*, Vol. 18, pp. 160–167, 2000.
11. R. León, M. Martín, J. Vigara, C. Vilche, and J. Vega, *Biomol Eng*, Vol. 20 , pp. 177–182, 2003.
12. M.K. Kim, J.W. Park, C.S. Park, S.J. Kim, K.H. Jeune, M.U. Chang, and J. Acreman, *BioresTechnol*, Vol. 98 , pp. 2220–2228, 2007.
13. J.F. Sánchez, J.M. Fernández, F.G. Acién, A. Rueda, J. Pérez-Parra, and E. Molina, *ProcBiochem*, Vol. 43 , pp. 398–405, 2008.
14. A. Parmar, N.K. Singh, A. Kaushal, S. Sonawala, and D. Madamwar, *Bioresour. Technol.*, Vol. 102, pp. 1795–1802, 2011.
15. D. Somma, H. Lobkowicz, and J.P. Deason, *Clean Technol. Environ. Policy*, Vol. 12, pp. 373–380, 2010.
16. C.S. Jones, S.P. Mayfield, *Curr. Opin. Biotechnol*, Vol. 23, pp. 346–351, 2011.
17. A.M.J. Kliphuis, L. de Winter, C. Vejrazka, D.E. Martens, M. Janssen, and R.H. Wijffels, *Biotechnol. Prog*, Vol. 26, pp. 687–696, 2010.
18. Malcata F. Xavier, *Trends in Biotechnology*, Vol. 29, Issue 11, pp 542–549, 2011.
19. H.M. Amaro, A.C. Guedes, and F.X. Malcata , *Appl Energ*, Vol. 88 , pp. 3402–3410, 2011.

20. T.M. Mata, A.A. Martins, and N.S. Caetano, *Renew Sust Energ Rev*, Vol. 14, pp. 217–232, 2010.
21. R.H. Wijffels and M.J. Barbosa, *Science*, Vol. 329, pp. 796–799, 2010.
22. K. Chojnacka and F.J. Marquez-Rocha, *Biotechnol*, Vol. 3, pp. 21–34, 2004.
23. G. Oron, G. Shelef, and A. Levi, *Biotechnol. Bioeng*, Vol. 21, pp. 2165–2173, 1979.
24. C.V. Seshadri and S. Thomas, *Biotechnol. Lett*, Vol. 1, pp. 287–291, 1979.
25. A. Vonshak, Z. Cohen, and A. Richmond, *Biomass*, Vol. 8, pp. 13–25, 1985.
26. M.A. Borowitzka, J Biotechnology, Vol. 70, pp.313–321, 1999.
27. C. Posten and G. Schaub, *J. Biotechnol*, Vol. 142, pp. 64–69, 2009.
28. G Roesijadi, S.B. Jones, L.J. Snowden-Swan, and Y. Zhu, *"Macroalgae as a Biomass Feedstock: A Preliminary Analysis"*, U.S. Department of Energy under contract DE-AC05–76RL01830 by Pacific Northwest National Laboratory, 2010.
29. C. Santillan, *Experientia*, Vol. 38, pp. 40–43, 1982.
30. R.M.A. el-Shishtawy, S. Kawasaki, and M Morimoto, *Biotechnol Tech*, Vol. 11 , pp. 403–407, 1997.
31. L. Metcalf and H.P. Eddy, *Waste Water Engineering*, (*second Ed.*) McGraw-Hill, San Francisco, 1980.
32. T. Matsunaga, H. Takeyama, H. Miyashita, and H. Yokouch, *Adv. Biochem. Eng. Biot*, Vol. 96, pp. 165–188, 2005.
33. R. M. Knuckey, M. R. Brown, R Robert, and D. M. F.Frampton, Aquacult. Eng, Vol. 35 *(3)*, *pp.*300–313, 2006,
34. J.A. Borchard, and C.R. Omelia, *J. Am. Water Works Assoc*, Vol. 53, pp. 1493–1502, 1961.
35. M.K. Danguah, L. And, N. Uduman, N. Moheimani, and G.M. Forde, *J. Chem. Technol. Biotechnol*, Vol. 84 , pp. 1078–1083, 2008.
36. C.G. Golueke and W.J. Oswald, *J. Water Pollut. Con. F*, Vol. 37, pp. 471–498, 1965.
37. D.S. Parker, "Performance of alternative algal removal systems". University of Texas Water Research Publication, no. 9, *Ponds as a Wastewater Treatment alternative*, University of Texas, Austin, 1975.
38. A. Sukenik, W. Schroder, J. Lauer, G. Shelef, and C.J. Soeder, *Water Res*, Vol. 19, pp. 127–129, 1985.
39. J. Prakash, B. Pushpuraj, P. Carbzzi, G. Torzillo, E. Montaini, and R Materassi, *Int. J. Solar Energy*, Vol. 18 , pp. 303–311, 1997.
40. H. Desmorieux, and N Decaen, *J. Food Eng*, Vol. 77, pp. 64–70, 2006.
41. N.K. Singh, A. Parmar, and D.. Madamwar, *Bioresour. Technol.*, Vol. 100, pp. 1663–1669, 2009.
42. A. Parmar, N.K. Singh, and D. Madamwar, *J. Phycol*, Vol. 86 , pp. 285–289, 2010.
43. B. Soni, U. Trivedi, and D Madamwar, *Bioresour. Technol.*, Vol. 99, pp. 188–194, 2008.
44. M Cartens, E. Moina Grima, A. Robels Medina, A. Gimenez, and J Ibanez Gonzalez , *J. Am. Oil Chem. Soc*, Vol. 73 , pp. 1025–1031, 1996.
45. P.K. Park, E.Y. Kima, and K.H. Chub, *Sep. Purif. Technol*, Vol. 53, pp. 148–152, 2007.
46. T. Lewis, P.D. Nichols, and T.A.McMeekin, *J. Microbial. Methods*, Vol. 43, pp. 107–116, 2000.

47. M. Herraro, A. Cifuentes, and E. Ibanez, *Food Chem*, Vol. 358, pp. 136–148, 2006.

48. B. Metting, W.J. Zimmerman, I. Crouch, and J van Staden, "Agronomic Uses of Seaweed and Microalgae," in I. Akatsuka ed., *Introduction to Applied Phycology*, SPB Academic Publishing, The Hague, Netherlands, pp. 589–627, 1990.

49. R.S. Ayala, and L. Castro, *Food Chem*, Vol. 75, pp. 109–113, 2001.

50. M.A. Hejazi, C. de Lamarlie, J.M.S. Rocha, M. Vermue, J. Tramper, and R.H. Wijffels, *Biotechnol. Bioeng*, Vol. 79, pp. 29–36, 2002.

51. A.P. Carvalho, and F.X. Malcata, *J. Agr. Food. Chem*, Vol. 53, pp. 5049–5059, 2005.

52. A. Richmond, *Handbook of microalgal culture: biotechnology and applied phycology*, Blackwell Science, 2004.

53. B. Hankamer, F. Lehr, J. Rupprecht, J.H. Mssgnug, C. Posten, and O Kruse, *Physiol. Plant*, Vol. 131 , pp. 10–21, 2007.

54. J. Tollefson, *Nature*, Vol. 451, pp. 880–883, 2008.

55. S. Elshahed Mostafa, *Journal of Advanced Research*, Vol. 1, (2), pp. 103–111, 2010.

56. P. Spolaore, Joannis- C. Cassan, E. Duran, and A. Isambert, *J. Biosci. Bioeng*, Vol. 101, pp. 87–96, 2006.

57. E.A. Ehimen, Z.F. Sun, and C.G. Carrington, *Fuel*, Vol. 89, pp. 677–684, 2010.

58. F.E. Rey, E.K. Heiniger, and C.S. Harwood, *Appl. Environ. Microbiol*, Vol. 73 (5), pp. 1665–1671, 2007.

59. J.L. Gosse, B.J. Engel, F.E. Rey, C.S. Harwood, L.E. Scriven, and M.C. Flickinger, *Biotechnol. Prog*, Vol. 23 (1), pp. 124–130, 2007.

60. P.C. Hallenbeck, and D Ghosh, *Trends Biotechnol*, Vol. 27 (5), pp. 287–297, 2009.

61. L.T. Angenent, K. Karim, M.H. Al Dahhan, B.A. Wrenn, and R Domíguez Espinosa, Trends *Biotechnol*, 22 (9), pp. 477–485, 2004.

62. C.N. Dasgupta, J.J.G ilbert, P. Lindblad, T. Heidorn, S.A. Borgvang, and K.D. Skjanes, *Int J Hydrogen Energ*, Vol. 35 , pp. 10218–10238, 2010.

63. O. Kruse, and B Hankamer, *Curr Opin Biotechnol*, Vol. 20 , pp. 257–263, 2010.

64. T. Happe, B. Mosler, and J.D. Naber, *J Biochem*, Vol. 222 , pp. 769–774, 1994.

65. A. Melis, L. Zhang, M. Forestier, M.L. Ghirardi, and M Seibert, *Plant Physiol*, 122, pp. 127–136, 2000.

66. S.A. Angermayr, K.J. Hellingwer, P. Lindblad, and M.J. Teixeira de Mattos, *Curr. Opin. Biotechnol*, Vol. 20, pp. 257–263, 2009.

67. D.B. Levin, P. Lawrence, and L Murry, *Int. J. Hydrogen Energy*, Vol. 29, pp. 173– 185, 2004.

68. H. Sakurai, and H. Masukawa, *Mar. Biotechnol*, Vol. 2, pp. 128–145, 2007.

69. C.A. Cardona, and O.J. Sánchez, *BioresourTechnol*, Vol. 98, pp. 2415–2457, 2007.

70. M. K. Lam, and T. K. Lee, *Biotechnology Advances*, Vol. 30 (3), pp. 673–690, 2012.

71. L.A. Demain, *J. Ind. Microbiol. Biotechnol*, Vol. 36, pp. 319–332, 2009.

72. R .Ueda, S. Hirayama, K Sugata, and H Nakayama, US Patent 5578,472 (1996)

73. R.A. Bush, and K.M. Hall, Process for the production of ethanol from algae. US Patent 7135,308, 2006.

74. R. Harun, M.K. Danquah, and M Forde Gareth, *J. Chem. Technol. Biotechnol*, Vol. 85 , pp. 199–203, 2010.

75. N. Youssef, D.R. Simpson, K.E. Duncan, M.J. McInerney, M Folmsbee, T Fincher, and R.M. Knapp, *Appl. Environ. Microbiol*, Vol. 73 (4), pp. 1239–1247, 2007.

76. B.P. Singh, M.R. Panigrahi, and H.S. Ray. *Energy Sources*, Vol. 22 (7), pp. 649–658, 2000.

77. T.H. Willke, and K.D. Vorlop, *Appl. Microbiol. Biot*, Vol. 66, pp. 131–142, 2004.

78. J.R. Beneman, *J. Ind. Microbiol*, Vol. 31, pp. 247–256, 1990.

79. M.A. Borowitzka, *Curr. Microbiol*, Vol. 3, pp. 372–375, 1986.

80. G.F. Wu, Q.Y. Wu, and Z.Y. Shen, *Bioresour. Technol.*, Vol. 76, pp. 85–90, 2001.

81. R.J. Radmer, and B.C. Parker, *J. Appl. Phycol*, Vol. 6, pp. 93–98, 1994.

82. S. Philip, T. Keshavarz, and I Roy, *J. Chem. Technol. Biot*, Vol. 82, pp. 233–247, 2007.

83. M. Olaizola, *Biomol. Eng*, Vol. 20, pp. 459–466, 2003.

84. J. Benemann, and W Oswald, "Systems and economic analysis of microalgae, ponds for conversion for CO_2 to biomass". Final Report to the US Department of Energy, Pittsburgh Energy Technology Centre, 1996.

85. E.B. Sydney, W. Sturm, J. Cesar de Carvalho, V. Thomas- Soccol, C. Larroche, A. Pandey, and C.R. Soccol, *Bioresour. Technol.*, Vol. 101, pp. 5892–5896, 2010.

86. J.N. Rosenberg, G.A. Oyler, L. Wilkinson, and M.J. Betenbaugh, *Curr. Opin. Biotechnol*, Vol. 19, pp. 430–436, 2008.

87. X. Lui, and Curtiss III, *Proc. Nat. Acad. Sci*, Vol. 106, pp. 21550–21554, 2009.

88. A. Leonard, C.J. Rooke, C.F. Meunier, H. Sarmento, J.P. Descy, and B.L. Su, *Energy Environ. Sci.*, Vol. 3, pp. 370–377, 2010.

89. L. Brennan, and P Owende, *Renew. Sust. Energy Rev*, Vol. 14, pp. 557–577, 2010.

90. D. Song, J. Fu, and D Shi, Chin. *J. Biotechnol*, Vol. 24, pp. 341–348, 2008.

91. Y. Asada, and J Miyake, *J. Biosci. Bioeng*, Vol. 88, pp. 1–9, 1999.

92. J. Dexter, and P. Fu, *Energy Environ. Sci*, Vol. 2, pp. 857–864, 2009.

93. D.P. Chynoweth, *HortScience*, Vol. 40, pp. 283–286, 2005.

94. S. Ostergaard, L. Olsson, M. Johnston, and J Nielsen, *Nat. Biotechnol*, Vol. 18, pp. 1283–1286, 2000.

95. S. Ostergaard, L. Olsson, and J. Nielsen, *Metabolic Microbiol. Mol. Biol. Rev*, 64, pp. 34–50, 2000.

96. N Wei, J. Quarterman, and J Yong-Su, *Trends in Biotechnology*, Vol. 31 (2), pp. 70–77, 2013.

97. C. Bro, S. Knudsen, B. Regenberg, L. Olsson, and J Nielsen, *Appl. Environ. Microbiol*, 71, pp. 6465–6472, 2005.

98. G. Torzillo, A. Scoma, C. Faraloni, A. Ena, and U Johanningmeier, *Int J Hydrogen Energy*, Vol. 34 , pp. 4529–4536, 2009.

99. J. Beckmann, F. Lehr, G. Finazzi, B. Hankamer, C. Posten, L. Wobbe, and O. Kruse, *J Biotechnol*, Vol. 142 , pp. 70–77, 2009.

100. J.H. Mussgnug, S. Thomas-Hall, J. Rupprecht, A. Foo, V. Klassen, A. McDowall, P.M. Schenk, O. Kruse, and B Hankamer, *Plant Biotechnol J*, Vol. 5 pp. 802–814, 2007.

101. A. Melis, *Plant Sci*, Vol. 177, pp. 272–280, 2009.

102. http://www.biomara.org/

103. http://arpae.energy.gov/ProgramsProjects/OtherProjects/BiomassEnergy/MacroAlgaeButanol.aspx

8

Bioethanol Production Processes

Mohammad J. Taherzadeh*,[1], Patrik R. Lennartsson[1],
Oliver Teichert[2] and Håkan Nordholm[2]

[1]*School of Engineering, University of Borås, Borås, Sweden*
[2]*Lantmännen Agroetanol AB, Norrköping, Sweden*

Abstract

Ethanol, with more than 86 billion liters produced in 2011, is the dominant biofuel in the global fuel market. Nowadays, ethanol is produced from starch- and sugar sources, so-called 1st generation ethanol, while lignocellulosic ethanol (2nd generation) is not yet applied for industrial applications. In general, the production includes pretreatment, eventual hydrolysis of sugar polymers by enzymes or acids, fermentation by baker's yeast, for example and downstream processes such as distillation, dehydration and evaporation. Furthermore, the global production and consumption of ethanol has also raised ethical and social concerns about, for example, food vs. fuel. This chapter covers the details of the processes in biofuel production.

Keywords: Ethanol, fermentation

8.1 Introduction

Ethanol (C_2H_5OH) is a clear, colorless, flammable liquid chemical. It has been produced and used in aqueous solutions as an alcoholic beverage for several thousand years. Ethanol also has several industrial applications (e.g., in detergents, toiletries, coatings and pharmaceuticals) and has been used as transportation fuel for more than a century. Nicholas Otto used ethanol in the internal combustion engine invented in 1897 [1]. However, ethanol did not have

---

Corresponding author: Mohammad.Taherzadeh@hb.se

Dr.Vikash Babu, Dr. Ashish Thapliyal & Dr. Girijesh Kumar Patel (eds.) Biofuels Production, (211–254) 2014 © Scrivener Publishing LLC

a major impact in the fuel market until the 1970s, when two oil crises occurred in 1973 and 1979. Since the 1980s, ethanol has been a major actor in the fuel market as an alternative fuel as well as an oxygenating compound for gasoline. Ethanol can be produced synthetically from oil and natural gas or biologically from sugar, starch and lignocellulosic materials. The biologically produced ethanol is sometimes called *fermentative ethanol* or *bioethanol*. Application of bioethanol as fuel has none or very limited net emissions of CO_2 [2] and is able to fulfill the Kyoto Climate Change Protocol (1997) to decrease the net emissions of CO_2 [3]. However, there has been plenty of criticism about the sustainability of sugar and grain based ethanol, so called "first generation ethanol." In this chapter, the global market and the production of bioethanol are briefly reviewed.

8.2   Global Market for Bioethanol and Future Prospects

Ethanol is produced from a variety of feedstocks. Fermentative ethanol is produced from grains, molasses, sugarcane juice, fruits, surplus wine, whey and some other similar sources, which contain simple sugars and their polymers. On the other hand, synthetic ethanol is produced from oil, e.g., through hydration of ethylene [4]:

$$\text{Oil} \longrightarrow CH_2CH_2 \xrightarrow{H_2O} CH_3CH_2OH \text{ (ethanol)} \quad (8.1)$$

However, bioethanol completely dominates the world market, accounting for more than 95% of the annual world production [5]. The majority of ethanol is used as fuel, which is also responsible for the majority of the world market expansion. In 2011, 86.1 billion liters (68 million tons) ethanol was produced, compared to 17.0 billion liters in 2000 (Fig. 8.1) [6].

The two dominant ethanol producing nations in the world are the United States and Brazil, accounting for 63% (54 billion liters) and 24% (21 billion liters) of the global ethanol production in 2011, respectively. However, compared to 2010, the ethanol production in Brazil has decreased by 18%, partly due to poor sugarcane harvests, while the production in the United States has increased. Other notable world producers are China (2.1 billion liters),

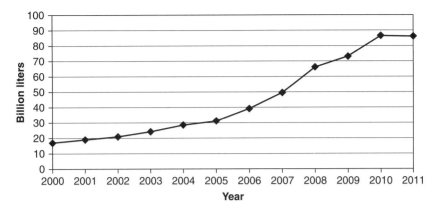

Figure 8.1 World fuel ethanol production since 2000 [6].

Canada (1.8 billion liters), France (1.1 billion liters) and Germany (0.8 billion liters) [6].

8.3 Overall Process of Bioethanol Production

The process of ethanol production depends on the raw materials used. A general simplified representation of these processes is shown in Fig. 8.2 and a brief description of different units of the process is presented in this chapter. It should be noted that if sugar substances, such as molasses and sugarcane juice, are used as raw materials, the milling, pretreatment, hydrolysis and detoxification are not necessary.

Milling, liquefaction and saccharification processes are usually necessary for the production of fermentable sugar from starchy materials, while milling, pretreatment and hydrolysis are typically used for ethanol production from lignocellulosic materials. Furthermore, a detoxification unit is not always considered, unless a toxic substrate is fed to the bioreactors.

8.4 Production of Sugars from Raw Materials

Sugar substances (such as sugarcane juice and molasses), starchy materials (such as wheat, corn, barley, potato and cassava) and lignocellulosic materials (such as forest residuals, straws and other

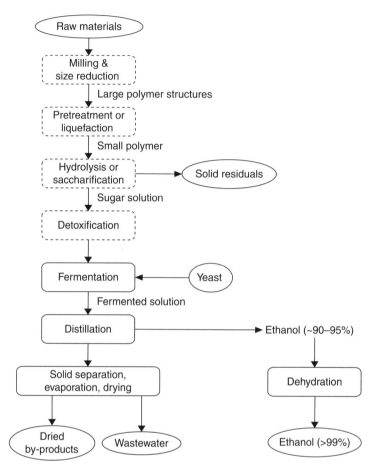

Figure 8.2 A general process scheme for ethanol production from different raw materials.

agricultural by-products) are considered the raw materials for ethanol production. The dominating sugars available or produced from these popular raw materials are:

- Glucose, fructose and sucrose in sugar substances
- Glucose in starchy materials
- Glucose from cellulose and either mannose or xylose from hemicellulose of lignocellulosic materials

Most ethanol-producing microorganisms can utilize a variety of hexoses such as glucose, fructose, galactose and mannose and even a limited number of disaccharides such as sucrose, lactose,

Table 8.1 Treatment for different types of substrates.

Substrate	Pretreatment or liquefaction	Hydrolysis or saccharification	Detoxification
Sugar materials	No	No	Typically no
Starchy materials	Yes	Yes	No
Lignocellulosic materials	Yes	Yes	Depends on the hydrolysis method

cellobiose and maltose, but only rarely their polymers. Therefore, it is necessary to convert the complex polysaccharides, such as cellulose and starch, to simple sugars or disaccharides. Different types of substrates that need treatment prior to fermentation are presented in Table 8.1.

In this section, sugar production from starchy materials is discussed; lignocellulosic materials are discussed in Sec. 8.5.

8.4.1 Sugar Solution from Starchy Materials

There are various raw materials that contain starch and are suitable for ethanol production. Corn is widely used industrially for this purpose. However, there are several other cereals, such as wheat, rye, barley, triticale, oats and sorghum; and crop roots such as potato and cassava, which are used as raw materials for ethanol production. The cereals contain about 50–70% starch, 8–12% proteins, 10–15% water and small amounts of fats and fibers. The compositions of the crop roots are almost identical to those of the cereals on a dry basis, but the water content of the roots is usually 70–80%. The exact composition of each raw material depends on the type and variety of materials used and can be found in literature (e.g., [2, 7]). The microorganisms use starch from these materials as a carbon and energy source, while part of their protein content is used as a nitrogen source.

Starch contains two fractions: amylose and amylopectin. Amylose, which typically constitutes about 20% of starch, is a straight-chain polymer of α-glucose subunits with a molecular weight that may vary from several thousands to half a million. Amylose is a water-insoluble polymer. The bulk of starch is amylopectin, which is also a polymer of glucose. Amylopectin contains a substantial number of branches in molecular chains. Branches occur from the ends of amylose segments, averaging 25 glucose units in length. Amylopectin molecules are typically larger than amylose, with molecular weights ranging up to 1–2 Mg/mol. Amylopectin is soluble in water and can form a gel by absorbing water at higher temperatures.

For ethanol production, hydrolysis is necessary for converting starch into fermentable sugar available to microorganisms. Traditional conversion of starch into sugar monomers requires a two-stage hydrolysis process: liquefaction of large starch molecules to oligomers and saccharification of the oligomers to sugar monomers. This hydrolysis may be catalyzed by acid or amylolytic enzymes.

8.4.2   Acid Hydrolysis of Starch

Acid hydrolysis is a well-known process still applied in some ethanol industries. Sulfuric acid is the most commonly applied acid in this process, where starch is converted into low molecular weight dextrins and glucose [8]. The main advantages of this process are the rapid hydrolysis and the lower cost for the catalyst, compared to enzymatic hydrolysis. However, the acid processes possess drawbacks, including (a) high capital cost for an acid-resistant hydrolysis reactor, (b) destruction of sensitive nutrients such as vitamins present in raw materials and (c) further degradation of sugar to hydroxymethylfurfural (HMF), levulinic acid and formic acid which lowers the ethanol yield and inhibits the fermentation process [9].

The acid hydrolysis process can be performed either in batches or in continuous systems. Dilute acid hydrolysis can also be used as a pretreatment for enzymatic hydrolysis. The starch or starchy material is soaked in the dilute acid prior to enzymatic hydrolysis, after this it is continuously passed through a steam-jet heater into a cooking tube (called a jet cooker or mash cooker) with a plug flow residence time for a couple of minutes and then it is subjected to enzymatic hydrolysis.

8.4.3   Enzymatic Hydrolysis of Starch

Enzymatic hydrolysis has several advantages compared to acid hydrolysis. First, the specificity of the enzymes allows for the production of sugar syrups with well-defined physical and chemical properties. Second, milder enzymatic hydrolysis results in fewer

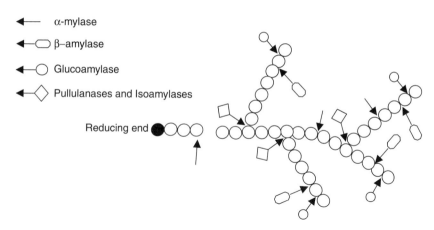

Figure 8.3 Mechanism of action of amylase on starch.

side reactions and less "browning" by Maillard-reactions [8]. The types of enzymes involved in the enzymatic hydrolysis of starch are α-amylase, β-amylase, glucoamylase, pullulanase and isoamylase. The mechanism of action for these enzymes is presented schematically in Fig. 8.3.

There are two popular industrial processes from starch materials: dry milling and wet milling. In the dry-milling process, grains are first ground and then processed as whole grain flour without a separation of the starch from the germ and fiber components. To be explicit, in this method the whole mixture of starch and other components is processed. Starch is converted into sugars in two stages: liquefaction and saccharification, by adding water, enzymes and heat (enzymatic hydrolysis). Dry-milling processes normally produce co-products, for example, distillers' dried grains with solubles (DDGS), which are used as animal-feed supplements. Revenues from co-products play an important part in making the overall process economically feasible [2].

In the wet-milling process, grain is steeped and can be separated into starch, germ and fiber components. Wet milling is capital intensive, but it generates numerous co-products that help to improve the overall production economics [2]. Wet mills can produce protein rich fractions and fiber enriched fractions as feed components or even germ-based products as germ oil. In this method, after the grain is cleaned, it is steeped and then ground to remove the germ. Further grinding, washing and filtering steps separate the fiber and gluten. The starch that remains is washed out in these separation steps and is then broken down into fermentable sugars by the addition of enzymes in the liquefaction and saccharification stages. The fermentable sugars produced are then subjected to fermentation for ethanol production.

8.5   Characterization of Lignocellulosic Materials

Lignocellulosic materials predominantly contain a mixture of carbohydrate polymers (cellulose and hemicellulose) and lignin. The carbohydrate polymers are tightly bound to lignin mainly through hydrogen bonding, but also through some covalent bonding. The contents of cellulose, hemicellulose and lignin in common lignocellulosic materials are listed in Table 8.2 [2, 10–13].

8.5.1   Cellulose

Cellulose is the main component of most lignocellulosic materials. Cellulose is a linear polymer of up to 27,000 glucosyl residues linked by β-1,4 bonds. However, each glucose residue is rotated 180° relative to its neighbors so that the basic repeating unit is cellobiose, a

Table 8.2 Contents of cellulose, hemicelluloses and lignin in common lignocellulosic materials.

lignocellulosic materials	Cellulose (%)	Hemicellulose (%)	Lignin (%)
Hardwood stems	40–75	10–40	15–25
Softwood stems	30–50	25–40	25–35
Corn cobs	45	35	15
Wheat straw	30	50	15
Rice straw	32–47	19–27	5–24
Sugarcane bagasse	40	24	25
Leaves	15–20	80–85	0
Paper	85–99	0	0–15
Newspaper	40–55	25–40	18–30
Waste paper from chemical pulps	60–70	10–20	5–10
Grasses	25–40	25–50	10–30

dimer of a two-glucose unit. As glucose units are linked together into polymer chains, a molecule of water is not used due to the bondings in-between, which makes the chemical formula $C_6H_{10}O_5$ for each monomer unit of "glucan." The parallel polyglucan chains form numerous intra- and intermolecular hydrogen bonds, which results in a highly ordered crystalline structure of native cellulose, interspersed with less-ordered amorphous regions [14–16].

8.5.2   Hemicellulose

Hemicelluloses are heterogeneous polymers of pentoses (e.g., xylose and arabinose), hexoses (e.g., mannose, glucose and galactose) and sugar acids. Unlike cellulose, hemicelluloses are not chemically homogeneous. Hemicelluloses are relatively easily hydrolyzed by acids to their monomer components consisting of glucose, mannose, galactose, xylose, arabinose and small amounts of rhamnose, glucuronic acid, methylglucuronic acid and galacturonic acid. Hardwood hemicelluloses contain mostly xylans, whereas softwood hemicelluloses contain mostly glucomannans. Xylans are the most abundant hemicelluloses. Xylans of many plant materials are heteropolysaccharides with homopolymeric backbone chains of 1,4-linked β-d-xylopyranose units. Xylans from different sources, such as grasses, cereals, softwood or hardwood, differ in composition. Besides xylose, xylans may contain arabinose, glucuronic acid or its 4-O-methyl ether, as well as acetic, ferulic and p-coumaric acids. The degree of polymerization of hardwood xylans (150–200) is higher than that of softwoods (70–130) [13, 14].

8.5.3   Lignin

Lignin is a very complex molecule. It is an aromatic polymer constructed of phenylpropane units linked in a three-dimensional structure. Generally, softwoods contain more lignin than hardwoods. Lignins are divided into two classes, namely, "guaiacyl lignins" and "guaiacyl-syringyl lignins." Although the principal structural elements in lignin have been largely clarified, many aspects of its chemistry remain unclear. Chemical bonds have been reported between lignin and hemicellulose and even cellulose. Lignins are extremely resistant to chemical and enzymatic degradation. Biological degradation can be achieved mainly by fungi, but also by certain actinomycetes (bacteria) [14, 17].

8.6   Sugar Solution from Lignocellulosic Materials

There are several possible ways to hydrolyze lignocellulose. The most commonly applied methods can be classified into two groups: chemical hydrolysis and enzymatic hydrolysis. In addition, there are some other hydrolysis methods in which no chemicals or enzymes are applied. For instance, lignocellulose may be hydrolyzed by γ-ray or electron-beam irradiation or microwave irradiation. However, those processes are commercially unfeasible [14].

8.6.1   Chemical Hydrolysis of Lignocellulosic Materials

Chemical hydrolysis involves the exposure of lignocellulosic materials to a chemical for a period of time, at a specific temperature and results in sugar monomers from cellulose and hemicellulose polymers. Predominantly, acids are applied in chemical hydrolysis. Sulfuric acid is the most investigated acid, although other acids such as hydrochloric acid (HCl) have also been used. Acid hydrolyses can be divided into two groups: concentrated-acid hydrolysis and dilute-acid hydrolysis [18].

8.6.1.1   *Concentrated Acid Hydrolysis*

Hydrolysis of lignocellulose by concentrated sulfuric or hydrochloric acids is a relatively old process. Concentrated-acid processes are generally reported to give higher sugar and ethanol yields, compared to dilute-acid processes. Furthermore, they do not need a very high pressure and temperature. Although this is a successful method for cellulose hydrolysis, concentrated acids are toxic, corrosive and hazardous and require that the material of construction used for reactors and other equipment is highly resistant to corrosion. High investment and maintenance costs have greatly reduced the commercial potential of this process. In addition, the concentrated acid must be recovered after hydrolysis to make the process economically feasible. Even, the environmental impact strongly limits the application of hydrochloric acid [11, 14].

8.6.1.2   *Dilute Acid Hydrolysis*

Dilute sulfuric acid hydrolysis is a method for either the pretreatment before enzymatic hydrolysis or the conversion of lignocellulose into sugars in a single step. As a pretreatment it gives high reaction rates and significantly improves enzymatic hydrolysis. Depending on the substrate used and the conditions applied, up to 95% of the hemicellulosic sugars can be recovered by dilute-acid hydrolysis from the lignocellulosic feedstock [2, 19]. Of all dilute acid processes, the processes using sulfuric acid have been the most extensively studied. Sulfuric acid is typically used in a 0.5–1.0% concentration. However, the time and temperature of the process can vary. It is common to use one of the following conditions in dilute-acid hydrolysis:

- mild conditions, i.e., low pressure and long retention time
- Severe conditions, i.e., high pressure and short retention time

In dilute acid hydrolysis, the hemicellulose fraction is depolymerized at lower temperatures than the cellulose fraction. If higher temperature or longer retention times are applied, the monosaccharides formed will be further hydrolyzed to other compounds. Therefore, it is suggested that the hydrolysis process be carried out in at least two stages. The first stage is carried out at relatively milder conditions during which the hemicellulose fraction is hydrolyzed and a second stage can be carried out by enzymatic hydrolysis or dilute acid hydrolysis, at higher temperatures, during which the cellulose is hydrolyzed [19]. These first and second stages are sometimes called "pretreatment" and "hydrolysis," respectively.

Hydrolyzates of first stage dilute acid hydrolysis usually consist of hemicellulosic carbohydrates. The dominant sugar in the first stage hydrolyzate of hardwoods (such as alder, aspen and birch) and most agricultural residues such as straws is xylose, whereas first-stage hydrolyzates of softwoods (e.g., pine and spruce) predominantly contain mannose. However, the dominant sugar in the second stage hydrolyzate of all lignocellulosic materials, either by enzymatic or dilute acid hydrolysis, is glucose, which is originated from cellulose.

8.6.1.3 Detoxification of Acid Hydrolyzates

In addition to sugars, several byproducts are formed or released in the acid hydrolysis process. The most important byproducts are carboxylic acids, furans and phenolic compounds (see Fig. 8.4).

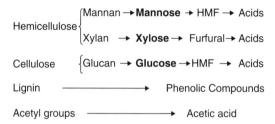

Figure 8.4 Formation of inhibitory compounds from lignocellulosic materials during acid hydrolysis.

Acetic acid, formic acid and levulinic acid are the most common carboxylic acids found in hydrolyzates. Acetic acid is mainly formed from acetylated sugars in the hemicellulose, which are already cleaved off at mild hydrolysis conditions. Since the acid is not further hydrolyzed, formation of acetic acid is dependent on the temperature and pressure of dilute-acid hydrolysis, until the acetyl groups are fully hydrolyzed. Therefore, the acetic acid yield in the hydrolysis does not significantly depend on the severity of the hydrolysis process [19, 20].

Furfural and HMF are the only furans usually found in hydrolyzates in significant amounts. They are hydrolysis products of pentoses and hexoses, respectively [19]. Formation of these byproducts is affected by the type and size of lignocellulose, as well as hydrolysis variables such as acid type and concentration, pressure and temperature and the retention time.

A large number of phenolic compounds have been found in hydrolyzates. However, reported concentrations are normally a few milligrams per liter. This could be due to the low water solubility of many of the phenolic compounds or a limited degradation of lignin during the hydrolysis process. Vanillin, syringaldehyde, hydroxybenzaldehyde, phenol, vanillic acid and 4-hydroxybenzoic acid are among the phenolic compounds found in dilute-acid hydrolyzates [21].

Biological (e.g., by using enzymes Mn peroxidase and laccase), physical (e.g., by evaporation of volatile fraction and by extraction of nonvolatile fraction by diethyl ether) and chemical (e.g., by alkali treatment) methods have been employed for detoxification of lignocellulosic hydrolyzates [22, 23].

Detoxification of lignocellulosic hydrolyzates by overliming is a common method used to improve fermentability [24–27]. In this method, $Ca(OH)_2$ is added to hydrolyzates to increase the pH (up to

9–11) and this condition is kept for a period of time (from 15 min up to several days), followed by decreasing the pH to the fermentation pH. It has been found that time, pH and temperature of overliming are the effective parameters in detoxification [28]. However, the drawback of this treatment is that part of the sugar is also degraded during the overliming process. Therefore, it is necessary to optimize the process to reduce the loss of fermentable sugars [23, 28]. Different methods to overcome the inhibitors have recently been reviewed [29].

8.6.2   Pretreatment Prior to Enzymatic Hydrolysis of Lignocellulosic Materials

Native (indigenous) cellulose fractions of cellulosic materials are recalcitrant to enzymatic breakdown, so a pretreatment step is required to render them amenable to enzymatic hydrolysis to glucose. A number of pretreatment processes have been developed in laboratories, including:

- *Physical pretreatment*—mechanical comminution, irradiation and pyrolysis
- *Physicochemical pretreatment*—steam explosion or autohydrolysis, ammonia fiber explosion (AFEX), SO_2 explosion and CO_2 explosion
- *Chemical pretreatment*—ozonolysis, dilute acid hydrolysis, alkaline hydrolysis organosolvent process and oxidative delignification
- *Biological pretreatment*, using lignin-degrading enzymes and white-rot fungi

However, not all of these methods may be technically or economically feasible for large-scale processes. In some cases, one method is used to increase the efficiency of another method. For instance, milling could be applied to achieve a better steam explosion by reducing the chip size. Furthermore, it should be noted that the selection of the pretreatment method should be compatible with the selection of hydrolysis. For example, if acid hydrolysis is to be applied, a pretreatment with alkali may not be beneficial [21]. Pretreatment methods were reviewed by Wyman [2], Sun and Cheng [11], Hendriks and Zeeman [30] and Aslanzadeh *et al* [31].

Among the different types of pretreatment methods, dilute acid, SO_2 and steam explosion methods have been successfully developed for the pretreatment of lignocellulosic materials. The methods show promising results for industrial application. Dilute sulfuric acid hydrolysis is a favorable method for either pretreatment before enzymatic hydrolysis or for the conversion of lignocellulose into sugars.

8.6.3   Enzymatic Hydrolysis of Lignocellulosic Materials

Enzymatic hydrolysis of cellulose and hemicellulose can be carried out by highly specific cellulase and hemicellulase enzymes (glycosyl hydrolases). This group includes at least 15 protein families and some subfamilies [14, 32]. Enzymatic degradation of cellulose to glucose is generally accomplished by synergistic action of three distinct classes of enzymes [2]:

- 1,4-β-d-glucan-4-glucanohydrolases   or   endo-1,4-β-glucanases, which are commonly measured by detecting the reducing groups released from carboxymethylcellulose (CMC).
- Exo-1,4-β-d-glucanases, including both 1,4-β-d-glucan hydrolases and 1,4-β-d-glucan cellobiohydrolases. 1,4-β-d-glucan hydrolases liberate d-glucose and 1,4-β-d-glucan cellobiohydrolases liberate d-cellobiose.
- β-d-glucoside glucohydrolases or β-d-glucosidases, which release d-glucose from cellobiose and soluble cellodextrins, as well as an array of glycosides.

There are synergies between exo-exo, exo-endo and endo-endo enzymes, which have been demonstrated several times.

Substrate properties, cellulase activity and hydrolysis conditions (e.g., temperature and pH) are the factors that affect the enzymatic hydrolysis of cellulose. To improve the yield and rate of enzymatic hydrolysis, there has been some research focused on optimizing the hydrolysis process and enhancing cellulase activity. Substrate concentration is one of the main factors that affect the yield and initial rate of enzymatic hydrolysis of cellulose. At low substrate levels, an increase of substrate concentration normally results in an increase of the yield and reaction rate of the hydrolysis. However, high substrate concentrations can cause substrate inhibition, which substantially lowers the rate of hydrolysis. The extent of substrate inhibition depends on the ratio of total substrate to total enzyme [11]. Increasing the dosage of cellulases in the process to a certain extent can enhance the yield and rate of hydrolysis, but would significantly increase the costs of the process. Cellulase loading of 10 FPU/g (filter paper units per gram) of cellulose is often used in laboratory studies because it provides a hydrolysis profile with high levels of glucose yield in a reasonable time (48–72 h) at a reasonable enzyme cost. Cellulase enzyme loadings in hydrolysis vary from 5 to 33 FPU/g substrate, depending on the type and concentration of substrates. β-glucosidase can act as a limiting agent in the enzymatic hydrolysis of cellulose. Adding supplemental β-glucosidase can enhance the saccharification yield [33, 34].

The enzymatic hydrolysis of cellulose consists of three steps: (1) adsorption of cellulase enzymes onto the surface of cellulose, (2) biodegradation of cellulose to simple sugars and (c) desorption of cellulose [11]. Cellulase activity decreases during hydrolysis. Irreversible adsorption of cellulase on cellulose is partially responsible for this deactivation. The addition of surfactants during hydrolysis is capable of modifying the cellulose surface property and minimizing the irreversible binding of cellulase on cellulose. Tween 20 and Tween 80 are the most efficient non-ionic surfactants in this regard. The addition of Tween 20 as an additive in simultaneous saccharification and fermentation (SSF) at 2.5 g/l has several positive effects on the process. It increases the ethanol yield, increases the enzyme activity in the liquid fraction at the end of the process, reduces the amount of enzyme loading and reduces the required time to attain maximum ethanol concentration [33].

8.7   Basic Concepts of Fermentation

The general reaction for ethanol production during fermentation is:

$$\text{Sugar (s)} \xrightarrow{\text{Microorganisms}} \text{Ethanol} + \text{Byproducts}$$

In this reaction, the microorganisms work as a catalyst.

8.8   Conversion of Simple Sugars to Ethanol

Conversion of simple hexose sugars, such as glucose and mannose, in fermentation into ethanol can take place anaerobically as follows:

$$C_6H_{12}O_6 \xrightarrow{\text{Microorganisms}} 2C_2H_5OH \text{ (ethanol)} + 2CO_2$$

If all of the sugar is converted into ethanol according to the above reaction, the yield of ethanol will be 0.51 g/g of the consumed sugars, meaning that from 1.0 g of glucose, 0.51 g of ethanol can be produced. This is the maximal theoretical yield for ethanol from hexoses. However, the ethanol yield obtained in fermentation does not usually exceed 90–95% of the maximal theoretical yield, since part of the carbon source in sugars is converted into biomass by growth of the production organism and to byproducts such as glycerol and acetic acid [9, 35].

A similar reaction for anaerobic conversion of pentoses, such as xylose to ethanol, might be considered. Xylose is generally converted first into xylulose by a one-step reaction catalyzed by xylose isomerase (XI) in many bacteria or by a two-step reaction through xylitol in yeasts and fungi. It can then be converted into ethanol anaerobically through pentose phosphate pathway (PPP) and glycolysis. The general reaction can be written as:

$$3C_5H_{10}O_{10} \xrightarrow{\text{Microorganisms}} 5C_2H_5OH \text{ (ethanol)} + 5CO_2$$

In this case, a theoretical ethanol yield of 0.51 g/g from xylose can be expected, as from glucose. However, the redox imbalance and slow rate of ATP formation are two major factors that make anaerobic ethanol production from xylose very difficult [36, 37]. A few anaerobic ethanol producing strains have been developed from xylose in research laboratories, but so far no strain is available for industrial scale process. Attempts have been made to overcome this problem of xylose assimilation by co-metabolization or working with microaerobic conditions, where oxygen is available at a low concentration. A number of microorganisms can produce ethanol aerobically from xylose, where the practical yield of ethanol from xylose and other pentoses is usually lower than its theoretical yield. The challenges in ethanol production from xylose were reviewed by van Maris et al. [38] and Madhavan *et al.* [39].

8.9   Biochemical Basis for Ethanol Production from Hexoses

A simplified central metabolic pathway for ethanol production in yeast and bacteria under anaerobic conditions is presented in Fig. 8.5 [14, 40–42].

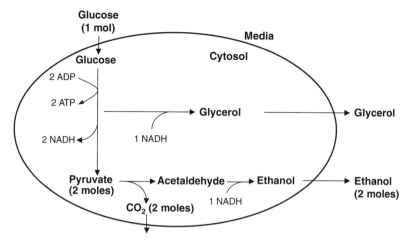

Figure 8.5 central metabolic pathway in yeast under anaerobic conditions.

The three major interrelated pathways that control catabolism of carbohydrate in most ethanol-producing organisms are:

- Embden-Meyerhof pathway (EMP) or glycolysis
- Pentose phosphate pathway (PPP)
- Krebs or tricarboxylic acid cycle (TCA)

In glycolysis, glucose is anaerobically converted into pyruvic acid and then into ethanol through acetaldehyde. This pathway provides energy in the form of ATP to the cells. The net yield in glycolysis is two moles of pyruvate (or ethanol) and two moles of ATP from each mole of glucose. This pathway is also the entrance for other hexoses such as fructose, mannose and galactose to metabolic pathways. With only two moles of ATP formed per glucose catabolized, large amounts of ethanol (at least 3.7 g of ethanol per gram of biomass) must be formed [14, 43].

The PPP handles pentoses and is important for nucleotide (ribose-5-phosphate) and fatty acid biosynthesis. The PPP is mainly used to reduce NADP$^+$. In *Saccharomyces cerevisiae*, 6–8% of the available glucose passes through the PPP under anaerobic conditions [8, 14].

The TCA cycle function is to convert pyruvic and lactic acids and ethanol aerobically into the end products CO_2 and H_2O. It is also a common channel for the ultimate oxidation of fatty acids and the carbon skeletons of many amino acids. In cells containing this additional aerobic pathways, the NADH that forms during glycolysis results in ATP generation in the TCA cycle [8].

Ethanol production from hexoses is redox-neutral, i.e., no net formation of NADH or NADPH occurs. However, biosynthesis of the cells results in the net formation of NADH and the consumption of NADPH. The PPP is mainly used to reduce $NADP^+$ to NADPH. The oxidation of surplus NADH under anaerobic conditions in *S. cerevisiae* is carried out through the glycerol pathway. Furthermore, there are other byproducts—mainly carboxylic acids: acetic acid, pyruvic acid and succinic acid—that add to the surplus NADH. Consequently, glycerol is also formed to compensate the NADH formation coupled with these carboxylic acids. Thus, formation of glycerol is coupled with biomass and carboxylic acid formation in anaerobic growth of *S. cerevisiae* [14, 44].

We should keep in mind that growth of the cells and increasing their biomass is the ultimate goal of the cells. They produce ethanol under anaerobic conditions in order to provide energy through catabolic reactions. Glycerol is formed to keep the redox balance of the cells and carboxylic acids may leak from the cells to the medium. Therefore, the ethanol-producing microorganisms produce ethanol as the major product under anaerobic conditions, while biomass, glycerol and some carboxylic acids are the byproducts.

8.10   Biochemical Basis for Ethanol Production from Pentoses

In general, yeast and filamentous fungi metabolize xylose through a two-step reaction before they enter the central metabolism (glycolysis) through the PPP. The first step is conversion of xylose into xylitol using xylose reductase (XR) and the second step is conversion of xylitol into xylulose using another enzyme, xylitol dehydrogenase (XDH) (Fig. 8.6) [45–47]. In total, the two steps are redox neutral; however, in most eukaryotic species, the two different

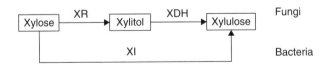

Figure 8.6 Xylose metabolism by bacteria and *S. cerevisiae*.

enzymes involved use different redox carriers, which leads to an imbalance in the cell [48, 49].

Wild strains of *S. cerevisiae* possess the enzymes XR and XDH, but their activities are too low to allow growth on xylose. Although *S. cerevisiae* cannot utilize xylose, it can utilize its isomer, xylulose. Thus, if *S. cerevisiae* is to be used for xylose fermentation, it requires a genetic modification to encode XR/XDH or xylose isomerase (XI) [46, 50].

Bacteria have a slightly different metabolic pathway for xylose utilization. They convert xylose into xylulose in one reaction using XI [51–54].

8.11 Microorganisms Related to Ethanol Fermentation

The criteria for an ideal ethanol producing microorganism are to have: (a) high growth and fermentation rate, (b) high ethanol yield, (c) high ethanol and glucose tolerance, (d) osmotolerance, (e) low optimum fermentation pH, (f) high optimum temperature, (g) general hardiness under physiological stress and (h) tolerance to potential inhibitors present in the substrate [35, 55]. Ethanol and sugar tolerance allows the conversion of concentrated feeds into concentrated products, reducing energy requirements for distillation and stillage handling. Osmotolerance allows handling of relatively dirty raw materials with high salt content. Low pH fermentation combats contamination by competing organisms. High temperature tolerance simplifies fermentation cooling. General hardiness allows microorganisms to survive stress such as that of handling (e.g., centrifugation) [55]. The microorganisms should also tolerate the inhibitors present in the medium. Microorganisms able to hydrolyze the (lignocellulosic) substrate and ferment the sugars into ethanol, referred to as consolidated bioprocessing, have also been sought. Consolidated bioprocessing has recently been reviewed [56].

8.11.1 Yeasts

Historically, yeasts have been the most commonly used microorganisms for ethanol production. Yeast strains are generally chosen among *S. cerevisiae, S. ellypsoideuse, S. fragilis, S. carlsbergensis,*

Schizosaccharomyces pombe, Torula cremoris and *Candida pseudotropicalis*. Yeast species which can produce ethanol as the main fermentation product are reviewed, for example by Lin and Tanaka [8].

Among the ethanol producing yeasts, the "industrial working horse" *S. cerevisiae* is by far the most well-known and most widely used yeast. This yeast can grow both on simple hexose sugars, such as glucose and on the disaccharide sucrose. *S. cerevisiae* is also generally recognized to be safe as a food additive for human consumption and is therefore ideal for producing alcoholic beverages and for leavening bread. However, it cannot ferment pentoses, such as xylose and arabinose into ethanol [13, 35]. There have been several research efforts to genetically modify *S. cerevisiae* to be able to consume xylose [36, 39, 57–59] . Several attempts have been made to clone and express various bacterial genes, which are necessary for fermentation of xylose into *S. cerevisiae* [60, 61]. Many have succeeded, but problems associated with pentose fermentation still persist.

Alternatively, xylose is converted into ethanol by some other naturally occurring recombinant. Among the wildtype xylose fermenting yeast strains for ethanol production, *Pichia stipitis* and *Candida shehatae* have reportedly shown promising results for industrial applications in terms of complete sugar utilization, minimal byproduct formation, low sensitivity to temperature and substrate concentration. Furthermore, *P. stipitis* is able to ferment a wide variety of sugars into ethanol and has no vitamin requirement for xylose fermentation [2].

Olsson and Hahn-Hägerdal [22] presented a list of bacteria, yeasts and filamentous fungi that produce ethanol from xylose. Certain species of the yeasts *Candida, Pichia, Kluyveromyces, Schizosaccharomyces and Pachysolen* are among the naturally occurring organisms. Jeffries and Kurtzman [62] reviewed the strain selection, taxonomy and genetics of xylose-fermenting yeasts.

Utilization of cellobiose is important in ethanol production from lignocellulosic materials by simultaneous saccharification and fermentation or SSF (*cf*. fig. 8.7). However, only a few ethanol-producing microorganisms are able to utilize cellobiose. Cellobiose utilization eliminates the need for one class of cellulase enzymes [2]. *Brettanomyces custersii* is one of the yeasts identified as a promising glucose and cellobiose fermenting microorganism for SSF of cellulose for ethanol production [63].

High temperature tolerance could be a good characterization for ethanol production, since it simplifies fermentation cooling. One of

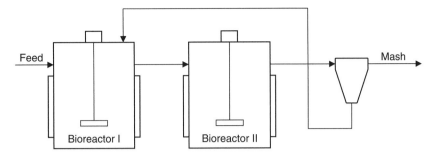

Figure 8.7 Two-stage continuous ethanol fermentation process with yeast recirculation [99, 100].

the problems associated with SSF is the different optimum temperature for saccharification and fermentation. Many attempts have been made to find thermotolerant yeasts for SSF. Sheikh Idris and Berryi [64] tested 58 yeast strains belonging to 12 different genera and capable of growing and fermenting sugars at temperatures of 40–46°C. They selected several strains belonging to the genera *Saccharomyces, Kluyveromyces and Fabospora*, in view of their capacity to ferment glucose, galactose and mannose at 40°C, 43°C and 46°C, respectively. *Kluyveromyces marxianus* has been found to be a suitable strain for SSF [65].

8.11.2   Bacteria

A great number of bacteria are able to produce ethanol, although many of them generate multiple end products in addition to ethanol. *Zymomonas mobilis* is an unusual Gram-negative bacterium that has several appealing properties as a fermenting microorganism for ethanol production. It has a homoethanol fermentation pathway and tolerates up to 12% ethanol. Its ethanol yield is comparable with *S. cerevisiae*, while it has much higher specific ethanol productivity (2.5×) than the yeast. However, the tolerance of *Z. mobilis* to ethanol is lower than that of *S. cerevisiae*, since some strains of *S. cerevisiae* can produce ethanol to result in a concentration as high as 18% of the fermentation broth. The tolerance of *Z. mobilis* to inhibitors and low pH is also low. Similar to *S. cerevisiae*, *Z. mobilis* cannot utilize pentoses [13, 41]. Several genetic modifications have been performed for utilization of arabinose and xylose by *Z. mobilis*. However, *S. cerevisiae* has been more welcomed for industrial application, probably because of the industrial problems that may arise in working with bacteria. Separation of *S. cerevisiae* from fermentation media is much easier than separation of *Z. mobilis*, which is an important characteristic for reuse of the microorganisms in ethanol production processes.

Using genetically engineered bacteria for ethanol production has also been applied in many studies. Ingram *et al.* [66] and Jojima *et al.* [67] have reviewed the metabolic engineering of bacteria for ethanol production. Recombinant *Escherichia coli* is a valuable bacterial resource for ethanol production. The construction of *E. coli* strains to selectively produce ethanol was one of the first successful applications of metabolic engineering. *E. coli* has several advantages as a biocatalyst for ethanol production, including the ability to ferment a wide spectrum of sugars, no requirements for complex growth factors and prior industrial use (e.g., for production of recombinant protein). The major disadvantages associated with using *E. coli* cultures are a narrow and neutral pH growth range (6.0–8.0), less hardy cultures compared to yeast and public perceptions regarding the danger of *E. coli* strains. Lack of data on the use of residual *E. coli* cell mass as an ingredient in animal feed is also an obstacle to its application [8].

8.11.3　Filamentous Fungi

A great number of molds are able to produce ethanol. The filamentous fungi *Fusarium, Mucor, Monilia, Rhizopus, Ryzypose and Paecilomyces* are among those that can ferment pentoses into ethanol [36]. Zygomycetes are saprophytic filamentous fungi, which are able to produce several metabolites including ethanol. Among the three genera *Mucor, Rhizopus* and *Rhizomucor, Mucor indicus* (formerly *M. rouxii*) and *Rhizopus oryzae* showed good performances on ethanol productivity from glucose, xylose and wood hydrolyzate [68]. *M. indicus* has several industrial advantages compared to baker's yeast for ethanol production, such as (a) capability of utilizing xylose, (b) having a valuable biomass, e.g., for production of chitosan and (c) high optimum temperature of 37°C [69]. Skory *et al.* [70] examined 19 *Aspergilli* and 10 *Rhizopus* strains for their ability to ferment simple sugars (glucose, xylose and arabinose) as well as complex substrates. An appreciable level of ethanol has been produced by *Aspergillus oryzae, R. oryzae and R. javanicus*. Recently, starter cultures for Indonesian tempe were evaluated for ethanol production, which resulted in 32 isolates of zygomycetes with ethanol yields up to 0.47 g/g [71].

The dimorphic organism *M. circinelloides* has also been investigated for production of ethanol from pentose and hexose sugars. Large amounts of ethanol were produced during aerobic growth on glucose by this mold. However, ethanol production on galactose or xylose was less significant [72]. Yields as high as 0.48 g/g ethanol from glucose by *M. indicus*, under anaerobic conditions, have been reported [73]. However, the yield and productivity of ethanol from xylose is lower than that of *P. stipitis* [74].

Although filamentous fungi have been industrially used for a long time for several purposes, a number of process engineering problems are associated with these organisms due to their filamentous growth. Problems can appear in mass and heat transfer. Furthermore, attachment and growth on the bioreactor walls, agitators, probes and baffles cause heterogeneity within the bioreactor and problems with measurement of controlling parameters and also cleaning of the bioreactor [75, 76].

8.12   Fermentation Processes

In this section, we discuss different fermentation processes applicable to ethanol production. Fermentation processes, as well as other chemical and biotechnological processes, can be classified into batch, fed-batch and continuous operation. All these methods are applicable in industrial fermentation of sugar substances and starch materials. These processes are well established; the fed-batch and continuous modes of operation being dominant in the ethanol market. When configuring the fermentation process, several parameters must be considered, including (a) high ethanol yield and productivity, (b) high conversion of sugars and (c) low equipment cost. The need for detoxification and choice of the microorganism must also be evaluated in relation to the fermentation configuration.

Presentation of a variety of inhibitors and their interaction effects, e.g., in lignocellulosic hydrolyzates, makes the fermentation process more complex than with other substrates for ethanol production [17, 23]. In fermentation of this hydrolyzate, the pentoses should be utilized in order to increase the overall yield of the process and to avoid problems with wastewater treatment. Therefore, it is still a challenge to use a hexose-fermenting organism such as *S. cerevisiae* for fermentation of the hydrolyzate.

When a mixture of hexoses and pentoses is present in the medium, microorganisms usually take up hexoses first and produce ethanol. As the hexose concentration decreases, they start to take up pentose. Fermentation of hexoses can be successfully performed under anaerobic or microaerobic conditions, with high ethanol yield and productivity. However, fermentation of pentoses is generally a slow and aerobic process. If one adds air to ferment pentoses, the microorganisms will start utilizing the ethanol as well. It makes the entire process complicated and demands a well-designed and controlled process.

8.12.1 Batch Processes

In batch processes, all nutrients required for fermentation are present in the medium prior to cultivation. Batch technology has been preferred in the past due to the ease of operation, low cost of controlling and monitoring system, low requirements for complete sterilization, use of unskilled labor, low risk of financial loss and easy management of feedstocks. However, the overall productivity of the process is very low because of long turnaround times and initial lag phase [9].

In order to improve traditional batch processes, cell recycling and the application of several bioreactors has been used. Reuse of produced cells can increase productivity of the process. Application of several bioreactors operated at staggered intervals can provide a continuous feed to the distillation system. One of the successful batch methods applied for the industrial production of ethanol is Melle-Boinot fermentation. This process achieves a reduced fermentation time and increased yield by recycling yeast and applying several bioreactors operated at staggered intervals. In this method, yeast cells from previous fermentation are separated from the media by centrifugation and up to 80% are recycled [9, 77]. Instead of centrifugation, the cells can be filtered, followed by the separation of yeast from the filter aid using hydrocyclones [78].

In well-detoxified or completely non-inhibiting acid hydrolyzates of lignocellulosic materials, exponential growth will be obtained after inoculation of the bioreactor. If the hydrolyzate is slightly inhibiting, there will be a relatively long lag phase during which part of the inhibitors are converted. However, if the hydrolyzate is severely inhibiting, no conversion of the inhibitors will occur and neither cell growth nor fermentation will occur. A slightly inhibiting hydrolyzate can thus be detoxified during batch fermentation. However, very high concentrations of the inhibitors will cause a complete inactivation of the metabolism [21].

Several strategies may be considered for fermentation of hydrolyzate to improve the *in situ* detoxification in batch fermentation and to obtain a higher yield and productivity of ethanol. Having high initial cell density, increasing the tolerance of microorganisms against the inhibitors by either the adaptation of cells to the medium or genetic modification of the microorganism and choosing optimal reactor conditions to minimize the effects of inhibitors are among these strategies.

Volumetric ethanol productivity is low in lignocellulosic hydrolyzates when low cell-mass inoculums are used due to poor cell growth. usually, high cell concentration, e.g., 10 g/L dry cells has been used in order to find a high yield and productivity of ethanol in different studies. In addition, a high initial cell density helps the process for *in situ* detoxification by the microorganisms; therefore, the demand for a detoxification unit decreases. *In situ* detoxification of the inhibitors may even lead to increased ethanol yield and productivity, due to uncoupling by the presence of weak acids or due to decreased glycerol production in the presence of furfural [23]. Adaptation of the cells to hydrolyzate or genetic modification of the microorganism can significantly improve the yield and productivity of ethanol. Optimization of reactor conditions can be used to minimize the effects of inhibitors. Among the different parameters, cell growth is found to be strongly dependent on pH [21, 23].

8.12.2   Fed-batch Processes

In fed-batch processes (or semi-continuous processes), the substrate and required nutrients are added continuously or intermittently to the initial medium after the start of cultivation or from the halfway point throughout the batch process. Fed-batch processes have been utilized to avoid problems associated with substrates that inhibit growth rate if present at high concentration, to overcome catabolic repression, to demand less initial biomass, to overcome the problem of contamination and to avoid mutation and plasmid instability found in continuous culture. Furthermore, fed-batch processes do not face the problem of washout, which can occur in continuous fermentation. A major disadvantage of a fed-batch process is the need for additional control instruments that require a substantial amount of operator skill. In addition, for systems without feedback control, where the feed is added on a predetermined fixed schedule, there can be difficulty in dealing with any deviation (i.e., time courses may not always follow the expected profiles) [79]. The fed-batch processes without feedback control can be classified as intermittent fed-batch, constant rate fed-batch, exponential fed-batch and optimized fed-batch. The fed-batch processes with feedback control have been classified as indirect control and direct control fed-batch processes [79, 80].

The fed-batch technique is one of the most promising methods for the fermentation of dilute acid hydrolyzates of lignocellulosic materials. The basic concept behind the success of this technique is the capability of *in situ* detoxification of hydrolyzates by the fermenting microorganisms. Since the yeast has a limited capacity for the conversion of the inhibitors, the achievement of a successful fermentation strongly depends on the feed rate of the hydrolyzate. By adding the substrate at a low rate in fed-batch fermentation, the concentrations of bioconvertible inhibitors such as furfural and HMF in the bioreactor remain low and the inhibiting effect therefore decreases. At a very high feed rate, using an inhibiting hydrolyzate, both ethanol production and cell growth can stop, whereas at a very low feed rate, the hydrolyzate may still be converted, but at a very low productivity rate, which was experimentally confirmed. Consequently, an optimum feed rate should exist [14, 21, 23].

Similar to batch operations, higher optimum dilution rates in fed-batch cultivation can be obtained by: (a) high initial cell concentration, (b) increasing the tolerance of microorganisms against the inhibitors and (c) choosing optimal reactor conditions to minimize the effects of inhibitors. Productivity in fed-batch fermentation is generally limited by the feed rate which, in turn, is limited by the cell mass concentration [23].

8.12.3   Continuous Processes

Process design studies of molasses fermentation have shown that the investment cost was considerably reduced when continuous rather than batch fermentation was employed and that the productivity of ethanol could be increased by more than 200%. Continuous operations can be classified into continuous fermentation with or without feedback control. In continuous fermentation without feedback control, called a *chemostat*, the feed medium containing all the nutrients is continuously fed at a constant rate (dilution rate D) and the cultured broth is simultaneously removed from the bioreactor at the same rate. The chemostat is quite useful in the optimization of media formulation and to investigate the physiological state of the microorganism [80]. Continuous fermentations with feedback control are turbidostat, phauxostat and nutristat. A turbidostat with feedback control is a continuous process to maintain the cell concentration at a constant level by controlling the medium feeding rate. A phauxostat is an extended nutristat, which maintains the pH value of the medium in the bioreactor at a preset value. A nutristat with feedback control is a cultivation technique to maintain nutrient concentration at a constant level [80].

When lignocellulosic hydrolyzates are added at a low feed rate in continuous fermentation, low concentration of bioconvertible inhibitors in the bioreactor is assured. In spite of a number of potential advantages in terms of productivity, this method has not yet been well developed in fermentation of acid hydrolyzates. One should consider the following points in the continuous cultivation of acid hydrolyzates of lignocelluloses:

- Cell growth is necessary at a rate equal to the dilution rate in order to avoid washout of the cells in continuous cultivation.
- growth rate is low in fermentation of hydrolyzates because of the presence of inhibitors.
- The cells should keep their viability and vitality for a long time.

The major drawback of continuous fermentation is that, in contrast to the situation in fed-batch fermentation, cell growth is necessary at a rate equal to the dilution rate, in order to avoid washout of the cells in continuous cultivation [23]. The productivity is a function of the dilution rate and since the growth rate is decreased by the inhibitors, the productivity in continuous fermentation of lignocellulosic hydrolyzates is low. Furthermore, at a very low dilution rate, the conversion rate of the inhibitors can be expected to decrease due to the decreased specific growth rate of the biomass. Thus, washout may occur even at very low dilution rates [21]. On the other hand, one of the major advantages of continuous cultivation is the possibility to run the process for a long time (e.g., several months), whereas the microorganisms usually lose their activity after facing the inhibitory conditions of the hydrolyzate. By employing cell-retention systems, the cell mass concentration in the bioreactor, the maximum dilution rate and consequently the maximum ethanol productivity increase. Different cell-retention systems have been investigated by cell immobilization, encapsulation, cell recirculation by filtration, settling and centrifugation. A relatively old study [81] shows that the investment cost for a continuous process with cell recirculation is less than that for continuous fermentation without cell recirculation.

The concept of partial recirculation of both yeast and wastewater is quite common in several industrial technologies for ethanol production. In these processes, the bioreactor works continuously; the cells are separated, e.g. by using a centrifuge or sedimentation (of flocculating yeasts) and the separated cells are returned to the bioreactor. Most of the ethanol-depleted beer, including residual sugars, is then recycled to the bioreactor. In this process, besides providing enough cell concentration in the bioreactor, less water is consumed and a more concentrated stillage is produced. Therefore, the process has a lower wastewater problem. If using centrifuges, the equipment has to be designed to avoid deactivation of the cells [55, 82].

Application of an encapsulated cell system in continuous cultivation has several advantages, compared to a free cell or traditionally entrapped cell system, e.g., in alginate matrix. Encapsulation provides higher cell concentrations than free cell systems in the medium, which leads to higher productivity per volume of the bioreactor in continuous cultivation. Furthermore, the biomass can easily be separated from the medium without centrifugation or filtration. The advantages of encapsulation, compared to cell entrapment, are less resistance to diffusion through beads/capsules, some degree of freedom in the movement of the encapsulated cells, no cell leakage from the capsules and higher cell concentration [83]. However, the encapsulation still needs improvement, for example, a more robust membrane in order to approach industrial application [84, 85] .

8.12.4   Series Arranged Continuous Flow Fermentation

Ethanol can be produced by using continuous flow bioreactors arranged in a series with complete sugar utilization or high ethanol concentration. With two bioreactors arranged in a series, the retention time can be chosen so that the sugar is only partially utilized in the first bioreactor, with fermentation completed in the second. Ethanol inhibition is reduced in the first bioreactor, allowing a faster throughput. The second, lower productivity bioreactor can now convert less sugar than if operated alone. For high product concentration, the productivity of a two stage system has been 2.3 times higher than that of a single stage [55, 86].

A two-stage continuous ethanol fermentation process with yeast recirculation is used industrially by Danish Distilleries Ltd., Grena, for molasses fermentation (see Fig. 8.7). Two bioreactors with 170,000 L volume produce 66 g/L ethanol in 21 h retention time [55]. A seven bioreactor series system (70,000 L volume each bioreactor) was also used in the Netherlands to produce 86 g/L ethanol in 8-h retention time [87]. A Japanese company used a six bioreactor series system (total volume 100,000 L) with 8.5 h retention time to produce 95 g/L ethanol [88].

8.12.5   Strategies for Fermentation of Enzymatic Lignocellulosic Hydrolyzates

The cellulose fraction of lignocelluloses can be converted into ethanol by either simultaneous saccharification and fermentation (SSF) or by separate enzymatic hydrolysis and fermentation (SHF) processes. A schematic of these processes is shown in fig. 8.8.

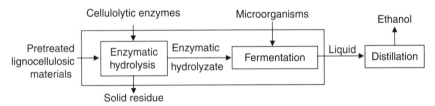

A: Separate enzymatic hydrolysis and fermentation

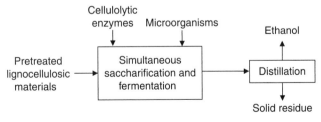

B: Simultaneous saccharification and fermentation

Figure 8.8 Main steps in SSF or SHF for ethanol production.

8.12.6 Separate Enzymatic Hydrolysis and Fermentation (SHF)

In SHF, enzymatic hydrolysis for the conversion of pretreated cellulose into glucose is the first step. Produced glucose is then converted into ethanol in the second step. enzymatic hydrolysis can be performed in the optimum conditions of the cellulase. The optimum temperature for hydrolysis by cellulase is usually between 45°C and 50°C, depending on the microorganism that produces the cellulase. The major disadvantage of SHF is that the released sugars severely inhibit cellulase activity [32, 89]. Removal of the solid residues can also result in a loss of fermentable sugars. Another possible problem in SHF is contamination. Hydrolysis is a lengthy process (one or possibly several days) and a dilute solution of sugar always has a risk of contamination, even at rather high temperatures such as 45–50°C.

8.12.7 Simultaneous Saccharification and Fermentation (SSF)

SSF combines the enzymatic hydrolysis of cellulose and fermentation into one step. As cellulose is converted into glucose, a fermenting microorganism is present in the medium and it immediately consumes the glucose produced. As mentioned, cellobiose and glucose significantly decrease the activity of cellulase. SSF results in higher reported ethanol yields and requires lower amounts of enzyme because end product inhibition from cellobiose and glucose formed during enzymatic hydrolysis is avoided by the yeast fermentation. SSF has the following advantages compared to SHF:

- Fewer vessels required for SSF, in comparison to SHF, resulting in capital cost savings
- Less contamination during enzymatic hydrolysis, since the presence of ethanol reduces the possibility of contamination
- Less contamination due to the fact that the fermentation organism can consume sugars directly as soon as they are released so that they are not available to other organisms
- Higher yield of ethanol
- Lower enzyme loading requirement

On the other hand, SSF has the following drawbacks compared to SHF:

- SSF requires that enzyme and culture conditions be compatible with respect to pH and temperature. In particular, the difference between optimum temperatures of the hydrolyzing enzymes and fermenting microorganisms is usually problematic. *Trichoderma reesei* cellulases, which constitute the most active preparations, have optimal activity between 45°C and 50°C, whereas *S. cerevisiae* has an optimum temperature between 30°C and 35°C. The optimal temperature for SSF is around 38°C, which is a compromise between the optimal temperatures for hydrolysis and fermentation. Hydrolysis is usually the rate limiting process in SSF [32]. Several thermotolerant yeasts (e.g., *C. acidothermophilum and K. marxianus*) and bacteria have been used in SSF to raise the temperature close to the optimal hydrolysis temperature.
- Cellulase is inhibited by ethanol. For instance, at 30 g/L ethanol, the enzyme activity was reduced by 25% [2]. Ethanol inhibition may be a limiting factor in the production of high ethanol concentration. However, there has been less attention paid to ethanol inhibition of cellulase, since practically it is not possible to work with very high substrate concentration in SSF, because of the problem with mechanical mixing.
- Another problem arises from the fact that most microorganisms used for converting cellulosic feedstock cannot utilize xylose, a hemicellulose hydrolysis product [8].

8.12.8 Comparison between Enzymatic and Acid Hydrolysis for Lignocellulosic Materials

The two most promising processes for the industrial production of ethanol from cellulosic materials are two stage dilute acid hydrolysis (a chemical process) and SSF (an enzymatic process). The advantages and disadvantages of dilute acid and enzymatic hydrolyses are summarized in Table 8.3. Enzymatic hydrolysis is carried out under mild conditions, whereas high temperature and low pH result in corrosive conditions for acid hydrolysis. While it is possible to

Table 8.3 Comparison between enzymatic and acid hydrolysis for ligno-cellulosic materials.

Parameters	Dilute-acid hydrolysis	Enzymatic hydrolysis
Rate of hydrolysis	Very high	Low
Overall yield of sugars	Low	High and depend upon pretreatment
Catalyst costs	Low	High
Conditions	Harsh reaction conditions, e.g., high pressure and temperature	Mild conditions (e.g., 50°C, atmospheric pressure, pH 4.8)
Inhibitors formation	Highly inhibitory hydrolyzate	Non-inhibitory hydrolyzate
Degradation of sensitive nutrients such as vitamins	High	Low

obtain a cellulose hydrolysis of close to 100% by enzymatic hydrolysis after a pretreatment, it is difficult to achieve such a high yield with acid hydrolysis. The yield of glucose out of cellulose by dilute acid hydrolysis is usually less than 60%. Furthermore, the previously mentioned inhibitory compounds are formed during acid hydrolysis, whereas this problem is not so severe for enzymatic hydrolysis. Acid hydrolysis conditions may destroy nutrients sensitive to acid and high temperatures such as vitamins, which may enter the process together with the lignocellulosic materials. On the other hand, enzymatic hydrolysis has some drawbacks in comparison to dilute-acid hydrolysis. Enzymatic hydrolysis can take up to several days, whereas acid hydrolysis can be conducted in a few minutes. Furthermore, the enzyme cost is still high, even if developments of new generations of enzyme solutions have improved the situation.

8.13   Ethanol Recovery

Fermented broth, also called "beer" or "mash" typically contains 2–12% ethanol. In addition, it contains a number of other materials that can be classified into microbial biomass, fusel oil and volatile components. Fusel oil is a mixture of primary methylbutanols and methylpropanols formed from α-ketoacids and derived from or leading to amino acids. Depending on the raw materials used, major components of fusel oil can be isoamyl alcohol, n-propyl alcohol, sec-butyl alcohol, isobutyl alcohol, n-butyl alcohol, active amyl alcohol and n-amyl alcohol. The amount of fusel oil in mash depends on the pH of the bioreactor. Fusel oil is used in solvents for paints, polymers, varnishes and essential oils. Acetaldehyde and trace amounts of other aldehydes and volatile esters are usually produced from grains and molasses. Typically, 1 L of acetaldehyde and 1–5 L of fusel oil are produced per 1000 L of ethanol [9, 55].

Stillage consists of the nonvolatile fraction of materials remaining after alcohol distillation. Its composition depends greatly on the type of feedstock used for fermentation. Stillage generally contains solids, residual sugars, residual ethanol, waxes, fats, fibers and mineral salts. The solids may be originated from feedstock proteins and spent microbial cells [9].

8.14 Distillation

Mash can be centrifuged or settled in order to separate the microbial biomass from the liquid and then sent to the ethanol recovery system. Distillation is typically used for the separation of ethanol, aldehydes, fusel oil and stillage [9]. Ethanol is readily concentrated from mash by distillation, since the volatility of ethanol in dilute solution is much higher than the volatility of water. However, ethanol and water form an azeotrope at 95.57%w/w ethanol (89 mol% ethanol) with a minimum boiling point of 78.15°C. This mixture behaves as a single component in a simple distillation and no further enrichment other than 95.57%w/w of ethanol can be achieved by simple distillation [9, 55, 90]. Various industrial distillation systems for ethanol purification are: (a) simple two column systems, (b) three or four column barbet systems, (c) three column othmer systems, (d) vacuum rectification, (e) vapor recompression, (f) Multi effect distillation and (g) six column reagent alcohol systems [9, 55]. These methods are reviewed by Kosaric [9]. The following parameters should be considered for the selection of industrial distillation systems:

- Energy consumption (e.g., steam consumption or cooling water consumption per unit of ethanol produced)
- Quality of ethanol (separation of fusel oil and light components to fit specifications)
- How to deal with the problems associated with the risk for clogging of distillation equipment as column internals and reboilers because of precipitation or formation of solids. Special design and use of a vacuum may be applied for overcoming the problem in the column. Using direct steam instead of using reboiler increases the amount of water to be taken care of in the downstream process
- Simplicity in controlling the system
- Simplicity in opening column parts and cleaning the columns

Of course, lower capital investment is also one of the main parameters in the selection of distillation systems.

Ethanol is present in the market with different degrees of purity. The majority of ethanol is 190 proof (95% or 92.4%, minimum) used for solvent, pharmaceutical, cosmetic and chemical applications. Technical grade ethanol, containing up to 5% volatile organic aldehyde, ester and sometimes methanol, is used for industrial solvents and chemical syntheses. High-purity 200 proof (99.85%) anhydrous ethanol is produced for special chemical applications. For fuel use in mixture with gasoline (gasohol), nearly anhydrous (99.2%) ethanol, but with higher available levels of organic impurities, is used [55].

A simple two column system is described here, while other systems are presented in the literature (e.g., [9, 55]). Simple one or two column systems with only a stripping and rectification section are usually used to produce lower quality industrial alcohol and azeotrope alcohol for further dehydration to fuel grade. The simplest continuous ethanol distillation system consists of stripping and rectification sections, either together in one column or separated into two columns (see Fig. 8.9).

The fermented mash is pumped into a continuous distillation process, where steam is used to heat the mash to its boiling point in the stripper column. The ethanol enriched vapors pass through a rectifying column where ethanol concentration increases from stage to stage. On top of the column the vapors are taken out of the column, a part is condensed and fed back into the column as reflux.

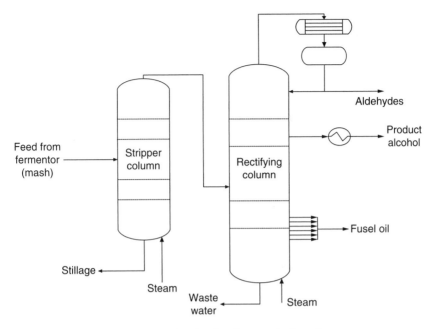

Figure 8.9 Two-column system for the distillation of ethanol.

The ethanol stripped stillage leaves the column at the bottom and is pumped to a stillage tank. Aldehydes are drawn from the head vapor, condensed and partly used as reflux. Fusel oil is taken out from several stages of the rectifying section [9, 55, 91].

With efficient distillation, the stillage should contain less than 0.1% ethanol since the presence of ethanol significantly increases the chemical oxygen demand (COD) of wastewater. For each 1% ethanol left in the stillage, the COD of the stillage is incremented by more than 20 g/L. Due to the potential impact of residual ethanol content, therefore, proper control over distillation can greatly affect the COD of stillage [91].

8.15 Alternative Processes for Ethanol Recovery and Purification

Since distillation is a highly energy consuming process, several processes have been developed for the purification of ethanol from fermentation broth, including, solvent extraction, membrane processes, CO_2 extraction, vapor recompression systems and low temperature blending with gasoline [9]. However, these processes are not established in the industrial production of ethanol.

8.16   Ethanol Dehydration

In order to allow the blending of alcohol with gasoline, the water content of ethanol must be reduced to less than 1% by volume, which is not possible by distillation. Higher water levels can result in phase separation of an alcohol-water mixture from the gasoline phase, which may cause engine malfunction. Removal of water beyond the last 5% is called *dehydration, absolutation* or *drying of ethanol*. Azeotropic distillation was previously employed to produce higher purity ethanol by adding a third component, such as benzene, cyclohexane or ether, to break the azeotrope and produce dry ethanol [91]. To avoid the illegal transfer of ethanol from the industrial market into the potable alcohol market, where it is highly regulated and taxed, dry alcohol usually requires the addition of denaturing agents that render it toxic for human consumption; the azeotropic reagents conveniently meet this requirement [91]. Except in the high purity reagent grade ethanol market, azeotropic drying has been supplanted by molecular sieve drying technology.

8.16.1   Molecular Sieve Adsorption

The molecular sieve is a more energy efficient method than azeotropic distillation. Furthermore, this method avoids the occupational hazards associated with azeotropic chemical admixtures. In molecular sieve drying, 95% ethanol is passed through a bed of synthetic zeolite with uniform pore sizes that preferentially adsorb water molecules. Approximately ¾ of adsorbed material is water and ¼ is ethanol. The bed becomes saturated after a few minutes and must be regenerated by heating or evacuation to drive out the adsorbed water. During the regeneration phase, a side stream of ethanol/water (often around 50%) is produced, which must be re-distilled before returning to the drying process [91].

8.16.2   Membrane Technology

Membranes can also be used for ethanol purification. Reverse osmosis (RO), which employs membranes that are impermeable to ethanol and permeable to water, can be used for the purification of ethanol from water. Using a membrane that is permeable to ethanol, but not to water is another approach [9]. Pervaporation, a promising membrane technique for separation of organic liquid mixtures, such as azeotropic mixtures or near boiling point mixtures, can also be used for separation of these azeotropes [90, 92]. It involves the separation of ethanol-water azeotrope or near azeotropic ethanol-water composition (from about 95 to 99.5 wt% ethanol) through water permeable (or water selective) membranes to remove the remaining water from the concentrated ethanol solution [93].

8.17   Distillers' Dried Grains with Solubles

In a dry-mill process, approximately one third of the initial grain substrates remain in the stillage after fermentation and distillation. The stillage is processed in a centrifuge to separate the solids from the liquids. The solid is a fibrous material, which is usually referred to as distillers dried grains (DDG), which can be dried and used directly as feed for ruminants. The remaining liquid, with a solid content of approximately 10%, is condensed to a solid content of 30%. It can be used directly and is sold in the feed market as a wet feed raw material, referred to as condensed distillers soluble (CDS) or syrup. For handling and logistical reasons, most ethanol producers mix the fibrous solids with the syrup. The mixture is then dried, pelletized and sold on the market as distillers' dried grain with solubles (DDGS).

Depending on the substrate, the coproducts will differ in nutrient composition. Typical nutritional composition for wheat DDGS produced in northern Europe, is approximately 32–36% protein, 6–7% fat, 10–12% water, 8–9% crude fiber, 4–6% ash, 1% phosphorus and 0,1% calcium. Coproducts from maize are usually lower in protein and higher in fat content than wheat coproducts. Approximately, 4–8 % of the biomass in the DDGS origin from spent yeast.

The ethanol process often means cooking at high temperatures and long retention times. It means that the ethanol production conditions are rough to sensitive proteins. The resulting Maillard reactions between sugars and amino acids or peptides decrease both the digestibility and the quality of the protein in the DDGS. The amino acid lysine is particularly sensitive. Since the lysine content of the raw material (wheat) is low, it therefore requires special attention when using DDGS in feed products. The sulfur content in wheat DDGS, which can vary between 0.3% and 1.1%, also requires attention as it could be high for some species. The phosphorous in DDGS are, due to enzyme treatment and yeast fermentation, considered more available to monogastric animals than other vegetable feed raw materials.

Wheat DDGS is an excellent feedstuff for ruminants, since the microbes in the rumen enable the animals to utilize feeds that are high in fiber content and they are able to utilize poor quality protein sources and non-protein nitrogen. Historically, ruminants have also been the main consumers of DDGS. If the drying/heating process in the ethanol plant is optimized according to protein quality, it is possible to produce a co-product with a considered amount of "RUP" or rumen protected protein. A practical inclusion level of DDGS in ruminant feed is approximately 25–40%, depending on the specific characteristics.

Wheat DDGS can be used in poultry feed with an inclusion level of 5–15 % and is often used to replace less sustainable protein sources such as soy meal. However, the relatively high amount of soluble and insoluble fiber in DDGS needs to be considered. The non-starch polysaccharides fraction can negatively influence gut health of the poultry and cause so called "sticky droppings". The appropriate type and amount of feed enzymes need to be added. Variations in protein quality and nutrient composition between different ethanol producers and even normal in-house variations in ethanol plants can also be a problem when composing poultry feed. DDGS can be used in pig feeds up to 25% inclusion with the same consideration of lysine content, feed enzyme addition and attention to nutrient variations.

There is limited information available on feeding wheat DDGS to fish in aquaculture diets. It is probably a very useful raw material in omnivorous species and it needs more careful consideration for carnivorous species, such as salmon. Product development and product improvement such as fractionating protein concentrate from stillage and separating wheat bran in different process stages need to be evaluated.

8.18   Sustainability of Bioethanol Production

The implementation of large scale 1st generation bioethanol production has caused debates regarding the sustainability of the process and ethical concerns have been raised [94–98]. Depending on how the raw materials are produced, how byproducts are utilized, what kind of energy source is used for process energy and, of course, how the comparisons are made, production of bioethanol from, e.g., corn may even lead to an increase in greenhouse gas emissions [97]. This is not considered the norm, however. Other than greenhouse gas emissions, other environmental factors have to be considered as well, such as soil erosion, water usage, local air and water pollution and loss of biodiversity [95]. Since the 1st generation bioethanol production is based on either starch or sugar, the raw materials could potentially be used as human food. Thus, the food security issue should also be considered [96]. This is also the main ethical issue; is it ethical to produce fuel from potential food sources? On the other, how ethical is it to deplete a limited resource like oil by using it for something simple such as a transportation fuel? Furthermore, the protein coproducts from the ethanol industry are an excellent sustainable protein source in the food chain.

Overall, 1st generation bioethanol should still be considered advantageous compared with fossil fuels. Furthermore, 2nd generation ethanol production promises to solve most of the environmental problems and ethical issues of the 1st generation process.

8.19   Concluding Remarks and Future Prospects

Ethanol has been well established in the fuel market, where its share from less than one billion liters in 1975 is expected to increase to 100 billion liters in 2015. Grains, sugarcane juice and molasses are the dominant raw materials for the time being, while lignocellulosic materials are expected to have a significant share in this market in the future. There have been great achievements in the development of ethanol production from lignocellulosic materials, but large-scale facilities are yet to be built. Several challenges still persist in the lignocellulosic ethanol, until the process is fully economically feasible.

References

1. H. Rothman, R. Greenshields and F.R. Calle, *The alcohol economy: fuel ethanol and the Brazilian experience*, London, Francis Printer, 1983.
2. C.E. Wyman, *Handbook on bioethanol: production and utilization*, Washington, DC, Taylor & Francis, 1996.
3. C. Breidenich, *et al.*, *American Journal of International Law*, Vol. 92, p. 315–331, 1998.

4. F.O. Licht, *World ethanol markets : the outlook to 2015*, Tunbridge Wells, F.O. Licht, 2006.
5. S.I. Mussatto, *et al.*, *Biotechnology Advances*, Vol. 28(6), p. 817–830, 2010.
6. REN21, *Renewables 2012 global status report*, Paris, REN21 Secretariat, 2012.
7. M. Roehr, *The biotechnology of ethanol: classical and future applications*, Weinheim, Wiley-VCH, 2001.
8. Y. Lin and S. Tanaka, *Applied Microbiology and Biotechnology*, Vol. 69, p. 627– 642, 2006.
9. N. Kosaric, *et al.*, "Ethanol fermentation", in G. Reed, eds., *Biotechnology: A comprehensive Treatise*, Verlag-Chemie, pp. 257–386, 1983.
10. J.E. Bailey and D.F. Ollis, *Biochemical engineering fundamentals*, Singapore, McGraw-Hill, 1986.
11. Y. Sun and J. Cheng, *Bioresource Technology*, Vol. 83(1), p. 1–11, 2002.
12. K. Karimi, G. Emtiazi and M.J. Taherzadeh, *Enzyme and Microbial Technology*, Vol. 40, p. 138–144, 2006.
13. B.C. Saha, *Journal of Industrial Microbiology and Biotechnology*, Vol. 30, p. 279– 291, 2003.
14. M.J. Taherzadeh, *Ethanol from lignocellulose: physiological effects of inhibitors and fermentation strategies*, Gothenburg, Chalmers University of Technology, 1999.
15. P. Béguin and J.-P. Aubert, *FEMS Microbiology Reviews*, Vol. 13, p. 25–58, 1994.
16. C.R. Soccol, *et al.*, "Lignocellulosic bioethanol: Current status and future perspectives", in A. Pandey, *et al.*, eds., *Biofuels*, Academic press, pp. 2011.
17. E. Palmqvist and B. Hahn-Hägerdal, *Bioresource Technology*, Vol. 74, p. 25–33, 2000.
18. M.J. Taherzadeh and K. Karimi, *BioResources*, Vol. 2(3), p. 472–499, 2007.
19. K. Karimi, S. Kheradmandinia and M.J. Taherzadeh, *Biomass and Bioenergy*, Vol. 30(3), p. 247–253, 2006.
20. M.J. Taherzadeh, C. Niklasson and G. Lidén, *Chemical Engineering Science*, Vol. 52(15), p. 2653–2659, 1997.
21. M.J. Taherzadeh and C. Niklasson, "Ethanol from Lignocellulosic Materials: Pretreatment, Acid and Enzymatic Hydrolyses and Fermentation", in K. Hayashi, eds., *Lignocellulose Biodegradation*, American Chemical Society, pp. 49–68, 2004.
22. L. Olsson and B. HahnHagerdal, *Enzyme and Microbial Technology*, Vol. 18(5), p. 312–331, 1996.
23. E. Palmqvist and B. Hahn-Hägerdal, *Bioresource Technology*, Vol. 74, p. 17–24, 2000.
24. P. Persson, *et al.*, *Journal of Agricultural and Food Chemistry*, Vol. 50(19), p. 5318– 5325, 2002.
25. R. Millati, C. Niklasson and M.J. Taherzadeh, *Process Biochemistry*, Vol. 38(4), p. 515–522, 2002.
26. A. Martinez, *et al.*, *Biotechnology Progress*, Vol. 17(2), p. 287–293, 2001.
27. S. Amartey and T. Jeffries, *World Journal of Microbiology & Biotechnology*, Vol. 12(3), p. 281–283, 1996.
28. R. Purwadi, C. Niklasson and M.J. Taherzadeh, *Journal of Biotechnology*, Vol. 114(1–2), p. 187–98, 2004.
29. M.J. Taherzadeh and K. Karimi, "Fermentation inhibitors in ethanol processes and different strategies to reduce their effects", in A. Pandey, *et al.*, eds., *Biofuels*, Academic press, pp. 2011.

30. A.T.W.M. Hendriks and G. Zeeman, *Bioresource Technology*, Vol. 100(1), p. 10–18, 2009.
31. S. Aslanzadeh, *et al.*, "An overview of existing individual unit operations in biological and thermal platforms of biorefineries", in N. Qureshi, D. Hodge and A.V. Vertes, eds., *Biorefineries: Integrated biochemical processces for liquid biofuels (ethanol and butanol)*, Elsevier, pp. in press, 2013.
32. G.P. Philippidis and T.K. Smith, *Applied Biochemistry and Biotechnology*, Vol. 51/52, p. 117–24, 1995.
33. M. Alkasrawi, *et al.*, *Enzyme and Microbial Technology*, Vol. 33(1), p. 71–78, 2003.
34. M. Linde, M. Galbe and G. Zacchi, *Enzyme and Microbial Technology*, Vol. 40(5), p. 1100–1107, 2007.
35. T. Brandberg, *Fermentation of undetoxified dilute acid lignocellulose hydrolyzate for fuel ethanol production*, Gothenburg, Chalmers University of Technology, 2005.
36. T.W. Jeffries, *Current Opinion in Biotechnology*, Vol. 17(3), p. 320–326, 2006.
37. M. Sonderegger, *et al.*, *Biotechnology and Bioengineering*, Vol. 87(1), p. 90–98, 2004.
38. A.J.A. van Maris, *et al.*, *Antonie van Leeuwenhoek*, Vol. 90(4), p. 391–418, 2006.
39. A. Madhavan, *et al.*, *Critical Reviews in Biotechnology*, Vol. 32(1), p. 22–48, 2012.
40. M. Desvaux, E. Guedon and H. Petitdemange, *Applied and Environmental Microbiology*, Vol. 66(6), p. 2461–2470, 2000.
41. I.S. Horvath, *Fermentation inhibitors in the production of bio-ethanol*, Gothenburg, Chalmers University of Technology, 2004.
42. M. Jeppsson, *et al.*, *Applied and Environmental Microbiology*, Vol. 68(4), p. 1604–9, 2002.
43. W.H. Kampen, "Nutritional requirements in fermentation processes", in C.L. Todaro, eds., *Fermentation and biochemical engineering handbook*, Noyes publications, pp. 122–160, 1997.
44. M.J. Taherzadeh, L. Adler and G. Lidén, *Enzyme and Microbial Technology*, Vol. 31(1–2), p. 53–66, 2002.
45. K. Karhumaa, *et al.*, *Applied Microbiology and Biotechnology*, Vol. 73(5), p. 1039–1046, 2006.
46. R. Millati, *Ethanol production from lignocellulosic materials*, Gothenburg, Chalmers University of Technology, 2005.
47. H. Schneider, *Critical Reviews in Biotechnology*, Vol. 9(1), p. 1–40, 1989.
48. P.M. Bruinenberg, J.P. van Dijken and W.A. Scheffers, *Journal of General Microbiology*, Vol. 129, p. 953–964, 1983.
49. S.S. Silva, M.G.A. Felipe and I.M. Manchilha, *Applied Biochemistry and Biotechnology*, Vol. 70–72(1), p. 331–339, 1998.
50. M. Jeppsson, *et al.*, *FEMS Yeast Research*, Vol. 3(2), p. 167–75, 2003.
51. Y.C. Bor, *et al.*, *Gene*, Vol. 114(1), p. 127–32, 1992.
52. M. Gulati, *et al.*, *Bioresource Technology*, Vol. 58(3), p. 253–264, 1996.
53. Y. Wang, *et al.*, *Chinese Journal of Biotechnology*, Vol. 10(2), p. 97–103, 1994.
54. X. Zhu, *et al.*, *Acta Crystallographica Section D: Biological Crystallography*, Vol. 56(Pt 2), p. 129–136, 2000.
55. B.L. Maiorella, "Ethanol", in M. Moo-Young, eds., *Comprehensive biotechnology: the principles, applications and regulations of biotechnology in industry, agriculture and medicine*, Pergamon press Ltd, pp. 861–914, 1985.

56. D.G. Olson, *et al.*, *Current Opinion in Biotechnology*, Vol. 23(3), p. 396–405, 2012.
57. S. Govindaswamy and L.M. Vane, *Bioresource Technology*, Vol. 98(3), p. 677–685, 2007.
58. C. Martin, *et al.*, *Bioresource Technology*, Vol. 98(9), p. 1767–1773, 2007.
59. L. Ruohonen, *et al.*, *Enzyme and Microbial Technology*, Vol. 39(1), p. 6–14, 2006.
60. N. Ho, Z. Chen and A. Brainard, *Applied and Environmental Microbiology*, Vol. 64(5), p. 1852–1859, 1998.
61. M.H. Toivari, *et al.*, *Metabolic Engineering*, Vol. 3(3), p. 236–249, 2001.
62. T.W. Jeffries and C.P. Kurtzman, *Enzyme and Microbial Technology*, Vol. 16(11), p. 922–932, 1994.
63. D.D. Spindler, *et al.*, *Biotechnology Letters*, Vol. 14(5), p. 403–407, 1992.
64. E.T.A. Sheikh Idris and D.R. Berry, *Biotechnology Letters*, Vol. 2(2), p. 61–66, 1980.
65. M. Ballesteros, *et al.*, *Process Biochemistry*, Vol. 39(12), p. 1843–1848, 2004.
66. L. Ingram, *et al.*, *Biotechnology and Bioengineering*, Vol. 58(2&3), p. 204–214, 1998.
67. T. Jojima, M. Inui and H. Yukawa, *Biofuels*, Vol. 2(3), p. 303–313, 2011.
68. R. Millati, L. Edebo and M.J. Taherzadeh, *Enzyme and Microbial Technology*, Vol. 36(2–3), p. 294–300, 2005.
69. K. Karimi, G. Emtiazi and M.J. Taherzadeh, *Process Biochemistry*, Vol. 41(3), p. 653–658, 2006.
70. C.D. Skory, S.N. Freer and R.J. Bothast, *Biotechnology Letters*, Vol. 19(3), p. 203–206, 1997.
71. R. Wikandari, *et al.*, *Applied Biochemistry and Biotechnology*, Vol. 167(6), p. 1501–1512, 2012.
72. T.L. Lubbehusen, J. Nielsen and M. McIntyre, *Applied Biochemistry and Biotechnology*, Vol. 63(5), p. 543–8, 2004.
73. A. Sues, *et al.*, *FEMS Yeast Research*, Vol. 5(6–7), p. 669–676, 2005.
74. K. Karimi, *et al.*, *Biotechnology Letters*, Vol. 6, p. 1395–1400, 2005.
75. P.A. Gibbs, R.J. Seviour and F. Schmid, *Critical Reviews in Biotechnology*, Vol. 20(1), p. 17–48, 2000.
76. M. Papagianni, *Biotechnology Advances*, Vol. 22(3), p. 189–259, 2004.
77. J.N. de Vasconcelos, C.E. Lopes and F.P. de França, *Brazilian Journal of Chemical Engineering*, Vol. 21(3), p. 357–365, 2004.
78. V.M. da Matta and A. Medronho Rde, *Bioseparation*, Vol. 9(1), p. 43–53, 2000.
79. B. McNiel and L.M. Harvey, "Fermentation a practical approach", in B.D. Hames, eds., *Practical Approach Series*, Oxford University press, pp. 113–120, 1990.
80. Y. Harada, *et al.*, *Fermentation Pilot Plant*, 1997.
81. G.R. Cysewski and C.R. Wilke, *Biotechnology and Bioengineering*, Vol. 20, p. 1421–1444, 1978.
82. B. Goggin and G. Thorsson, *Alfa-Laval, Tumba, Sweden*, 1982.
83. F. Talebnia and M.J. Taherzadeh, *Journal of Biotechnology*, Vol. 125(3), p. 377–384, 2006.
84. P. Ylitervo, C.J. Franzén and M.J. Taherzadeh, *Journal*, DOI: 10.1002/jctb.3944, 2012
85. J.O. Westman, *et al.*, *Applied Microbiology and Biotechnology*, Vol. 96(6), p. 1441–1454, 2012.

86. T.K. Ghose and R.D. Tyagi, *Biotechnology and Bioengineering*, Vol. 21, p. 1401– 1420, 1979.

87. C.S. Chen, W. Gibson and K.I. Mashima, *Hawaii ethanol from molasses project phase 1, Final report*, Manoa, Hawaii Natural Energy Institute, 1980.

88. I. Karaki, *et al.*, *Hakko KoyoKaishi*, Vol. 30, p. 106, 1972.

89. G.P. Philippidis, T.K. Smith and Charles E. Wyman, *Biotechnology and Bioengineering*, Vol. 41(9), p. 846–853, 1993.

90. T. Uragami, *et al.*, *Macromolecules*, Vol. 35, p. 9156–9163, 2002.

91. A.C. Wilkie, K.J. Riedesel and J.M. Owens, *Biomass and Bioenergy*, Vol. 19, p. 63–102, 2000.

92. V. Dubey, L.K. Pandey and C. Saxena, *Journal of Membrane Science*, Vol. 251, p. 131–136, 2005.

93. Z. Changluo, L. Moe and X. We, *Desalination*, Vol. 62, p. 299–313, 1987.

94. H. Azadi, *et al.*, *Renewable and Sustainable Energy Reviews*, Vol. 16(6), p. 3599– 3606, 2012.

95. M.E. Dias De Oliveira, B.E. Vaughan and E.J. Rykiel Jr, *BioScience*, Vol. 55(7), p. 593–602, 2005.

96. J.C. Escobar, *et al.*, *Renewable and Sustainable Energy Reviews*, Vol. 13(6–7), p. 1275–1287, 2009.

97. M.S.d.P. Gomes and M.S. Muylaert de Araújo, *Renewable and Sustainable Energy Reviews*, Vol. 13(8), p. 2201–2204, 2009.

98. R.K. Niven, *Renewable and Sustainable Energy Reviews*, Vol. 9(6), p. 535–555, 2005.

99. R. Purwadi, *Continuous ethanol production from pilute-acid hydrolyzates: Detoxification and fermentationstrategy*, Gothenburg, Chalmers University of Technology, 2006.

100. K. Rosen, *Process Biochemistry*, Vol. 13(5), p. 25, 1978.

9

Production of Butanol: A Biofuel

Sapna Jain, Mukesh Kumar Yadav and Ajay Kumar*

Department of Biotechnology, Institute of Biomedical Education and Research, Mangalayatan University, Aligarh 202145, India

Abstract

Butanol is a four carbon, primary, saturated alcohol with the molecular formula C_4H_9OH. Butanol is well known for multiple industrial uses as well as a potential fuel. The focus on butanol as a biofuel is due to the growing demand for non-fossil biofuels. When compared to the traditional biofuel ethanol, butanol has many advantages because of its low vapour pressure and low hygroscopicity; it can be mixed with fossil fuels or can be used alone, allowing an alternative to gasoline. Butanol is widely produced through petrochemical methods but increasing oil prices and a growing awareness of global warming has brought significant attention to the production of butanol from biomass using Acetone-butanol-ethanol (ABE) fermentation utilizing ABE-producing clostridia. In spite of a number of advantages, ABE fermentation runs short due to low productivity, solvent toxicity, difficulty in controlling clostridial metabolism and inefficient as well as costly methods of product recovery. To date, a considerable amount of research has been conducted in various fields such as microbial technology, metabolic engineering and systems biology to overcome the above problems. This chapter provides a detailed account of properties, applications, chemical and biological production techniques, metabolic engineering approaches and *in situ* separation methods of butanol.

Keywords: Butanol, petrochemical methods, ABE-producing clostridia, solvent toxicity, metabolic engineering approaches, *in situ* purification methods

Corresponding author: ajaymtech@gmail.com

Dr.Vikash Babu, Dr. Ashish Thapliyal & Dr. Girijesh Kumar Patel (eds.) Biofuels Production, (255–284) 2014 © Scrivener Publishing LLC

9.1 Introduction

In response to the continuous increase in the cost of crude oil as a main energy source, intensive research for the development of sustainable, economical and environmental alternatives to fossil fuel is becoming more and more prominent. One of the most promising alternatives to petroleum-derived fuels is biofuels. Biofuels usually work as fuel additives rather than petroleum substitutes [1]. Until now, bioethanol has been the primary biofuel, because it is economically favorable and easy to manufacture. However, biobutanol has proved to be much more advantageous than bioethanol.

Commercially, the extensive production of butanol is exclusively done from petrochemical methods but nowadays, there has been a resurgence in butanol production via Acetone-Butanol-Ethanol (ABE) fermentation. The butanol produced biologically from any sort of biomass is considered biobutanol. The production of biobutanol through ABE fermentation is not a recent matter. ABE fermentation, promoted by bacteria of the genus *Clostridium* sp., is in fact one of the oldest known anaerobic industrial fermentation [2]. Until the 20th century, 66% of the butanol consumed worldwide was usually produced from biotechnological means i.e. ABE fermentation [3]. However, with the increasing development of the petrochemical industry and the higher costs of carbohydrate sources as feed substrate, butanol production from biotechnological means started to cease rapidly and led to its complete eradication by 1960s.

It was in the beginning of the 1970s that oil prices started to escalate due to the "oil crisis". Also, ever emerging environmental awareness and the uncertainty of petroleum supplies in energy driven worldwide societies have led to the revival and interest in this bio industry [4]. Since then, research has been conducted with modest improvements in the ABE fermentation process, to sufficiently lower the operating costs to make biobutanol an economically viable advanced biofuel.

This chapter highlights the different chemical and biological methods for the production of butanol. It also entails certain metabolic engineering approaches towards modifying microbial strains to increase the quantity of biobutanol produced and to improve the quality of the process. This chapter also focuses on commonly used techniques such as gas stripping, perstraction, pervaporation, vacuum distillation etc. for the *in situ* separation of biobutanol from the fermentation broth.

9.2   Butanol and its Properties

Butanol (butyl alcohol or 1-butanol) is a four carbon, aliphatic alcohol that has the molecular formula C_4H_9OH (MW 74.12 g. mol^{-1}). Butanol is a colorless liquid with a distinct banana like aroma and a strong alcoholic odor. It is highly flammable and slightly hydrophobic. In case of direct contact, it may irritate the eyes and skin. It shows a narcotic effect when inhaled in high concentrations and its vapor may cause irritation to the mucous membranes [5].

Butanol exists in four isomeric forms viz- (normal) n-butanol, *iso*-butanol, *sec*-butanol and *tert*-butanol (Fig. 9.1). The four isomeric forms have slightly different properties and a variety of distinctive applications. Some of the important physical and chemical properties of the four isomeric forms are summarized in Table 9.1.

Butanol is completely soluble with organic solvents, but only partially soluble in water [6]. Other characteristics of butanol are summarized in Table 9.2.

9.3   Butanol as Fuel

One of the major surpassing roles of butanol is its application as a next generation motor-fuel. Some of the distinctive properties of butanol over other fuels such as methanol, ethanol and gasoline are compiled in Table 9.3. Butanol can be used as a direct

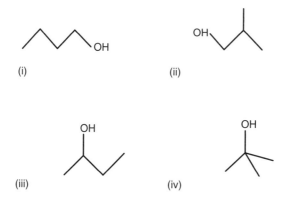

(i)   (ii)

(iii)   (iv)

Figure 9.1 Structures of the four isomeric forms of butanol (i) n-butanol (ii) iso-butanol (iii) sec-butanol (iv) tert-butanol.

Table 9.1 Physical and chemical properties of the four isomers of butanol.

Properties	*n*-butanol	*iso*-butanol	*sec*-butanol	*tert*-butanol
Physical	Liquid, Banana-like odour	Liquid, Sweet-musty odour	Liquid, Fruity odour	Solid at room temperature, Camphor-like odour
Density at 20°C (g/cm³)	0.810	0.802	0.806	0.781
Melting point (°C)	−90	−101.9	−115	Slightly above 25
Boiling point (°C)	118	108	99	82
Water solubility (g/100mL)	7.7	8.0	12.5	Miscible
Flash point (°C)	35	28	24	11

Table 9.2 Properties of Butanol.

Properties	Butanol
Specific gravity	0.810–0.812
Ignition temperature (°C)	35–37
Auto-ignition temperature (°C)	343–345
Relative density (water: 1.0)	0.81
Critical pressure (hPa)	48.4
Critical temperature (°C)	287
Explosive limits (Vol. % in air)	1.4–11.3
Vapor pressure (kPa at 20°C)	0.58

Table 9.3 Comparison of Butanol with other fuels.

Properties	Butanol	Gasoline	Ethanol	Methanol
Boiling point (°C)	117–118	127–221	78	64.7
Density at 20°C (g/ml)	0.809	0.7–0.8	0.785	0.786
Solubility in 100 g of water	immiscible	immiscible	miscible	miscible
Energy content/ value (BTU/gal)	110000	115000	84000	76000
Air-fuel ratio	11.2	14.6	9	6.5
Energy density (MJ/L)	27–29.2	32	19.6	16
Research Octane number	96	91–99	129	136
Motor Octane number	78	81–89	102	104
Dipole moment (polarity)	1.66	N.A	1.7	1.6
Viscosity (10^{-3} Pa.s)	2.593	0.24	1.708	0.5445
Heat of vaporization (MJ/kg)	0.43	0.36	0.92	1.2
Liquid Heat capacity at STP (kJ/k-mol. °K)	178	160–300	112.3	81.1

replacement of gasoline (petrol) or as a fuel additive. Due to its similar characteristics to gasoline (Table 9.3), it can be used efficiently in any gasoline engine without any modification or substitution. Branched chain 4-carbon alcohols including *iso*-butanol, *sec*-butanol and *tert*-butanol possess higher octane numbers than *n*-butanol and thus can be eminently used as fuel additives [7].

Butanol, when used as a fuel, leads to drastically reduced emissions of CO, hydrocarbons and NO_x pollutants. Thus, it has no detrimental effects on the environment.

Butanol is a better fuel additive option than ethanol in various aspects including: (a) it is less corrosive [8]; (b) less volatile, hence less explosive (c) it has approximately 30% more energy/ BTU per gallon (Table 9.3); (d) less hydroscopic i.e. it does not adsorb moisture readily, so it is less affected by weather changes; (e) safer than ethanol because of its high flash point. The biggest advantage of butanol over ethanol is that it can be dispersed through existing pipelines and filling stations [9]. This is primarily due to two main reasons, firstly, butanol is completely miscible with gasoline and diesel fuels. Secondly, it can be added to gasoline without considering evaporation emissions and the consequent problems of the vapor pressure of butanol (4mmHg at 20°C) is 11 times lower than that of ethanol (45 mmHg at 20°C).

9.4   Industrial applications of Butanol and its Derivatives

In 2007, Donaldson *et al.* estimated that 10–12 billion pounds of butanol, produced annually, accounted for a 7–8.4 billion dollar market [10]. It has also been estimated that butanol has a projected market expansion of 3% per year [11]. Besides the expected role of butanol as an engine-biofuel, it has a broad range of industrial applications. It is used in the manufacturing of safety glass, flotation aids (e.g. Butyl xanthate), detergents, de-icing fluids and cosmetics. Butanol also finds application in the food and flavor industries as an extracting agent [6].

Approximately half of the total butanol produced worldwide is used in the form of butyl acrylate and methacrylate esters which are used in the production of latex surface coatings, enamels, lacquers, super absorbants, flocculants, fibers and textiles [11]. Butyl glycol ether, butyl acetate, plasticizers, butyl amines and amino resins are among the other important butanol derived compounds. These are excellent when used as diluting agents in paint thinners and brake fluid formulations. Butanol is also employed as a solvent in the perfume industry and in the manufacturing of antibiotics, vitamins and hormones.

9.5   Methods for Production of Butanol

In the early days, the industrial production of butanol was mainly based on the fermentation of the bacterium *Clostridium acetobutylicum* which acts on carbohydrates and produces butanol and acetone [6]. However, the escalating demand for butanol and the dramatic growth of the petrochemical industry led to the replacement of biological processes by more efficient chemical processes. This section describes in detail both the chemical and biological methods of butanol production.

9.5.1   Chemical Method

There are three major chemical processes- Oxo synthesis, Reppe synthesis and crotonaldehyde hydrogenation for the production of butanol in large-scale in industries. These three processes are described as below.

9.5.1.1   Oxo Synthesis

Oxo synthesis, also called hydroformylation, was discovered by O. Roelen in 1938 [12]. It is the most widespread means of producing butanol. In this process, carbon monoxide (CO) and hydrogen (H_2), collectively called as syngas, react with propylene in the presence of metal catalysts such as Co, Rh, or Ru substituted hydrocarbonyls [13]. Although both aldehydes and ketones are produced in the process, only the aldehyde mixture undergoes hydrogenation for the production of butanol [Fig. 9.2a]. Different isomeric ratios of butanol are obtained depending upon reaction conditions such as pressure, temperature and the type of catalyst. For example, a process operating under a pressure of 200–300 kg/cm^2, temperature 100–180°C and using Co as a catalyst produces 75% butanol and 25% isobutanol, while a process that is operating under much lower pressure (7–20 kg/cm^2) and using Rh as a catalyst yields 92% butanol and 8% isobutanol. However, it is not possible to produce secondary and tertiary butyl alcohol by using this process.

9.5.1.2   Reppe Synthesis

Reppe process was developed in 1942. In this process, propylene, carbon monoxide and water are made to react under pressure in

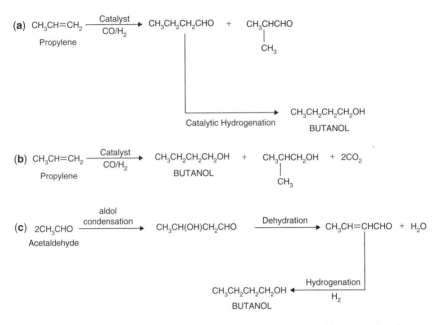

Figure 9.2 Chemical synthesis of butanol: (a) Oxo synthesis (b) Reppe Synthesis (c) Crotonaldehyde hydrogenation [6].

the presence of a catalyst [14]. This generates a mixture of n-butar-aldehyde and isobutaraldehyde [Fig. 9.2b], and then the former is reduced to n-butanol [15]. This process leads to the direct production of butanol at low temperatures and pressure. This process has not been commercially successful because it requires expensive processing technology.

9.5.1.3   Crotonaldehyde hydrogenation

Crotonaldehyde hydrogenation was the common route for butanol synthesis from acetaldehyde until a few decades ago [14]. It consists of aldol condensation, dehydration and hydrogenation [Fig. 9.2c]. Currently, this process is rarely utilized but it may become significant in the future. While other processes completely rely on petroleum, the crotonaldehyde hydrogenation process provides an alternative route for the production of butanol from ethanol, which can be ubiquitously produced from biomass. In this case, ethanol is dehydrogenated to form acetaldehyde from which the synthesis of butanol can proceed [16].

9.5.2   Biological Method

The biological production of butanol by means of ABE fermentation using different strains of *Clostridium* sp. was one of the first large-scale industrial fermentation productions. In the early 20th century, ABE fermentation was ranked the second largest industrial fermentation in the world, after ethanol fermentation. The reasons for the total demise of this fermentation process in 1960s were its inability to compete with the petrochemical method of butanol production and the higher costs of carbohydrate sources as feed substrates. However, in South Africa and Russia, the ABE fermentation processes continued to operate until the late 1980s to early 1990s [17]. Due to emerging environmental awareness and a recent increase in demand for the use of renewable resources as feedstock for the production of chemicals, combined with the advances in the field of biotechnology, there has been a renewal of interest in fermentative butanol production.

Louis Pasteur first reported the production of butanol in a microbial fermentation in 1861. In 1986, Jones and Woods documented a detailed account of the complete history of the development of ABE fermentation in a review article. [18]. Some of the more important events from the development of ABE fermentation are summarized in Table 9.4.

Butanol is produced by fermentation using different strains of *Clostridium* sp. The fermentation process by these microorganisms is a biphasic process i.e. they undergo two growth phases: an acidogenic phase followed by a solventogenic phase (Fig. 9.3). In the acidogenic phase, the production of acids such as acetic acid, butyric acid and lactic acid take place while the solventogenic phase accounts for the production of acetone, butanol and ethanol in a mass ratio of 3:6:1, respectively, from the previously produced acids [19]. Substrates such as glucose, starch, molasses and whey permeate have been used to produce butanol in the past decades. Since substrate cost affects the price of butanol production, the use of economically favorable substrates such as agricultural residues, wastes and energy crops are also being investigated [18].

9.5.2.1   *Microorganisms and metabolic pathway for Butanol production*

A number of different species of butanol-producing Clostridia have been recognized. They differ from each other only in the

Table 9.4 Important events during the development, popularity and decline of ABE fermentation.

Year	Event	Reference
1861	Production of butanol through microbial fermentation firstly observed by Louis Pasteur	[20]
1911	Fernbach isolated a culture which could ferment potatoes to produce butanol	[21, 22]
1912–1914	Weizmann isolated and studied *Clostridium acetobutylicum*	[21]
1916	ABE fermentation was transferred to USA and Canada	[23]
1935	First industrial fermentation of solvents (acetone and butanol)from molasses	[24]
End of 1935	Hydrogen from the ABE fermentation was used for hydrogenation of edible oils	[24]
After 1936	Industrial ABE fermentation started in Japan, Brazil, India, Australia, South Africa and Egypt	[25, 26]
1937	Maize mash was utilized as substrate for production of acetone and butanol at Germiston in South africa	[27]
1960	The production of axetone and butanol ceased virtually in United States and Britain. (However, last factory operated till 1986 in South Africa)	[19]

type and ratio of solvents produced. In addition to the production of butanol, Clostridia can produce chiral products that are difficult to make by chemical synthesis [28] and can degrade a number of toxic chemicals [29, 30]. The genus *Clostridium* was named in 1880 by Prazmowski [31]. Clostridia are rod-shaped, strictly anaerobic, spore-forming, gram-positive bacteria and their isolation is relatively easy as they have very simple nutrient requirements. Generally, complex nitrogen sources such as yeast

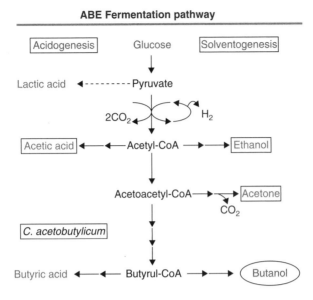

Figure 9.3 Two-phase ABE fermentation pathway [79].

extract are also required for good growth [32]. Solventogenic (solvent producing) Clostridia can utilize a variety of substrates from monosaccharides such as pentoses and hexoses to polysaccharides [18].

Among many solventogenic Clostridia, *Clostridium acetobutylicum*, *C. beijerinckii*, *C. pasteurianum*, *C. saccharobutylicum* and *C. saccharoperbutylaccetonicum* are the primary solvent producers. *C. acetobutylicum* stains as Gram-positive when growing cultures whereas it stains as Gram-negative during the stationary growth phase when forming spores (Fig. 9.4) [33, 34]. *C. beijerinckii* produces a solvent at approximately the same ratio as *C. acetobutylicum* but isopropanol is produced in the place of acetone. On the contrary, *C. aurantibutyricum* produces both acetone and isopropanol in addition to butanol [35] and *C. tetanomorphum* produces only butanol and ethanol in equal concentrations [36]. Thus, the choice of strains to be used in an industrial fermentation depends on the ratio of the end products required and the nature of the raw material used [21, 23].

The metabolic pathways of clostridia that contribute to butanol production are comprised of two phases viz- acidogenesis and solventogenesis. In the acidogenic phase, which usually occurs during the exponential growth phase, acid forming pathways are

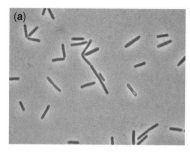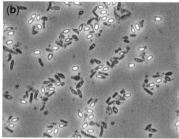

Figure 9.4 *Clostridium* bacteria: (a) Exponential growth phase during which acidogenesis takes place (b) Stationary growth phase during which sporulation and solventogenesis occur [80].

activated and acetate, butyrate, hydrogen and carbon dioxide are produced as major products [37]. In the solventogenesis phase, acids are reassimilated by the bacteria and used in the production of acetone, butanol and ethanol. The solventogenic phase is coupled with the sporulation process [38]. A detailed representation of the metabolic pathways in *C. acetobutylicum* along with different enzymes and genes involved in the reactions has been shown in Fig. 9.5.

Certain dramatic changes in the gene expression pattern are responsible for the transition from the acidogenic to the solventogenic phase [39]. The transcription factor *spo0A* is responsible for the initiation of sporulation and solvent production in *C. acetobutylicum* [40] by activating the transcription of acetoacetate decarboxylase (*adc*) CoA trasferase (*ctfAB*) and alcohol dehydrogenase (*adhE*) coding genes [41]. *Spo0A* is also involved in solvent formation and sporulation in *C. beijerinckii,* however, the mechanism for control of the solvent formation differs from that of *C. acetobutylicum* [42]. According to Scotcher *et al.* (2005), *spo0A* is a multifunctional regulator, which is activated when it is phosphorylated by the phospho-relay two-component signal transduction system [43]. They also proposed that transcription factors encoded by *sinR* and *abrB* play an important role in regulating the sporulation initiation and *abr310* might act as a regulator at the transition between the acidogenic and solventogenic phases. For the enhancement of solvent production, it is required to selectively utilize the positive effects of *spo0A* on solvent formation. The reason behind this is that the effects of *spo0A* on solvent formation act as a balance in regulating gene expression for sporulation versus solvent formation [42].

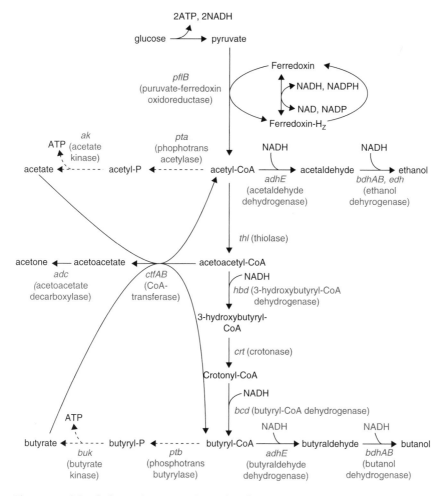

Figure 9.5 Metabolic pathways in *C.acetobutylicum*. Reactions which predominate during acidogenesis and solventogenesis are indicated by dotted and solid arrows respectively. Green letters indicate genes and enzymes involved in the reactions [6].

9.5.2.2   Fermentation Substrates

Butanol has been produced from a number of substrates such as molasses, starch, whey permeate and corn [44, 45]. The high cost of conventional sugar (molasses) or starch (maize, wheat, millet, rye, etc.) substrates is one of the major factors affecting the economic viability of the ABE fermentation. In addition to this, the ability of saccharolytic clostridia to use different carbohydrates has stimulated research in the use of alternative and cheaper substrates.

Various studies on the fermentation of various carbohydrates for the production of acetone and butanol have concluded that fructose, glucose, mannose, sucrose, lactose and dextrin are completely consumed during fermentation, while galactose, xylose, arabinose, melezitose, inulin, and mannitol are partially utilized [18]. Noncellulosic substrates such as Jerusalem artichokes, cheese whey, apple pomace and algal biomass have also been investigated as potential substrates for ABE fermentation. The use of non-traditional substrates such as agricultural residues and wastes including rice straw, wheat straw, barley straw hydrolysate, switchgrass hydrolysate, wood (hardwood), corn fibers, annual and perennial crops and waste paper is also being investigated [46, 47, 48]. These substrates need to be hydrolyzed using a combination of alkali/acid pretreatment and enzymes, prior to using them for fermentation.

9.5.2.3   Industrial Fermentation Process and Advanced Techniques

The initial industrial fermentation process was carried out in a batch process, using fermenters that lacked mechanical agitation systems and had a capacity of 50,000 to 200,000 gallons. The most common substrates were maize mash and beet molasses, which were first cooked for 60 to 90 min at 130 to 133°C and 15 to 60 min at 107 to 120°C, respectively. It was a normal practice to supplement the molasses with additional sources of organic and inorganic nitrogen, phosphorus and a buffering agent. However, no further nutritional additions were necessary while using maize mash. The fermenters were filled to 90 to 95% of their capacity under a blanket of carbon dioxide, and sterile carbon dioxide was often bubbled through the fermentation broth before and after inoculation to facilitate mixing. Cultures were normally kept as spores in sterile sand or soil. Inocula were prepared by heat-activating spores at 65 to 100°C for 1 to 3 min. After two to four buildup stages, the cells were inoculated into the fermenter. Fermenters using maize mash were run at 34 to 39°C for 30 to 60 hours giving a product yield of 25 to 26% based on dry-weight corn equivalents while those utilizing molasses were run at 29 to 35°C providing a solvent yield of 29 to 33% based on the fermentable sugars. The carbon dioxide and hydrogen produced during the fermentation process were recovered, separated and used for a variety of purposes [18]. The important parameters for ABE fermentation at the industrial level are summarized in Table 9.5.

Table 9.5 ABE fermentation parameters at industrial scale.

Micro-organism
Clostridium acetobutylicum *Clostridium beijerinckii*
Feedstock-Initial concentration
Maize mash – 8% wt/volume Starch feedstocks such as corn, cassava, potato – 8 to 10% wt/volume Beet molasses – 5 to 7.5% wt/volume
Yield
25 – 26% based on dry-weight corn equivalents 35 -37% of starch
Fermentation time
30 – 60 hours
Products/ Solvents ratios
Acetone: Butanol: Ethanol – 3:6:1 CO_2: H_2 – 60:40
Type of fermentation / capacity
Batch fermentation – 90 to 750 m³ Cycle of continuous fermentation – 1000 to 2000 m³

The pH of the medium plays an important role in the fermentation process. In the acidogenic phase, the rapid formation of acetic and butyric acids causes a decrease in pH. Solventogenesis starts when pH reaches a critical point, beyond which acids are reassimilated and butanol and acetone are produced. Therefore, it can be easily recognized that low pH is a prerequisite for solvent production. Another point of consideration is that before choosing the initial concentration of substrate, it is important to realize that the final concentration of butanol is not likely to be higher than 13g/l as butanol is highly toxic to biological systems at quite low concentrations, a result of which, clostridia can lose the ability to produce solvents during the cultivation. This degeneration phenomenon was found to be due to the loss of the megaplasmid pSOL1 in *C. acetobutylicum* [49].

Due to their simple operation and reduced risk of contamination, batch reactors are generally preferred in the bioprocessing industry. However, in a batch reactor, the productivity is low due to the long lag phase and product inhibition. In order to eliminate the lag phase, continuous reactors can be used. Continuous reactor systems also prevent solvent accumulation and cell inhibition since the concentrations of the solvents in the reactor never exceed 6g/l, which is far from the bacteria's tolerance level. A fed-batch reactor integrated with an *in situ* product removal system has proved to be quite efficient in solving the problem of product inhibition. For example, in an integrated fed-batch fermentation-gas stripping product recovery system using *C. beijerinckii* BA101, with H_2 and CO_2 as the carrier gases, the solvent productivities were improved to 400% of the control batch fermentation productivities [50].

Furthermore, fibrous bed reactor and cell cycle reactors have also been applied to butanol production, in order to increase productivity. The fibrous bed reactor is essentially an immobilized bed reactor, where the cells are immobilized on a roll of fibers such as cotton fibers. The orientation of the fibers creates many channels through which the reactor contents flow up. Fresh medium is constantly supplied to the reactor and solvents that normally inhibit cell growth are constantly removed. This bioreactor is also self-renewing in the sense that as one cell dies; more cells grow in its place because of the continuous supply of nutrients. Therefore, no new cells need to be introduced into the bioreactors, once the initially inoculated biomass adheres to the fibrous surface during the immobilization period. The bioreactor has a very large void space as well as large surface area, which allow for both greater cell density (40–100 g/l) and greater productivity, leading to much faster reactions. The design of this reactor is such that the dead cells fall to the bottom of the vessel and the gases that are produced during the process (CO_2 and H_2) easily flow up through the channels, helping to mix the contents of the reactors. The greatest advantage of this bioreactor is that it can run continuously over one year without the need for maintenance or downstream processing [51].

In membrane cell cycle reactors, a hollow-fiber ultrafilter is applied to separate and recycle cells in continuous fermentation [52]. With this bioreactor a solvent productivity of 6.5 g/l/h can be achieved at a dilution rate of 0.5 h^{-1}. A major obstacle of this system is the clogging of the membrane with the fermentation broth. However, this problem can be overcome by allowing only the fermentation broth to undergo filtration by using the immobilized cell system [53].

9.5.2.4 Solvent Toxicity

During the solventogenic phase, clostridial cellular metabolism usually continues until the concentration of the solvent reaches around 20 g/l, after this level, further cell metabolism ceases. Of all the solvents produced, the lipophilic solvent butanol is the most toxic and is the only solvent produced to the level that becomes toxic to the cells during fermentation. It has been demonstrated that an addition of 7 to 13 g/l of butanol to culture medium results in a 50% inhibition of growth [18]. Butanol disrupts the phospholipids' components of the cell membrane causing an increase in membrane fluidity [54]. Increased membrane fluidity leads to the destabilization of the membrane and the disruption of membrane-associated functions such as glucose uptake, transport processes and membrane bound ATPase activity. Butanol toxicity is one of the major limiting factors to the amount of solvent that can be produced during ABE fermentation. It has been hypothesized that enhancing the butanol tolerance of the cells could help in eradicating the problem of solvent toxicity. One important example of a butanol-tolerant strain is the *C.beijerinckii* BA101 strain derived from the *C.beijerinckii* NCIMB 8052 by random mutagenesis [55]. Antisense RNA technology has also been used to develop a butanol-tolerant mutant strain of *C.beijerinckii* NCIMB 8052 by targeting the *gld A* gene which encodes glycerol dehydrogenase [56]. The observation that membrane fluidity increases with increasing temperature suggests that a decrease in temperature during the solvent producing phase might enhance butanol tolerance. Thus, butanol production can be enhanced by decreasing the temperature from 30°C during the acidogenesis phase to 24°C during the solventogenic phase. Another approach towards overcoming the solvent toxicity problem is the integration of an *in situ* solvent recovery process with fermentation. It is also expected that metabolic and cellular engineering using the systems biology approach will help to develop a more solvent tolerant strain of clostridia [57].

9.5.2.5   Metabolic Engineering

The main purposes behind the metabolic engineering of butanol producing clostridia involve increased butanol production with respect to final concentration and productivity, increased butanol (solvent) tolerance, extended substrate utilization range, modifying the metabolic network for better medium design and the selective production of butanol instead of mixed acid/solvent production. These purposes can be achieved by utilizing various genetic engineering tools such as transformation techniques, shuttle vectors and gene knockout systems.

C. acetobutylicum ATCC 824 possesses a strong restriction system encoded by Cac824I (restriction endonuclease) that recognizes 5'-GCNGC-3'. This restriction system prevents efficient transformation of recombinant plasmid prepared in E. coli. Thus, Mermelstein et al. (1992) developed a B. subtilis – C. acetobutylicum shuttle vector pFNK1, which allowed higher transformation efficiency [58]. Using this shuttle vector, the acetoacetate decarboxylase (adc), and the phosphotransbutyrylase (ptb) genes can be successfully expressed at elevated levels in C. acetobutylicum ATCC 824. The copy number of commonly used plasmids in C. acetobutylicum is around 7 – 20 copies per cell, which seem to be suitable for metabolic engineering purposes [59].

Due to the lack of an efficient gene knockout system, metabolic engineering of clostridia is hindered. Only five genes viz- buk, pta, adhE, solR, spo0A have been knocked-out in C. acetobutylicum with very low reproducibility [42, 60]. Heap et al. (2007) has developed a gene knockout system called the clostron system using Lactobacillus lactis [61]. It is the most efficient gene knockout system for clostridia known to date.

The first successful example of metabolic engineering was the amplification of the acetone formation pathway in C. acetobutylicum. In the fermentation process, using recombinant C. acetobutylicum with amplified adc and ctfAB genes (encoding acetoacetate decarboxylase and CoA transferase respectively), the acetone forming genes become active earlier, resulting in earlier induction of acetone formation. As compared to fermentation using its parental strain, this process resulted in an increased final concentration of acetone, butanol and ethanol by 95%, 37% and 90% respectively [62]. In another study, buk and pta genes, encoding butyrate kinase and phosphor transacetylase respectively, were deactivated to redirect carbon flow from acid formation to solvent formation. The PJC4BK strain, having inactivated buk gene, produced 10% more butanol than the parental strain [63]. Anti-sense RNA (asRNA) technology has also been found to be effective in reducing the activities of butyrate forming enzymes in C. acetobutylicum. Desai and

Papoutsakis (1999) designed two different asRNAs, one for *buk* and the other for *ptb* and successfully reduced the activities of butyrate kinase and phosphotransbutyrylase by 85% and 70% respectively. The strain transformed with *buk*-asRNA produced 50% and 35% higher final concentrations of acetone and butanol, respectively.

Solvent tolerance and stress response can be improved in *C. aceto-butylicum* by the overexpression of many genes including molecular pumps, chaperones (e.g. *groES, dnaKJ, hsp18*) and genes involved in sporulation, fatty acid synthesis and transcriptional regulators [64]. The overexpression of heat shock proteins GroES and GroEL also increase the final solvent concentration by stabilizing solventogenic enzymes [65].

Before 2007, the complete genome sequence of *C. acetobutylicum* was unavailable, as a result of which the engineering targets were limited to those genes that were directly related to acid or solvent formation. Now, as the complete genome sequence is available, it is possible to carry out more systematic strain improvement process. With advances in systems biology, *in silico* modeling and simulation can also be used as efficient tools for metabolic pathway engineering of clostridia. In fact, Papoutsakis (1984) was the first to develop such stoichiometric models for the quantitative analysis of metabolic fluxes in *C. acetobutylicum* [66].

Efforts have also been made to develop tools for the metabolic engineering of *E. coli* for butanol production. A mutant *E. coli* BW25113 strain overexpressing the *crt, bcd, etfAB, hbd* and *adhE2* genes of *C. acetobutylicum*, and *atoB* gene of *E. coli* was able to produce 552 mg/l butanol using 2% (w/v) glycerol as a carbon source [67]. By overexpressing the *thl, hbd, crt, bcd, ald* and *bdhA B* genes of *C. acetobutylicum* involved in butanol biosynthesis, 0.8 mM, 0.19 mM, and 0.01 mM butanol could be produced by engineered *E. coli, B. subtilis* and *S. cerevisiae*, respectively [68].

9.6   *In situ* Separation Techniques for Butanol

The major problem in the production of butanol is the high cost of product recovery. Distillation is one of the oldest methods for butanol recovery from the fermentation broth. However, the traditional recovery systems using distillation suffer from a high cost of product recovery due to the low concentration of butanol in the fermentation broth [4]. In order to resolve the solvent toxicity problem and to reduce the operation costs, several *in situ* recovery systems such as gas stripping, adsorption, liquid-liquid adsorption, pervaporation, perstraction and vacuum distillation have been employed. There are certain advantages and disadvantages of each system that need to be examined thoroughly before incorporating any of these systems with the production systems at the industrial level. A comparison between distillation, gas stripping, pervaporation, perstraction and adsorption techniques has been documented in Table 9.6.

The recovery and purification processes are greatly influenced by the performance of the fermentation process, which in turn is affected by the strain characteristics. For instance, when a strain is metabolically engineered to produce butanol without or with a very low production of acetone and ethanol, the purification process will be extremely simplified. When a butanol-tolerant strain is employed in the fermentation process, it facilitates the recovery process as higher concentration of butanol can be achieved during the fermentation.

9.6.1   Gas stripping

Gas stripping is one of the simplest techniques to recover butanol from fermentation broth because it does not employ any expensive equipment (Fig. 9.6). In this technique, a fermentation gas such as nitrogen (N_2) or carbon-dioxide (CO_2) is sparged through the fermentation broth during fermentation, which causes the volatile butanol to vaporize and go out with the gas stream in the top of the reactor. The fermentation gas is then passed through a condenser for solvent recovery and the stripped gas is recycled back to the fermentor. This process continues until all the sugar in the fermentor is utilized.

Gas stripping can be efficiently integrated with batch, continuous and fed-batch reactor systems to recover the solvents from the fermentation broth [50, 69]. Qureshi and Blaschek have demonstrated that by using gas stripping the productivity can be increased by 41% and the bacteria use 50% more substrate than normal [70].

9.6.2   Adsorption

In this technique, a stream of fermentation broth is firstly made to run through a membrane filter or centrifuge to separate the biomass from it. The solids, so obtained, can be recirculated into the fermenter while the liquids are passed through an adsorption column

Table 9.6 Comparison between different *in situ* separation techniques.

Technology	Efficiency	Capacity	Selectivity	State of Development	Scale	Operating cost
Gas stripping	Medium	Moderate	Low	Research	Laboratory	High
Pervaporation	High	Moderate	Moderate	Developed	Pilot	High
Perstraction	High	High	High	Research	Laboratory	Medium
Adsorption	High	Low	Low	Research	Commercial	Low
Distillation	High	Moderate	High	Complete	Commercial	High

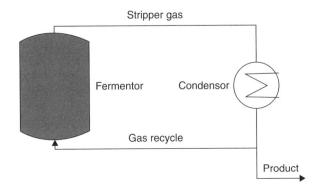

Figure 9.6 Flow diagram for Gas stripping [77].

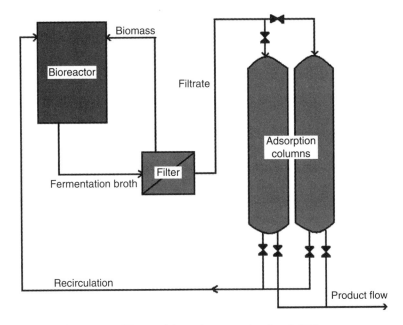

Figure 9.7 Adsorption of Butanol from fermentation broth [77].

(Fig. 9.7). The columns are filled with hydrophobic adsorbents such as silica over which butanol can be adsorbed easily. Researchers have shown that some zeolites can also favor adsorption of butanol [71]. After the columns have adsorbed butanol, they can be desorbed by increasing the temperature to around 200°C [72].

This technique is more advantageous than ordinary distillation in that it require only 8.2 MJ/kg butanol to be present in fermentation broth to operate, whereas distillation require the presence of at least 73.3 MJ/kg butanol to function properly.

9.6.3   Liquid-liquid extraction (Perstraction)

Liquid-liquid extraction is considered one of the best techniques for butanol recovery. This approach takes advantage of the differences in the distribution coefficients of the chemicals. Since butanol is more soluble in the extractant (organic phase) than in the fermentation broth (aqueous broth), it is selectively concentrated in the extractant [6]. Two commonly used extractants are oleyl alcohol and n-decanol. These extractants are applied in fermentation process with *C. beijerinckii* BA101 and *C. acetobutylicum* cultures respectively [73].

Liquid-liquid extraction also suffers from certain shortcomings due to the toxicity of the extractant to the cells and emulsion formation. However, these problems can be overcome by separating the fermentation broth and the extractant with the help of a membrane that provides surface area for butanol exchange between the two immiscible phases (Fig. 9.8). When the process of liquid-liquid extraction is carried out in presence of a membrane, it is referred to as "Perstraction" [46].

9.6.4   Pervaporation

Pervaporation is a membrane-based separation process that allows the selective removal of volatile compounds from fermentation broth. In this process, the membrane is placed in contact

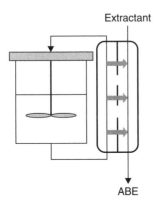

Figure 9.8 Fermentation reactor integrated with perstraction system [6].

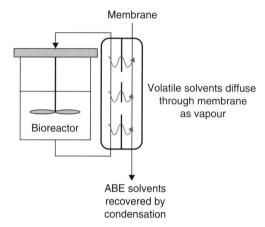

Figure 9.9 Fermentation reactor integrated with pervaporation system [6].

with the fermentation broth and the volatile solvents diffuse through the membrane as vapor that can be recovered by condensation (Fig. 9.9). Pervaporation processes may deploy both solid and liquid membranes. A liquid membrane containing oleyl alcohol has been illustrated to be the best membrane for liquid medium in batch and fed-batch reactors [74]. This membrane has very high specificity as compared to a solid silicone rubber membrane. In order to increase the selectivity of silicone rubber membranes, Qureshi *et al.* have synthesized a silicon-silicalite-1 composite membrane that possesses a 2.2-fold improvement in selectivity [75]. The solvent productivity in continuous fermentation integrated with pervaporation system can be enhanced by using an ionic liquid polydimethylsiloxane ultrafiltration membrane [76].

9.6.5   Vacuum Distillation

The main idea behind the vacuum distillation process is to lower the pressure in the distillation columns, as the decreased pressure leads to a lower boiling point so that less energy is needed to boil any component [77]. The set up for vacuum distillation does not need to be very different from the regular distillation set up. The only difference is that the pressure in the column has to be controlled somehow. A flow diagram showing where the vacuum distillation should be implemented in the ABE process is shown in Figure 9.10.

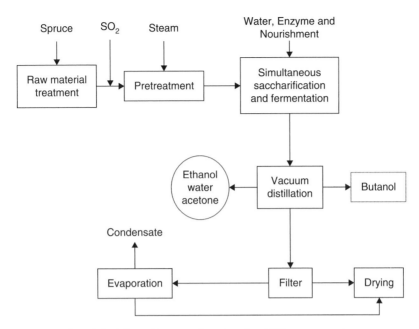

Figure 9.10 Simplified flow diagram showing the ABE process using vacuum distillation [77].

9.7   Future Prospects

Butanol has a promising future as a biofuel; still, more improvements are still needed to make the biobutanol process economically competitive. The main stress should be on research for developing organisms with higher butanol tolerance, better end product specificity, efficient substrate pretreatment and product recovery methods. From the last 30 years, there have been numerous reports on the ABE fermentation process carried out with different modifications. All these modifications together contributed to drastic advances in butanol production. The overall competitiveness of the bioprocess highly depends on the performance of the strain; therefore, successful metabolic engineering is crucial to enhance butanol production. Specific and volumetric productivities can be increased by optimizing metabolic fluxes. This goal can be achieved by identifying the target genes to be engineered by genome-scale metabolic flux analysis and omics studies [57]. As the complete genome sequences for both *C. beijerinckii* and *C. acetobutylicum* are available, it is just a matter of time and interest to construct a genome-scale metabolic network of clostridia. Gene knockout models can also be prepared by using the genome-scale metabolic networks to identify combinatorial target genes to be manipulated. Another tool for metabolic engineering such as genome breeding can be used to compare the genome of *C. beijerinckii* BA101 and its parental strain *C. beijerinckii* NCIMB 8052, with the motive of identifying mutated genes of *C. beijerinckii* BA101 involved in enhanced butanol production. This knowledge can be used to further enhance butanol production [78]. Transcriptome and proteome profiling can also be done under different environmental conditions to identify target genes and pathways to be manipulated. In addition to metabolic engineering, there is good scope in research field. Different transcription factors, chaperones, proteins can be engineered to increase butanol production and butanol tolerance. An efficient bioprocess for butanol production can also be achieved by integrating fermentation process with downstream process. The ongoing research in the field of metabolic engineering, microbial technology, bioprocess and downstream processing and systems biology is surely going to make bio-based butanol production, an economically feasible process in the near future.

References

1. S.E. David and S.A. Morton III, *Separation Science and Technology*, Vol. 43, p. 2460, 2008.
2. Y.L. Lin and H.P. Blaschek, *Applied Environmental Microbiology*, Vol. 45, p. 996, 1983.
3. P. Dürre, *Annals of the New York Academy of Science*, Vol. 1125, p. 353, 2008.
4. P. Dürre, *Applied Microbiology and Biotechnology*, Vol. 49, p. 639, 1998.
5. L. Gholizadeh, "Enhanced butanol production by free and immobilized *Clostridium* sp. cells using butyric acid as co-substrate" (Report), University of Boras, 2009.
6. S.Y. Lee, J.H. Park, S.H. Jang, L.K. Nielsen, J. Kim, and K.S. Jung, *Biotechnology and Bioengineering*, Vol. 101(2), p. 209, 2008.
7. S. Atsumi, T. Hanai, and J.C. Liao, *Nature*, Vol. 451, p. 86, 2008.
8. P. Dürre, *Biotechnology Journal;* Vol. 2, p. 1525, 2007.
9. Shang-Tian Yang, Methods for Producing Butanol. Patent: US 20080248540A1, 2008.
10. G.K. Donaldson, L.L. Huang, L.A. Maggio-Hall, V. Nagarajan, C.E. Nakamura, and W. Suh, Fermentative production of four carbon alcohols. International Patent WO2007/041269, 2007.
11. M. Kirschner, n-Butanol, Chemical Market Reporter ABI/INFORM Global, p. 42, 2006.

12. H. Weber, and J. Falbe, *Industrial and Engineering Chemistry Research*, Vol. 62(4), p. 33, 1970.
13. J. Falbe, *Carbon Monoxide in Organic Synthesis*. Berlin-Heidelberg-New York; Springer Verlag, 1970.
14. M. Bochman, F.A. Cotton, C.A. Murillo, and G. Wilkinson, *Advanced inorganic chemistry*. USA: John Wiley & Sons, Inc, 1999.
15. L.P. Wackett, *Current Opinion in Chemical Biology*, Vol. 12, p. 187, 2008.
16. W. Swodenk, Chemie-Ingenieur-Technik, Vol. 55, p. 683, 1983.
17. V.V. Zverlov, O. Berezina, G.A. Velikodvorskaya, and W.H. Schwarz, *Applied Microbiology and Biotechnology*, Vol. 71, p. 587, 2006.
18. D.T. Jones, and D.R. Woods. *Microbiology Review*. Vol. 50, p. 484, 1986.
19. T.C. Ezeji, N. Qureshi, P. Karcher, and H.P. Blaschek, "Production of butanol from corn", In: Minteer S (ed) Alcoholic fuels. Taylor and Francis Group, LLC, CRC Press, Boca Raton, pp. 99–122, 2006.
20. C. L. Gabriel, *Industrial and Engineering Chemistry Research*, Vol. 20, p. 1063, 1928.
21. C. L. Gabriel, and F.M. Crawford, *Industrial and Engineering Chemistry Research*, Vol. 22, p. 1163–1165, 1930.
22. J. H. Hastings, "Acetone-butyl alcohol fermentation", In A. H. Rose (ed.), Economic microbiology, Primary products of metabolism. Academic Press, Inc., New York. Vol. 2, pp. 31–45, 1978.
23. J. H. Hastings, *Advances in Applied Microbiology*, Academic Press, Inc., New York. Vol. 14, p. 1 1971.
24. D. Ross, *Progress in Industrial Microbiology*, Vol. 3, p. 73, 1961.
25. M. J. Spivey, *Process Biochemistry*, Vol. 13, p. 2–5, 1978.
26. P. M. Robson and D.T. Jones, "Production of acetone butanol by industrial fermentation", In 0. Chaude and G. Durand (ed.), Industrielle et Biotechnologie. Production d'intermediares industriels par culture anaerobie. *Societe Francaise de Microbiologie*, Paris, pp. 169–214. 1982.
27. A. H. Rose, *Industrial microbiology*, Butterworths, London, p. 160–166, 1961.
28. P. Rogers, *Advanced Applied Microbiology*, Vol. 31, p. 1, 1986.
29. A.J. Francis, C.J. Dodge, F. Lu, G.P. Halada, and C.R. Clayton, *Environmental Science Technology*, Vol. 28, p. 636, 1994.
30. J.C. Spain, *Annual Review of Microbiology*, Vol. 49, p. 523, 1995.
31. Jan et al. in chapter 2 Nigel 1989.
32. F. Monot, J.R. Martin, H. Petitdemange, and R. Gay, *Applied Environmental Microbiology*, Vol. 44, p. 1318, 1982.
33. P. Dürre, "Sporulation I clostridia", In: P. Dürre, (ed.). *Handbook on clostridia*. Boca Raton: CRC Press. pp. 659–699, 2005.
24. H. Biebl, *Journal of Industrial Microbiology and Biotechnology*, Vol. 27, p. 18, 2001.
35. H. A. George, and J.S. Chen, *Applied Environmental Microbiology*, Vol. 46, p. 321, 1983.
36. M. Gottwald, H. Hippe, and G. Gottschalk, *Applied Environmental Microbiology*, Vol. 48, p. 573, 1984.
37. M.G.N. Hartmanis, and S. Gatenbeck, *Applied Microbiology and Biotechnology*, Vol. 47, p. 1277, 1984.
38. C. Paredes, K.V. Alsaker, and E.T. Papoutsakis, *Nature Review*, Vol. 3, p. 969, 2005.

39. P. Dürre, A. Kuhn, M. Gottwald, and G. Gottschalk, *Applied Microbiology and Biotechnology*, Vol. 26, p. 268, 1987.

40. H. Bahl, H. Muller, H. Behrens, H. Joseph, and F. Narberhaus, *FEMS Microbiology Review*, Vol. 7, p. 341, 1995.

41. L. Sullivan, and G.N. Bennett, *Journal of Industrial Microbiology and Biotechnology*, Vol. 33, p. 298, 2006.

42. L.M. Harris, N.E. Welker, and E.T. Papoutsakis, *Journal of Bacteriology*, Vol. 184, p. 3586, 2002.

43. M.C. Scotcher, and G.N. Bennett, *Journal of Bacteriology*, Vol. 187, p. 1930, 2005.

44. T.C. Ezeji, N. Qureshi, and H.P Blaschek, *Current Opinions in Biotechnology*, Vol. 18, p. 220, 2007.

45. N. Qureshi, T.C. Ezeji, J. Ebener, B.S. Dien, M.A. Cotta, and H.P. Blaschek, *Bioresource Technology*, Vol. 99, p. 5915, 2008.

46. N. Qureshi, B.C. Saha, B. Dien, R.E. Hector, and M.E. Cotta, *Biomass and Bioenergy*, Vol. 34, p. 559, 2010.

47. N. Qureshi, T.C. Ezeji, J. Ebener, B.S. Dien, M.A. Cotta, and H.P. Blaschek, *Bioresource Technology*, Vol. 99, p. 5915, 2008.

48. N. Qureshi, B.C. Saha, R.E. Hector, B. Dien, S. Hughes, Liu Siqing, Iten Loren, M.J. Bowman, G. Sarath, and M.A. Cotta, *Biomass and Bioenergy*, Vol. 34, p. 566, 2010.

49. T.C. Ezeji, N. Qureshi, and H.P. Blaschek, "Industrially relevant fermentations", In: Dürre P, editor. *Handbook on clostridia*. Boca Raton: CRC Press. pp. 797–812, 2005.

5. T.C. Ezeji, N. Qureshi, and H.P. Blaschek, Applied Microbiology and Biotechnology. Vol. 63(6), p. 653, 2004.

51. Yang, Shang-Tian, Extractive Fermentation Using Convoluted Fibrous Bed Bioreactor. The Ohio State University Research Foundation, assignee, Patent 5,563,069, 1996.

52. P. Pierrot, M. Fick, and J.M. Engasser, *Biotechnology Letters*, Vol. 8, p. 253, 1986.

53. F. Lipnizki, S. Hausmanns, G. Laufenberg, R. Field, and B. Kunz, *Chemical Engineering Technology*, Vol. 23, p. 569, 2000.

54. L.K. Bowles, and W.L. Ellefson, *Applied Environmental Microbiology*, Vol. 50, p. 1165, 1985.

55. N. Qureshi, and H.P. Blaschek, *Journal of Industrial Microbiology and Biotechnology*, Vol. 27, p. 287, 2001.

56. H. Lyanage, M. Young, and E.R. Kashket, *Journal of Molecular Microbiology and Biotechnology*, Vol. 2, p. 87, 2000.

57. S.Y. Lee, D.Y. Lee, and T.Y. Kim, *Trends in Biotechnology*, Vol. 23, p. 349, 2005.

58. L.D. Mermelstein, N.E. Welker, G.N. Bennett, and E.T. Papoutsakis, *Bio/Technology*, Vol. 10, p. 190, 1992.

59. S.Y. Lee, L.D. Mermelstein, and ET. Papoutsakis, *FEMS Microbiology Letters*, Vol. 108, p. 319, 1993.

60. R.V. Nair, E.M. Green, D.E. Watson, G.N. Bennett, and E.T. Papoutsakis, *Journal of Bacteriology*, Vol. 181, p. 319, 1999.

61. J.T. Heap, O.J. Pennington, S.T. Cartman, G.P. Carter, and N.P. Minton, *Journal of Microbiology Methods*, Vol. 70, p. 452, 2007.

62. L.D. Mermelstein, E.T. Papoutsakis, D.J. Petersen, and G.N. Bennett, *Biotechnology and Bioengineering*, Vol. 42, p. 1053, 1993.

63. E.M. Green, Z.L. Boynton, L.M. Harris, F.B. Rudolph, E.T. Papoutsakis, and G.N. Bennett *Microbiology (UK)*, Vol. 142, p. 2079, 1996.
64. C. Tomas, J. Beamish, and E.T. Papoutsakis, *Journal of Bacteriology*, Vol. 186, p. 2006, 2004.
65. C.A. Tomas, N.E. Welker, and E.T. Papoutsakis, *Applied Environmental Microbiology*, Vol. 69, p. 4951, 2003.
66. E.T. Papoutsakis, *Biotechnology and Bioengineering*, Vol. 26, p. 174, 1984.
67. S. Atsumi, A.F. Cann, M.R. Connor, C.R. Shen, K.M. Smith, M.P. Brynildsen, K.J.Y. Chou, T. Hanai, and J.C. Liao, *Metabolic Engineering*, Vol. 10, p. 1016, 2007.
68. G.K. Donaldson, L.L. Huang, L.A. Maggio-Hall, V. Nagarajan, C.E. Nakamura, and W. Suh, Fermentative production of four carbon alcohols. International Patent WO2007/041269, 2007.
69. T.C. Ezeji, N. Qureshi, and H.P. Blaschek, *World Journal of Microbiology and Biotechnology*, Vol. 19, p. 595, 2003.
70. N. Qureshi, and H.P. Blaschek, *Renewable Energy*; Vol. 22, p. 557, 2001.
71. A. Oudshoorn, L.A.M. van der Wielen, and A.J.J. Straatho, *Biochemical Engineering Journal*, Vol. 48, p. 99, 2009.
72. E. Larsson, M. Max-Hansen, A. Pålsson, and R. Studeny, "A feasibility study on conversion of an ethanol plant to a butanol plant", 2008.
73. C. Thirmal and Y. Dahman, *The Canadian Journal of Chemical Engineering*; Vol. 90, p. 745, 2012.
74. M. Matsumura, H. Kataoka, M. Sueki, and K. Araki, *Bioprocess Engineering*, Vol. 3, p. 93, 1988.
75. N. Qureshi, H.P. Blaschek, *Biotechnology Progress*, Vol. 15, p. 594, 1999.
76. P. Iza´k, K. Schwarz, W. Ruth, H. Bahl, and U. Kragl, *Applied Microbiology and Biotechnology*, Vol. 78, p. 597, 2008.
77. S. Dahlbom, H. Landgren, and P. Fransson, "Alternatives for Bio-Butanol Production". Feasibility study presented to Statoil, 2011.
78. J. Ohnishi, S. Mitsuhashi, M. Hayashi, S. Ando, H. Yokoi, K. Ochiai, and M. Ikeda, *Applied Microbiology and Biotechnology*, Vol. 58, p. 217, 2002.
79. D. Ramey and S.-T. Yang, "Production of Butyric Acid and Butanol from Biomass-Final Report". *U.S. Department of Energy*, Morgantown, West Virginia, 2004.
80. K. Melzoch, P. Patáková, M. Linhová, J. Lipovský, P. Fribert, M.S.S. Toure, M. Rychtera, M.Pospíšil and G. Šebor "Experiences in the production and use of butanol as biofuel". *Institute of Chemical Technology Prague*, Prague 6, Czech Republic, 2010.

10

Production of Biodiesel from Various Sources

Komal Saxena[1], Avinash Kumar Sharma[1], Lalit Agrawal[2] and Ashish Deep Gupta*[,1]

[1]Department of Biotechnology, Institute of Biomedical Education and Research, Mangalayatan University, Aligarh, 202 145, Uttar Pradesh, India. [2]CSIR-National Botanical Research Institute, Rana Pratap Marg, Lucknow, 226 001, Uttar Pradesh, India.

Abstract

We all are aware that the depletion of the fossil fuel reservoirs has lead to an urgent search for alternative forms of present day fuels. The possible solution could come from biofuels, especially biodiesel. Biodiesel could be produced from plants, microorganisms, fungi, wastewater, microalgae etc., but these sources suffer from one or more drawbacks. However, microalgae could serve as the best source for biodiesel production since it does not lead to a land versus fuel conflict. This chapter is mainly focused on the sources for the production of biodiesel, biodiesel production processes, catalysis involved and the factors affecting biodiesel production. We have to understand that for the best utilization of biodiesel in diesel engines, we have to incorporate technical modifications in traditional diesel engines so that biodiesel production could be promoted commercially.

Keywords: Biodiesel feedstock, biodiesel production, biofuels, homogeneous catalysts, heterogeneous catalysts

10.1   Introduction

Due to the depletion of fossil fuels reserves, the rising of the crude oil prices, political instability in countries and the exhaustion of CO_2 in the environment, there is a need for an environmental

Corresponding author: ashishdeepgupta@rediffmail.com

Dr.Vikash Babu, Dr. Ashish Thapliyal & Dr. Girijesh Kumar Patel (eds.) Biofuels Production, (285–308) 2014 © Scrivener Publishing LLC

Oil + Methanol/Ethanol $\xrightarrow{\text{catalyst}}$ Fatty acid methyl esters / Fatty acid ethylesters + Glycerol

Figure 10.1 The trans-esterification reaction for biodiesel production.

sustainable, renewable, low pollution emitting, non-toxic, economically sustainable and carbon neutral fuel for transport systems and other purposes. A detailed description of existing and future fossil fuels which are capable of replacing diesel has been explored in this chapter, keeping in mind, that a source for the production of biodiesel should be one which has got the potential to replace the fossil fuel demand [1–4].

Biodiesel is one of the biofuels that has been produced on an industrial scale. The catalytic trans-esterification of oil with alcohol (methanol/ethanol) produces biodiesel that is a clear amber-yellow liquid, which is either fatty acids mono alkyl ester (FAME) or fatty acid ethyl ester (FAEE) (Fig. 10.1) [1, 2, 5]. The lipid profile of FAME is made mainly of saturated fatty acids such as palmitic acid, stearic acid and unsaturated fatty acids like oleic acid, linolenic acid and linoleic acid [6–10]. The lipid profile of FAEE is made of ethyl palmitate, ethyl oleate, ethyl myristate and ethyl palmitoleate [3, 4].

Biodiesel could be defined as the fatty acid methyl esters / fatty acid ethyl esters produced by the transesterification of triacylglycerols present in oil in the presence of methanol/ethanol respectively.

10.2   Sources/Feedstocks for the Production of Biodiesel

The prominent sources for biodiesel production are soybean oil, rapeseed oil, palm oil, corn oil, animal fat, non-edible oilseed crops and waste cooking oil, although these sources do not tend to satisfy the upcoming or existing biodiesel demands. The use of cheap feedstocks such as non-edible oils, animal fats, waste food oil and by-products of the refining of vegetable oils for biodiesel production could lead to a cost reduction [11, 12]. However, various algae have emerged as the preferred source for biodiesel production due to many reasons [11]. A detailed description of biodiesel production from different sources is listed below.

10.2.1 Production of Biodiesel from Plants

About 350 crops are known that could be used for the production of biodiesel. However, sunflower oil, safflower oil, soybean oil, cottonseed oil, rapeseed oil and peanut oil are the existing potential sources because they could be used in diesel engines without alteration. The different plant sources that are used for biodiesel production in a country depends upon the availability and cost of the plant oil produced from plant sources grown in that country [7]. However, using edible oils as feedstock for biodiesel production is very costly and therefore could not be commercialized [13].

Non-edible oils like *Mathuca indica, Shorea robusta, Pongamia glabra, Mesua* spp., *Mallotus philillines, Garcinea indica, Jatropha curcas, Salvadova* spp., *Nicotiana tabacum*, rice bran oil, *Azadirachta indica, Hevea brasiliensis* and castor oil are the prominent sources for biodiesel production. *Jatropha curcas* oil is a preferred feedstock for India [14, 15]. This plant can thrive well in adverse conditions. Therefore, it is grown in marginal soils, although the quantity of seeds obtained by the plant is very low [16, 17]. The fatty acid profile of *Jatropha curcas* consists of both saturated and unsaturated fatty acids. The saturated fatty acids are mainly palmitic acid (14.2%), stearic acid (7.0%) and the unsaturated fatty acids are mainly oleic acid (44.7%) and linoleic acid (32.8%) [6, 7, 8, 9]. *Pongamia pinnata* is also able to grow rapidly on marginal lands and has a fatty acid profile that consists of oleic acid (51.8%), linoleic acid (17.7%), palmitic acid (10.2%), stearic acid (7.0%) and linolenic acid (3.6%) [10].

10.2.2 Producton of Biodiesel using Bacteria

A future approach to biodiesel production is its production from bacteria which is known as micro diesel [18, 19]. The FAEE could be produced by genetically engineered *Escherichia coli* with ethanol producing pathway genes from *Zymomonas mobilis*. This increases the amount of fatty acids obtained by alterating the enzymes in various metabolic pathways. For example, higher amounts of fatty acids are obtained by the heterologous expression of acyl-CoA in *E. coli*, deletion of the *fadE* gene which encodes acyl-CoA dehydrogenase, overexpression of thioesterases, and the heterologous introduction of *fadD* gene from *Saccharomyces cerevisiae* in *E. coli* [4].

10.2.3 Production of Biodiesel from Fungi

Several filamentous fungi and oleaginous yeasts are capable of accumulating more than 20% of the lipid content in their cells namely, *Yarrowia* spp., *Candida* spp., *Rhodotorulla* spp., *Rhodosporidium* spp., *Cryptococcus* spp., *Trichosporon* spp., *and Lipomyces* spp. These yeast species are capable of utilizing agricultural residues for lipid production. The fatty acid profile of the lipid accumulated in a single cell of oleaginous fungi mainly contains palmitic acid (18.51%), oleic acid (67.29%), myristic acid (1.11%), stearic acid (1.25%) and linoleic acid (4.76%) which is quiet similar to that of the plant feedstock derived oil utilized for biodiesel production. Amongst them, *Rhodotorulla* 110 utilizes glucose as a sole carbon source, yeast extract and ammonium sulphate as a nitrogen source to give high lipid content. Although rice bran is the best carbon source which could be used for lipid accumulation in these yeast cells due to its low cost, other sources which could be used as carbon sources are glucose, sucrose, lactose and soluble starch [3, 20].

Y. lipolytica is a model oleaginous yeast for biodiesel production. It is grown on glucose and inexpensive wastes like cooking oil waste and waste motor oil. *Y. lipolytica* NCIM 3589 shows the highest lipid/biomass coefficient by using glucose as a carbon source whereas, the use of waste cooking oil results in a high accumulation of both saturated fatty acids and monounsaturated fatty acids in this yeast [3]. The single cell oil accumulation in a fungi contains high amounts of saturated and monounsaturated fatty acids like palmitic (C16:0), stearic (C18:0) and oleic (C18:1) acids. Yeast such as *Saccharomyces cerevisiae* and *Pichia stipitis*, fungi such as *Aspergillus niger* and *Trichoderma reesei* utilize lignocellulosic biomass hydrolysate for the accumulation of lipid in their biomass. Single cell oil accumulation in *Aspergillus terreus* IBB M1 from mangrove ecosystems has been reported. This fungus also contains lignocellulosic biomass. Other carbon sources in use could be sugarcane bagasse, grape stalk, groundnut shells, cheese whey and glucose. Though the use of glucose results in high lipid accumulation and high biomass growth, it is not preferred as a source due to its high cost [21].

The enzyme wax ester synthases obtained from *S. cerevisiae*, *Acinetobacter baylyi* ADP1, *Marinobacter hydrocarbonoclasticus* DSM 8798, *Rhodococcus opacus* PD630, *Mus musculus* C57BL/6 and *Psychrobacter arcticus* 273–4 can synthesize wax esters from alcohols and fatty acyl coenzyme. A thioesters can produce FAEE, which could be used as biodiesel. The production of FAEE could be enhanced further by up-regulation of acetyl coenzyme A carboxylase [22].

10.2.4 Production of Biodiesel from Waste Water

Beef tallow contains about 50% saturated fatty acid from its total fatty acid content, which could be utilized for biodiesel production. Another cheap source for biodiesel production is used domestic waste oil (UDWO). Like many plant oils, the problem with UDWO is its high amount of free fatty acid that limit its trans-esterification reaction because the physiochemical properties of biodiesel produced from UDWO are almost the same as that of the potential plant feedstocks in use [23]. Other wastes like municipal solid waste, agricultural waste and industrial waste cannot be used as sources for biodiesel production due to the presence of many hazardous contaminants in them [23].

Biodiesel could be produced from waste cooking oil using a mixture of TiO_2-MgO oxides having different Mg/Ti molar ratios obtained by the sol–gel method. The advantage of using titanium is that it improves the catalytic stability by substituting Ti ions in place of Mg ions in the defective magnesia lattice [24]. *Boettcherisca peregrine* larvae are capable of growing on solid organic wastes and could be used for biodiesel production [13]. *Hermetia illucens* larva and *Chrysomya megacephala* are capable of growing on organic wastes like pig manure and restaurant garbage and they could also be used for the production of biodiesel [13].

10.2.5 Production of Biodiesel from Microalgae

Microalgae have emerged as a sustainable and promising source for biodiesel production because of their rapid growth rate and high intracellular content of lipids [23]. The reasons for the choice of microalgae as the most suitable source for biodiesel production are its easy availability, a higher oil yield than traditional feedstocks and that it provides a feasible solution to the land versus fuel conflict since it can be grown away from farmlands and forests, thus providing minimal harm to the ecosystem and food chains. Microalgae are capable of growing in wastewater, fresh water and marine water too. *Isochrysis zhangjiangensis* can grow on waste water [25], *Chlorella pyrenoidosa* on the rice straw hydrolysate, *Botryococcus* spp., *Botryococcus* spp., *Botryococcus* spp. and *Botryococcus* spp. can grow in the fresh water ponds. Another fresh water microalgae *Scenedesmus* spp. LX1 can also accumulate high amounts of lipids in its biomass [26–29]. Other important species for biodiesel production are *Nannochloropsis* spp. [30] *Scenedesmus obliquus* SJTU-3, *Chlorella pyrenoidosa* SJTU-2 [31, 25] *Chlorella vulgaris, Neochloris oleoabundans, Cylindrotheca closterium* [32] *Chlorococcum* spp. [33] *Dunaliella tertiolecta, Nannochloropsis oculata* and *Dunaliella salina teodoresco* [34]. *Dunaliella salina teodoresco* is the most industrially important green halophytic microalgae because it can produce β-carotene in huge amounts. Therefore, this microalga could be used as a promising feedstock for the production of biofuels [34].

10.3   Various Processes of Biodiesel Production

10.3.1   Trans-esterification

This is the most commonly used method for the production of biodiesel from various sources because the oil obtained from plants is 10 to 20 times more viscous than petroleum diesel. Trans-esterification is an essential step which lowers the viscosity of the biodiesel oil. This is a process of reacting a triglyceride present in biodiesel feedstock oil with an alcohol (methanol/ethanol) to produce fatty acid methyl esters/fatty acid ethyl esters and glycerol by virtue of different catalysts [7, 35–37]. Herein, the triglycerides get converted into diglycerides, monoglycerides and glycerol in steps. In each step alkyl ester is produced as a by-product along with the production of one mole of FAME [38–41]. During the production of biodiesel from various sources, glycerol can be obtained as a major by-product [1, 2, 5].

The two-step trans-esterification process is essential to reduce the free fatty acid content of oil to less than 1% wherever high content of fatty acid is found in the feedstock. An example is *Madhuca indica* oil, which contains about 20% fatty acid [42, 43]. In the two-step catalysis of the feedstock oil, the first step, the pre-treatment of the oil, is done to reduce the free fatty acid content and in the second step, trans-esterification is performed [15]. By using this method, the high FFA level of *Madhuca indica* oil could be reduced to less than 1% by the two-step catalytic trans-esterification process

[15]. The feedstock oils, methanol/ethanol and FAME are observed as partially miscible two phases during trans-esterification i.e. the methanol/ethanol and oil phase [41, 44]. Based on the products of the trans-esterification reaction, the biphasic system formsthe FAME-rich phase which is the product and glycerol-rich phase which is the major by product [41, 44].

10.3.1.1 Types of Catalysis Involved in Trans-esterification

The selection among types of catalysts depends on the amount of free fatty acid content in the feedstock [45].They can be grouped into the following categories.

a. *Homogeneous acid catalysis and base catalysis*
   This method is commonly used during trans-esterification reactions but base catalysis is preferred for trans-esterification. Homogeneous base catalysis utilizes KOH, NaOH and alkali metal alkoxides such as sodium methoxide and potassium methoxide for trans-esterification reactions [35–37, 43, 46]. Although the reaction in the acid and base catalysis is very similar, they differ in reaction temperature range. The trans-esterification reaction by basic catalysis takes place efficiently between a temperature range of 25°C–125°C [35–37, 46] while in acid catalysis the trans-esterification takes place between 55°C–80°C [47–49].

   The base catalysis suffers from an undesirable reaction resulting in soponification. This happens from the reaction of the catalyst with the free fatty acids present in waste or frying oils [24]. Sometimes, together with soponification, emulsion formation also takes place leading to a lesser downstream recovery and a less purified biodiesel [24]. To solve this problem, two-step acidic catalysis using sulphuric acid & phosphoric acid and two-step basic catalysis using NaOH, KOH & $NaOCH_3$ is preferred [50]. During homogeneous catalytic trans-esterification, the glycerol produced is of low quality which increases the cost of production and the purification of the product and by-products. Basic homogenous catalysis could not be used with multiple feedstocks. Moreover, the basic homogenous catalysts like NaOH and KOH are hygroscopic in nature and contribute to environmental hazards [45].

b. Heterogeneous catalysis

Heterogeneous catalysts have higher activity, selectivity and water tolerance, although these properties depend exclusively on the amount and strength of active acid or base sites present in the catalyst to be used. It is advantageous because it provides recycling and thus allows the reuse of catalysts. In heterogeneous catalysis, solid Mg–Al hydrotalcite shows the best activity in mild reaction conditions. However, this catalyst is unstable due to the leaching out of the alkali cations from the fatty acid methyl ester phase [24, 45]. The heterogeneous catalysts in use for industrial applications of biodiesel production are depicted below [45].

The types of heterogeneous catalysts are mentioned in fig. 10.2. The heterogeneous catalysts that show good stability and good activity are Al–Zn spinel and Zr–La [24]. Alkaline earth metal oxide like CaO is the most widely used catalysts because of its capability to provide 98% FAME yield during the first cycle of reaction. The reactivity of CaO is further determined by its calcination temperature. The CaO derived from waste egg shell can be reused for 17 cycles and $Ca(OCH_3)/Ca(C_3H_7O_3)_2$ which could be reused, gives about 93% FAME yield even after 20 cycles. Moreover, $Ca(C_3H_7O_3)_2/CaCO$ could also be used for at least 5 cycles giving about 95%

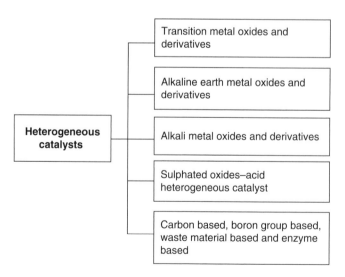

Figure 10.2 Heterogeneous catalysts in use for industrial application.

FAME yield [45]. The best catalyst in terms of yield is Nano-MgO, capable of providing 99% yield of FAME, although it cannot be reused. The industrially applicable catalyst is SrO, which has also been extensively used at the lab scale [45]. The use of activated carbon along with KOH has got 94% FAME yield and reusability for 3 cycles of reaction [45]. Na_2PEG dimethyl carbonate gives 99% FAME yield but cannot be reused. Others are boron group oxides- Chloro aluminate ionic liquid, S-ZrO_2 catalyst, ZrO_2 supported with La_2O_3 and CaO supported on mesoporous silica. During heterogeneous catalysis, the glycerol by-product is easier to separate and is of a high quality even though the rate of conversion of the triglycerides into biodiesel is slow [45].

c. *Using supercritical or subcritical fluids*
Supercritical fluids are those substances that exist as fluids above their critical temperature and pressure [7, 35–37, 46]. The catalysis of rapeseed oil using supercritical methanol has a profound effect on the rate of conversion of the oil into biodiesel at a reaction temperature of 35°C [14]. This approach for biodiesel production using supercritical or subcritical methanol is used during the trans-esterification of rapeseed oil involving the metal oxide catalysts - SrO, CaO, ZnO, TiO_2 and ZrO_2. The most efficient catalysis together with the highest catalytic activity and a minimum dissolution has been reported with ZnO whereas, SrO and CaO show high amounts of dissolution due to the conversion of the respective oxides to methoxides, namely, strontium methoxide and calcium methoxide [51]. Similarly, calcium oxide dissolution in products is seen when soybean oil is trans-esterified with refluxing methanol. When KOH/Al_2O_3 and KOH/NaY are used as catalysts, potassium gets leached out [51]. The trans-esterification of linseed oil involves supercritical fluids i.e. methanol and ethanol [7]. The alkali metal alkoxides and hydroxides for example, sodium methoxide and potassium methoxide / hydroxides of sodium and potassium are better base catalysts than the acid catalysts [35–37, 46].

d. *Bio-catalysis*
Due to the various drawbacks of the chemical trans-esterification, enzymatic trans-esterification has gained focus.

This is the most widely accepted catalysis that involves the use of immobilized lipase for the trans-esterification process. Immobilization offers an easy biodiesel separation, minimal wastewater treatment after biodiesel production, easy glycerol recovery, which is the major by-product obtained during trans-esterification, and no unwanted side reactions when pure immobilized lipase is used [52, 53]. The non-edible oil from *Jatropha carcus* undergoes solvent free methanolysis in batches and continuous bioreactors loaded with *Burkholderia cepacia* lipase-immobilized in n-butyl substituted hydrophobic silica monoliths prepared by sol-gel method. This approach makes the reaction mixture homogeneous and a very high yield of biodiesel is obtained at an optimal molar ratio of methanol:oil to be 3.3:3.5. This technology enables the continuous production of biodiesel using packed-bed bioreactors. Similarly, immobilized *Candida antarctica* B lipase is also used. The immobilized supports used for immobilization are diatomaceous earth polypropylene, mesoporous silica, kaolinite and silica-polyvinyl alcohol composite [53]. Lipase immobilized on a polymer based monolithic support in both organic solvents and their biphasic mixtures with water is also used [54]. The enzymatic production of biodiesel has been proposed to overcome the drawbacks of conventional chemically catalyzed processes. The problem with the use of lipase mediated trans-esterification is the high cost of the enzyme that is commercially available. Therefore, reuse of lipase is essential from an economic point of view. Hence, this enzyme is immobilized on an inert support for enabling its reuse together with increased enzyme stability and activity. For this purpose, lipase obtained from *Candida rugosa* DSM 70761 and *Yarrowia lipolytica* ATCC 8661 has been immobilized. Furthermore, the adsorption of lipase on celite increases the activity of *Candida rugosa* DSM 70761 lipase, more than that of *Yarrowia lipolytica* ATCC 8661 [55]. The semi-organic fluids consisting entirely of ions in liquid below 100°C are known as ionic liquids. They provide fast reaction rates and high yields in enzyme based trans-esterification processes. These eutectic solvents show physiochemical properties like low viscosity, high biodegradability and compatibility with NovozymR 435, a commercial immobilized *Candida antarctica* lipase [56].

e. Catalyst free trans-esterification
This technique is a modified form of the trans-esterification reaction. In this technique, supercritical methanol mediated trans-esterification involves catalyst free trans-esterification so that the purification and separation of catalysts from the product is quickly achieved. This method also takes less time than involving alkali in catalysis, although this reaction requires harsh reaction conditions [51]. Cottonseed is used for the production of biodiesel by using non-catalytic supercritical fluids [7].

Companies like Pacific Biodiesel, Lurgi and Desmet Ballestra are using alkaline catalysis. Pacific Biodiesel has come up with a solution to low yield due to free fatty acids. They have used feedstock with approximately 15% free fatty acid content without affecting the yield of the biodiesel produced by this process [57, 58]. Besides commonly used alkaline catalysis, Axens Technologies have commercialized a heterogeneous catalysis based process that was developed by the French Institute of Petroleum (IFP). In Axens Esterfip-H, a spinal mixed oxide of zinc and aluminium (zinc aluminate) acts as a heterogeneous catalyst for trans-esterification such that a high quality biodiesel is produced and the glycerol gets separated from FAME with ease. However, virgin oil or refined vegetable can be used in this process [59, 60]. The advantages and disadvantages of various types of catalysis involved in transesterification are described in table 10.1.

10.3.1.2   Factors Affecting Biodiesel Production by Trans-esterification

The factors which affect the rate of trans-esterification are reaction condition, molar ratio of alcohol (methanol/ethanol) to oil, catalyst used, physical parameters such as temperature and pressure, free fatty acid content and the amount of water present in the feedstock. While producing biodiesel, the alcohols i.e. methanol and to a lesser extent ethanol, work as acyl acceptors. Other alcohols such as propanol, butanol, isopropanol, tert-butanol, branched alcohols

Table 10.1 Advantages and disadvantages of different types of catalysis.

Type of catalysis	Advantages	Disadvantages	References
Homogeneous acid catalysis and base catalysis	Trans-esterification few hours	Difficulty in removal of catalyst from product during purification	[51]
Heterogeneous catalysis	Provides recycling and thus reuse of the catalyst, lower cost than homogeneous catalyst	Leaching out of the alkali cations from FAME	[24]
Supercritical/ Subcritical fluids with metal oxides	Suitable for biodiesel production from waste cocking oil	Dissolution of catalyst	[51]
Bio catalysis	Reuse of catalyst	Costly	[52]
Catalyst free trans esterification	Fast purification and separation	Requires harsh reaction conditions	[51, 7, 61]

and octanol could also be used but are not preferred due to their high costs. The choice of using methanol over ethanol is that FAME produced by use of methanol is more volatile, more viscous, exhibits a lower cloud point and pour point [62, 63, 64]. Karanja oil is also used for the production of biodiesel by trans-esterification using basic a catalyst i.e. KOH (0.25 – 1.5%) and with a molar ratio of MeOH/oil being 6:1 [64]. *Boettcherisca peregrine larva*, capable of growing on solid organic wastes, could be used for biodiesel production by the two-step trans-esterification process using a methanol to oil molar ratio of 12:1 in the presence of H_2SO_4 (1.5% w/w) as an acid catalyst. In the next step, triglycerides are converted to esters by alkaline trans-esterification [13]. Processed *Jatropha curcas* seed oil that has a high content of free fatty acids could be reduced by a treatment of oil with methanol to oil ratio (0.60 w/w) in the presence of acid catalysis H_2SO_4 (1% w/w) [12]. *Raphanus sativus,* which is a perennial plant of the Brassicaceae family, is used for the production of biodiesel by using a molar ratio of ethanol and oil of 6:1. Herein, ethanol is used as the alcoholysis agent during trans-esterification [65]. These eutectic solvents show physiochemical properties like low viscosity, high biodegradability and compatibility with NovozymR 435, a commercial immobilized *Candida antarctica* lipase when trans-esterification is performed with a molar ratio of choline acetate/glycerol of 1:1.5 [56].

The oils are trans-esterified so that their viscosity gets reduced, when the triglycerides of oils are converted into esters [64] however, the trans-esterification process does not occur if the free fatty acid content of oil is more than 3% [48, 66]. Although biodiesel has more lubricity due to the presence of free fatty acids, monoglycerides and diglycerides persist with the problem of oxidation and corrosion of the diesel engine [14, 67, 68, 69]. However, this problem could be solved by blend of 2% diesel oil [68]. Furthermore, the blending of methanol improves its viscosity whereas the addition of ethanol in blends reduces fuel consumption. It is known that biodiesel reduces emissions of unburnt hydrocarbons, carbon monoxide and particulates. In contrast, NOx emissions increase which affect the diesel engine. This could be removed by modification in the diesel engines i.e. retarding the injection timing of fuel and increasing exhaust gas recirculation [69, 70].

The advantages of biodiesel as a preferred source for fuel are its higher combustion efficiency, greater flash point, less sulphur content and less aromatic content and higher cetane number than that of the diesel fuel [71, 72]. The cetane number, which is the measure of auto-ignition of the fuel when injected into the engine, is used for the determination of the ignition quality of diesel fuel. The cetane number and the ignition of biodiesel depend upon the FAME component. Since it is known that biodiesel is more viscous than diesel, it leads to a poorer atomization of the fuel spray in the fuel injection system [73, 74]. Palm oil has appeared as a potential source for the production of biodiesel at a low temperature of 30°C. Since there are problems with mass transfer at 30°C due to the viscosity of oil, still, the optimization of enzyme loading, agitation speed, reaction time and methanol to oil substrate molar ratio of 6.5:1 could produce a maximum of 89.29% FAME yield. T-Butanol, methanol and hydrophobic oil form a homogeneous reaction mixture which reduces interphasic mass transfer resistance and enhances the viscosity of the mixture, whereas the external mass transfer resistance observed could be reduced by increasing the agitation speed in the batch reactor [63]. Biodiesel exhibits higher heating values (HHVs) as 39–41 MJ/kg, which is lesser than the higher heating values of petroleum diesel - 43 MJ/kg [49]. The cetane number of biodiesel varies according to the feedstock utilized for the biodiesel production. For example, the cetane number of grape biodiesel is 48 whereas that of palm biodiesel is 61 [75].

It has been described that water content affects the trans-esterification process and yields of biodiesel [76]. It has been suggested that the higher the biodiesel cetane number, the more readily the fuel is ignited. This results in increased NOx emissions in biodiesel because biodiesel is a fuel that has higher oxygen content than diesel but leads to a lesser calorific value of biodiesel than diesel. This is coupled with power losses and more fuel consumption, which increases the water and carbon dioxide content [77–79]. The blending of soybean oil biodiesel could be done with short chain methyl esters having fatty acids - caprilic (C8:0) and capric (C10:0). The presence of these short chain methyl esters decreases the cold flow properties and NOx emissions. Furthermore, the iodine numbers increase by converting linoleic acid (C18:2) and linolenic (C18:3) acids to oleic acid (C18:1) [80]. Biodiesel is used as a blend B20 having 20% biodiesel and 80% petroleum derived diesel rather than B100 having 100% biodiesel. Others are B30, B5 and lubricate additive B2. Amongst them the B5 blend does not require modification in conventional diesel engines [80].

The presence of metallic particles like iron, copper as well as the presence of water/moisture in the biodiesel leads to the corrosion of diesel engines. This is because the auto-oxidation of biodiesel forms formic acid, acetic acid, propionic acid and caproic acid, which are responsible for the enhanced corrosion and degradation of fuel properties as well as an alteration in fuel properties like density and the viscosity of biodiesel [81–84]. Biodiesel from *Jatropha curcas* and *Salvadora* contain ferrous and non-ferrous metals that also corrode biodiesel engines [68, 83, 85, 86].

10.3.2   Micro-emulsion

It has been reported that the micro-emulsion of methanol with vegetable oils can be used as biodiesel [87] however, micro-emulsions give a low quality biodiesel [88–90]. The disadvantage associated with it, is that the carbon particles present in it provide irregular injector needle sticking [88].

10.3.3   Thermal Cracking (Pyrolysis)

Pyrolysis is the thermal decomposition of any organic material in the absence of oxygen. For biodiesel production, pyrolysis is carried out by the thermal decomposition of triglycerides present in vegetable oil, animal fat and natural fatty acids, which could be used to produce biodiesel [91, 92]. Pyrolysis also gives a low quality biodiesel like micro emulsion [88–90]. Biodiesel produced by this method has a high cetane number, and an accepted amount of sulphur is present in its exhausts. However, water, sediment and metal contaminants of the biodiesel can corrode conventional diesel engines and affect the cloud point of fuel [88].

10.3.4   Strategies for the Production of Biodiesel from Microalgae

Microalgae based biodiesel production provides a solution for CO_2 fixation during the mass cultivation of microalgae in high rate raceway ponds and photo-bioreactors [93]. The raceway ponds are open systems whereas the photo-bioreactors are closed systems [23] although, microalgae need to be metabolically engineered to produce molecules with high energy content [94]. *Chlorella* strains like *Chlorella protothecoides, Chlorella vulgaris, Chlorella emersonii and Chlorella sorokiniana* can grow well in fresh water [11]. The increased productivity of *Chlorella vulgaris* derived biodiesel is possible only when the biomass productivity as well as an increased amount of storage lipids could be possible. In this respect, using chelated $FeCl_3$ in the growth media in the late growth phase increases final cell density but it does not show any induction of lipid accumulation in the algal cells [11]. *Isochrysis zhangjiangensis*, which grows in wastewater, shows 53% lipid at a high nitrate concentration (9g/l) after 24 hrs of culture [25]. The optimization of the reaction conditions for higher lipid productivity, higher growth rate, and oil content shows 50% of lipid accumulation together with a high biomass generation of (124 mg/l) in the microalgae in nitrogen deprived culture condition having fatty acid profile to be methyl palmitate (C16:0), methyl oleate (C18:1), methyl linoleate (C18:20) and methyl linolenate (18:3) [25]. The *Nannochloropsis* spp. of microalgae can grow best in the combined conditions of 13 g/l NaCl and light intensity of 700 µmol photons/m^2s to produce the highest biomass (47% DW) and average lipid productivity of 360 mg total fatty acid/

day. Hence, these species of microalgae are another potential source for the production of biodiesel [25]. *Chlorella pyrenoidosa* is a species of microalgae that can grow well on the rice straw hydrolysate and could be a lignocellulosic source for the growth of microalgae to obtain a maximum biomass concentration of 2.83 g/l after 48 hrs of growth with a high lipid content of 56.3%. Further, after the optimization of reaction conditions with 6 ml of n-hexane and 4ml of methanol with 0.5M sulphuric acid at 90°C after 2 hours of reaction, 99% of methyl ester and 95.5% biodiesel yield were obtained. Another medium suitable for the production of microalgae biomass was a single stage culture of *Chlorella vulgaris* ESP-31 on basal medium with low nitrogen source in medium (0.313 g/l KNO_3) to have a highest lipid productivity of 78 mg/l/d. The lipid productivity was further enhanced by the growth of the algae in a tubular photo-bioreactor to have a 6.36 g CO_2 fixation in 10 days of culturing with a CO_2 fixation rate of 430 mg/l/d having 65% saturated fatty acids as palmitic acid (C16:0), stearic acid (C18:0) and mono-unsaturated fatty acid as oleic acid (C18:1). Four green microalgae *Botryococcus* spp. TRG, *Botryococcus* spp. KB, *Botryococcus* spp. SK and *Botryococcus* spp. PSU were isolated from freshwater lakes of southern Thailand and possess a high lipid content of 35.9%, 30.2%, 28.4% and 14.7% respectively, when grown with acceptable light intensity and a high iron concentration (0.74 mM). Another fresh water microalgae *Scenedesmus sp.* LX1 can accumulate 30–50% lipids in the nitrogen or phosphorous depleted conditions of 2.5 mg/l or 0.1 mg/l respectively. On the other hand, this species shows a high biomass and lipid productivity of 0.47 g/l and 139 mg/l, respectively, with 30% lipid content per biomass. The TAGs content per lipid (~20% w/w) and TAGs productivity (~23 mg/l) were increased by 79% and 40%, respectively, when the culture was fed with ethyl-2-methyl acetoacetate. It could be suggested that the use of rice straw and KNO_3 or a high concentration of iron (0.74 mM) along with the feeding of a culture with ethyl-2-methyl acetoacetate together with balanced light intensity is beneficial for the production of high biomass from microalgae together with high lipid content [24, 25, 28, 29].

The use of flu gas from coal utilizing power stations (to reduce the cost of production) and direct trans-esterification (use of SrO as catalyst) using microwave and ultrasound radiations (to harvest biomass by the simplest and efficient process) are the strategies to produce biodiesel from *Nannochloropsis* spp. [30]. The microalgae *Scenedesmus obliquus* SJTU-3 and *Chlorella pyrenoidosa* SJTU-2 show the best growth in a 10% CO_2 concentration where the CO_2 bio-fixation rate is 84 g/l and 0.288 g/l/d for *S. obliquus* SJTU-3 and 1.55 g/l and 0.260 g/l/d for *C. pyrenoidosa* SJTU-2. More than 94% of the fatty acid constitutes, mainly, C16-C18,Herein the fatty acid profile is favorable for biodiesel production. A much higher level of CO_2 is favored for the accumulation of lipids and polyunsaturated fatty acids in the organism. Hence, these two microalgae species favor bio-mitigation in the production of biodiesel [25, 31]. The *Nannochloropsis* spp., can produce biodiesel on an industrial scale [95]. The mixed cultivation of microalgae obtained from various domestic wastewater bodies, when used for biodiesel production through acid catalyzed trans-esterification reactions, shows high lipid accumulation with a high portion of palmitic acid (C16:0) together with 33 different types of saturated and unsaturated fatty acids [96]. *Chlorella vulgaris*, *Neochloris oleoabundans* and a sea water diatom strain *Cylindrotheca closterium* can grow in photobioreactors under nitrogen depleted culture condition keeping in consideration the acceptable light intensity and the need of nitrates, phosphates and sulphates in the medium. The photo-bioreactor shows highest lipid content favorable for biodiesel production by *Neochloris oleoabundans* to be 25%–37% of DW. On the other hand, the highest TAG content was seen in *Chlorella vulgaris*, 11%–14% of DW [32]. The use of super critical carbon dioxide (SCCO$_2$) and hexane to extract lipids from the marine microalgae *Chlorococcum* spp. for the lab scale production of biodiesel shows a low lipid yield, but the fatty acid profile of the microalgae is suitable for biodiesel production. The fatty acid profile of this microalga is C18:1 (~63 %), C16:0 (~19%), C18:2 (~4%), C16:1 (~4%) and C18:0 (~3%). On the other hand, a decrease in the temperature and an increase in the pressure result in an increase in the yield of lipids. Whereas, the use of isopropanol as a co-solvent enhances the lipid yield when hexane is used as super critical fluid for extraction. The mass transfer coefficient for the extraction of lipids is also found to increase with the dielectric constant of the fluid as well as with the density of the fluid [33].

10.3.5 Production of Biodiesel by Heterotrophic Cultivation

A high quality biodiesel can be obtained by the heterotrophic cultivation of *C. protothecoides*, *Nitzschia communis*, *Botryococcus braunii*, *Dunaliella* and diatom *Chaetoceros muelleri* by about 50%–60%

of lipid accumulation in their biomass. This lipid content of the algal species could be enhanced further by silicon deficiency, phosphate limitation, high salinity and some heavy metal stress, such as cadmium. It has been observed that the co-immobilization of *Azospirillum brasilense* on the alginate beads results in a significant increase in its lipid content [11].

Another approach to producing high quality biodiesel by heterotrophic cultivation of microalgae *Chlorella protothecoids* is the use of corn powder hydrolysate as a carbon source instead of glucose. The extraction of the algal oil from the heterotrophic cultivated cells could be achieved by the extraction of oil with an *n*-hexane as a solvent and then acidic trans-esterification of the extracted oil with acid base catalysis [96, 97].

10.4   Determination of Yield, Process Optimization and Biodiesel Standardization

The optimal conditions for maximum biodiesel production depend on the type of feedstock i.e. the vegetable oil or animal fat in use. Response surface methodology is used to determine the yield as well as the optimal operational conditions for transesterification reactions to have the highest biodiesel yields. In this method, trans-esterification using KOH and the desired concentration of methanol is carried out by using different vegetable oils covering a wide range of fatty acid compositions and iodine values i.e. sunflower oil, maize oil, olive-pomace oil, linseed oil and palm oil [98]. Similarly, response surface methodology is used for the optimization of conditions for transesterification by ethanolysis of *R. sativus* crude oil for biodiesel production at moderate temperatures [65].

There are two international standards for biodiesel standardization, namely, the American standard - ASTM D6751 and the European Standard - EN14214. Although the parameters for both of them are quite similar, the difference lies in their probable application and the test methods used for them according to those given in European Committee for Standardization (CEN) and American Society for Testing and Materials, respectively. EN 14214 is used to specify the test methods for biodiesel for running diesel engines. This standard is further confirmed by the European Automotive Diesel Standard-EN-590, whereas, the American Standard - US

ASTM D6751 specifies the standards of biodiesel B100 for biodiesel blending with diesel fuel [99].

The biodiesel manufacturing cost includes the cost of feedstock, purification, operation and production [100]. However, the cost of biodiesel production accounts for about 75% of the total production cost [3]. The prolonged use of biodiesel from food sources, such as those that are produced from plants, has led to a food versus fuel conflict [48]. The use of microalga offers advantages over the use of plant sources due to its ease of growth in bioreactors, short life cycles, displays of rapid growth rates, non existent effect on space, ease of scaling up and the fact that it does not contribute to climate change [21]. *Pseudallescheria boydii* has been isolated from sewage sludge and from clinical sources. It is able to grow well on biodiesel agar but the production of biodiesel for such sources can result in the contamination of the environment with such pathogenic organisms [101].

Diesel cars utilize a biodiesel–diesel blend of up to 5% v/v. Furthermore, the conversion of a conventional diesel engine for pure biodiesel utilization and efficient usage costs 1,500€ per car in Germany. Hence, whether biodiesel driven cars can run on conventional diesel engines without any modification to the diesel engine is still another issue from the commercialization point of view [12]. In lieu of it, though, it has been observed that specific volume, isothermal compressibility, and cubic expansion coefficients of refined sunflower methyl ester oil and unrefined sunflower methyl ester oil as compared to diesel oil do not show much difference. Furthermore, plant oils obtained from the oilseed of *Lesquerella* spp. pose a unique fatty acid profile, having high levels of the hydroxy fatty acid - ricinoleic acid contributing to an increased lubricity of biodiesel obtained, hence it is a solution to the wear and tear on the injection system in a diesel engine because of the decreased lubricacy of the diesel fuel due to its less sulphur content [102, 103]. However, there is a need for further study to find an accurate cause and not the probable cause of wear and tear by the biodiesel in the diesel engine [97]. The biodiesel produced by growing palm oil grown on dried peat marsh is found to increase the amount of greenhouse gases in the atmosphere [104]. The *Nannochloropsis* spp., can produce biodiesel on an industrial scale when estimated to fulfill the industrial scale set up for biodiesel production together with greenhouse gases emission [95]. Biodiesel refineries show methanol to be the most difficult contaminant in production areas. Similarly, the palm oil industry in Malaysia is consuming more than 13 million tons of crude palm oil per annum to manufacture biodiesel. The biodiesel manufacturing waste effluent discharges and biodiesel spillage near the places of biodiesel production, increases biological oxygen demand and chemical oxygen demand in those areas. however, palm oil effluent could also serve as a source for the production of renewable bio-energies like bio ethanol and bio hydrogen [105, 106].

10.5   Conclusion

The production of biodiesel from sustainable feedstock is still an issue. The feasible solution to commercial and cost effective bio diesel production globally seems to be microalgal based biodiesel production in marine waters, as per the easy availability and suitable conditions for growth by either autotrophic or heterotrophic cultivation. Other sources such as plant oil, fungi and bacteria, which are in use for biodiesel production, suffer many drawbacks. However, the production of biodiesel from wastewater must be promoted. Moreover, we have to explore possible solutions for the wear and tear on the traditional diesel engine when using biodiesel as the main fuel. We need to incorporate proper modification of diesel engines in order to prevent them from any wear and tear. See section 10.4 where the need for such a modification is explained.

References

1. Y. Chisti, *Biotechnology Advances*, Vol. 25, p. 294–306, 2006.
2. A. Singh, P.S. Nigam, and J.D. Murphy, *Bioresource Technology*, Vol. 102, p. 10–16, 2011.
3. G.,Katre, C. Joshi, M. Khot, S. Zinjarde, and A.R. Kumar, *AMB Express*, Vol. 2, p. 36, 2012.
4. Y. Duan, Z. Zhu, K. Cai, X. Tan, and X. Lu, *PLoS ONE*, Vol. 6, p. e20265, 2011.
5. M. Hajek, F. Skopal, J. Kwiecien, and M. Cernoch, *Talanta*, Vol. 82, p. 283–285, 2010.
6. C.C. Akoh, S.W. Chang, G.C. Lee, and J.F. Shaw, *Journal of Agriculture and Food Chemistry*, Vol. 55, p. 8995–9005, 2007.
7. A. Demirbas, *Bioresource Technology*, Vol. 99, p. 1125–1130, 2008.
8. A. Demirbas, *Energy Conversion and Management*, Vol. 44, p. 2093–2109, 2003.
9. D.O. Edem, *Plant Foods for Human Nutrition*, Vol. 57, p. 319–341, 2002.

10. H. Raheman, and A.G. Phadatare, *Biomass and Bioenergy*, Vol. 27, p. 393–397, 2004.
11. Z.Y. Liu, G.C. Wang, and B.C. Zhou, *Bioresource Technology*, Vol. 99, p. 4717– 4722, 2008.
12. H.J. Berchmans, and S. Hirata, *Bioresource Technology*, Vol. 99, p. 1716–1721, 2008.
13. S. Yang, Q. Li, Q. Zeng, J. Zhang, Z. Yu, *et al. PLoS ONE*, Vol. 7, p. e45940, 2012.
14. S. Jain, and M.P. Sharma, *Bioresource Technology*, Vol. 101, p. 7701–7706, 2010.
15. M. Balat, *Energy Conversion and Management*, Vol. 52, p. 1479–1492, 2011.
16. D. Agarwal, and A.K. Agarwal, *Applied Thermal Engineering*, Vol. 27, p. 2314– 2323, 2007.
17. M. Debnath, and H.N. Verma, *Africal Journal of Biotechnology*, Vol. 7, p. 613– 616, 2008.
18. A. Yousuf, F. Sannino, V. Addorisio, and D. Pirozzi, *Journal of Agricultural and Food chemistry*, Vol. 58, p. 8630–8635, 2010.
19. D. Antoni, V.V. Zverlov, and W.H. Schwarz, *Applied Microbiology and Biotechnology*, Vol. 77, p. 23–35, 2007.
20. M. Enshaeieh, A. Abdoli, I. Nahir, and M. Madani, *Journal of Biology and today's world*, Vol. 1, p. 82–92, 2012.
21. M. Khot, S. Kamat, S. Zinjarde, A. Pant, B. Chopade, *et al.*, *Microbial Cell Factories*, Vol. **11**, p. 71, 2012.
22. S. Shi, J.O. Valle-Rodríguez, S. Khoomrung, V. Siewers, and J. Nielsen, *Biotechnology for Biofuels*, Vol. 5, p. 7, 2012.
23. J. Liu, J. Huang, Z. Sun, Y. Zhong, Y. Jiang, *et al.*, *Bioresource Technology*, Vol. 102, p. 106–110, 2011.
24. Z. Wen, X. Yu, S.T. Tu, J. Yan, and E. Dahlquist, *Bioresource Technology*, Vol. 101, p. 9570–9576, 2010.
25. D. Tang, W. Han, P. Li, X. Miao, and J. Zhong, *Bioresource Technology*, Vol. 102, p. 3071–3076, 2011.
26. K.L. Yeh, and J.S. Chang, *Biotechnology Journal*, Vol. 6, p. 1358–1366, 2011.
27. P. Li, X. Miao, R. Li, and J. Zhong, *Journal of Biomedicine and Biotechnology*, Vol. 2011, p. 141207, 2011.
28. C. Yeesang, and B. Cheirsilp, *Bioresource Technology*, Vol. 102, p. 3034–3040, 2011.
29. L. Xin, H. Hong-Ying, Y. Jia, and W. Yin-Hu, *Bioresource Technology*, Vol. 101, p. 9819–9821, 2010.
30. M. Koberg, M. Cohen, A. Ben-Amotz, and A. Gedanken, *Bioresource Technology*, Vol. 102, p. 4265–4269, 2011.
31. P.J McGinn, K.E. Dickinson, S. Bhatti, J.C. Frigon, S.R. Guiot, *et al.*, *Photosyntheis Reearch*, Vol. 109, p. 231–247, 2011.
32. J. Pruvost, G. Van Vooren, B. Le Gouic, A. Couzinet-Mossion, and J. Legrand, *Bioresource Technology*, Vol. 102, p. 150–158, 2011.
33. R. Halim, B. Gladman, M.K. Danquah, and P.A. Webley, *Bioresource Technology*, Vol. 102, p. 178–185, 2011.
34. Smith, D.R. Lee, R.W., Cushman, J.C., Magnuson, J.K., Tran, D., *et al.*, *BMC Plant Biology*, Vol. 10, p. 83, 2010.
35. F. Ma, and M.A. Hanna, *Bioresource Technology*, Vol. 70, p. 1–15, 1999.

36. A. Duran, M. Lapuerta, and J. Rodriuez-Fernandez, *Fuel*, Vol. 84, p. 2080–2085, 2005.

37. K. Maeda, H. Kuramochi, T. Fujimoto, Y. Asakuma, K. Fukui, *et al.*, *Journal of Chemical Engineering Data*, Vol. 53, p. 973–977, 2008.

38. A.W. Schwab, M.O. Bagby, and B. Freedman, *Fuel*, Vol. 66, p. 1372–1378, 1987.

39. F. Ferella, G. MazziottiDiCelso, I. DeMichelis, V. Stanisci, and F. Vegli, *Fuel*, Vol. 89, p. 36–42, 2010.

40. G. Vicente, M. Martinez, and J. Aracil, *Bioresource Technology*, Vol. 92, p. 297–305, 2004.

41. C. Stavarache, M. Vinatoru, R. Nishimura, and Y. Maeda, *Ultrasonics Sonochemistry*, Vol. 12, p. 367–372, 2005.

42. S.V. Ghadge, and H. Raheman, *Biomass Bioenergy*, Vol. 28, p. 601–605, 2005.

43. H. Raheman, and S.V. Ghadge, *Fuel*, Vol. 86, p. 2568–2573, 2007.

44. W. Zhou, and D.G. Boocock, *Journal of the American Oil Chemists Society*, Vol. 83, p. 1041–1045, 2006.

45. A.P.S. Chouhan, and A.K. Sarma, *Renewable and Sustainable Energy Reviews*, Vol. 15, p. 4378–4399, 2011.

46. A.S. Ramadhas, S. Jayaraj, and K.L.N. Rao, Experimental investigation on non-edible vegetable oil operation in diesel engine for improved performance. In: National conference on advances in mechanical engineering, J.N.T.U., Anantapur, India, 2002.

47. A.H. West, D. Posarac, and N. Ellis, *Journal of Bioresource Technology*, Vol. 99, p. 6587–6601, 2008.

48. J.M. Marchetti, V.U. Miguel, and A.F. Errazu, *Renewable and Sustainable Energy Reviews*, Vol. 11, p. 1300–1311, 2005.

49. N.N.A.N. Yusuf, S.K. Kamarudin, and Z. Yaakub, *Energy Conversion and Management*, Vol. 52, p. 2741–2751, 2011.

50. F.L. Pua, Z. Fang, S. Zakaria, F. Guo, and C.H. Chia, *Biotechnology for Biofuels*, Vol. 4, p. 56, 2011.

51. S.J. Yoo, H.S. Lee, B. Veriansyah, J. Kim, J.D. Kim, *et al.*, *Bioresource Technology*, Vol. 101, p. 8686–8689, 2010.

52. A. Karmakar, S. Karmakar, and S. Mukherjee, *Bioresource Technology*, Vol. 101, p. 7201–7210, 2010.

53. K. Kawakami, Y. Oda, and R. Takashashi, *Biotechnology for Biofuels*, Vol. 4, p. 42, 2011.

54. J. Urban, F. Svec, and J.M. Fréchet, *Biotechnology and Bioengineering*, Vol. 109, p. 371–380, 2012.

55. L. Tcacenco, A.A. Chirvase, C. Ungureanu, and E. Berteanu, *Romanian Biotechnological Letters*, Vol. 15, p. 5631–5639, 2010.

56. H. Zhao, G.A. Bakerb, and S. Holmesa, *Organic and Biomolecular Chemistry*, Vol. 9, p. 1908–1916, 2011.

57. Pacific biodiesel, http://www.biodiesel.com/index.php/technologies/biodieselprocess technology, 2013.

58. B. Freedman, E.H. Pryde, and T.L. Mounts, *Journal of the American Oil Chemists Society*, Vol. 61, p. 1638–1643, 1984.

59. M. DiSerio, M. Cozzolino, M. Giordano, R. Tesser, P. Patrono, *et al.*, *Industrial and Engineering Chemistry Research*, Vol. 46, p. 6379–6384, 2007.

60. J.A. Melero, J. Iglesias, and G. Morales, *Green Chemistry*, Vol. 11, p. 1285–1308, 2009.
61. A. Demirbas, *Bioresource Technology*, Vol. 99, p. 1125–1130, 2008.
62. K. Bozbas, *Renewable & Sustainable Energy Reviews*, Vol. 12, p. 542–552, 2008.
63. J.H. Sim, A.H. Kamaruddin, and S. Bhatia, *Bioresource Technology*, Vol. 101, p. 8948–8954, 2010.
64. L.C. Meher, V.S.S. Dharmagadda, and S.N. Naik, *Bioresource Technology*, Vol. 97, p. 1392–1397, 2006.
65. A.K. Domingos, E.B. Saad, H.M. Wilhelm, and L.P. Ramos, *Bioresource Technology*, Vol. 99, p. 1837–1845, 2008.
66. A.A. Refaat, N.K. Attia, H.A. Sibak, S.T.E. Sheltawy, and G.I.E. Diwani, *International Journal of Environmental Science and Technology*, Vol. 5, p. 75–82, 2008.
67. M. Sgroi, G. Bollito, G. Saracco, and S. Specchia, *Journal of Power Sources*, Vol. 149, p. 8–14, 2005.
68. T. Tsuchiya, H. Shiotani, S. Goto, G. Sugiyama, A. and Maeda, "Japanese standards for diesel fuel containing 5% FAME blended diesel fuels and its impact on corrosion". *SAE Technical Paper No.* 2006-01-3303, 2006.
69. J. Hu, Z. Du, C. Li, and E. Min, *Fuel*, Vol. 84, p. 1601–1606, 2005.
70. S.K. Hoekman, and C. Robbins, *Fuel Processing Technology*, Vol. 96, p. 237–249, 2012.
71. A. Demirbas, *Energy Source Part B: Economics, Planning, and Policy*, Vol. 4, p. 310–314, 2009.
72. M.F. Demirbas, and M. Balat, *Energy Conversion and Management*, Vol. 47, p. 2371–2381, 2006.
73. A.I. Bamgboye, and A.C. Hansen, *International Agrophysics*, Vol. 22, p. 21–29, 2008.
74. M. Balat, *Energy Source Part A: Recovery, Utilization, and Environmental Effects*, Vol. 30, p. 1856–1869, 2008.
75. M.J. Ramos, C.M. Fernandez, A. Casas, L. Rodriguez, and A. Perez, *Bioresource Technology*, Vol. 100, p. 261–268, 2009.
76. I.M. Atadashi, M.K. Aroua, A.R. Abdul Aziz, and N.M.N. Sulaiman, *Renewable and Sustainable Energy Reviews*, Vol. 16, p. 3456–3470, 2012.
77. G.A. Ban-Weiss, J.Y. Chen, B.A. Buchholz, and R.W. Dibble, *Fuel Processing Technology*, Vol. 88, p. 659–667, 2007.
78. D. Lance, C. Goodfellow, J. Williams, W. Bunting, I. Sakata, *et al.*, "The impact of diesel and biodiesel fuel composition on a euro V HSDI engine with advanced DPNR emissions control". SAE Technical Paper No. 2009-01-1903, 2009.
79. A. Tsolakis, A. Megaritis, M.L. Wyszynski, and K. Theinnoi, *Energy*, Vol. 32, p. 2072–2080, 2007.
80. G. Knothe, C.A. Sharp, and T.W. Ryan, *Energy Fuels*, Vol. 20, p. 403–408, 2006.
81. M.A. Jakab, S.R. Westbrook, and S.A. Hutzler, Testing for compatibility of steel with biodiesel. Southwest research institute Project No. 08.13070, 2008.
82. L. Diaz-Ballote, J.F. Lopez-Sansores, L. Maldonado-Lopez, and L.F. Garfias-Mesias, *Electrochemistry Communications*, Vol. 11, p. 41–44, 2009.
83. J. Kaminski, and K.J. Kurzydlowski, *Journal of Corrosion Measurements*, Vol. 6, p. 1–5, 2008.

84. M.N. Nabi, M.M. Rahman, and M.S. Akhter, *Applied Thermal Energy,* Vol. 29, p. 2265–2270, 2009.
85. M.A. Fazal, A.S.M.A. Haseeb, and H.H. Masjuki, *Fuel Processing Technology,* Vol. 91, p. 1308–1315, 2010.
86. A.S.M.A. Haseeb, H.H. Masjuki, L.J. Ann, and M.A. Fazal, *Fuel Processing Technology,* Vol. 91, p. 329–334, 2010.
87. A. Srivasta, and R. Prasad, *Renew Sustain Energy Reviews,* Vol. 4, p. 111–133, 2000.
88. H. Fukuda, A. Kondo, H. Noda, *Journal of Bioscience and Bioengineering,* Vol. 92, p. 405–416, 2001.
89. Z. Helwani, M.R. Othman, N. Aziz, W.J.N. Fernando, and J. Kim, *Fuel Processing Technology,* Vol. 90, p. 1502–1514, 2009.
90. H. Taher, S. Al-Zuhair, A.H. Al-Marzouqi, Y. Haik, and M.M.A. Farid, *Enzyme Research,* p. 1–25, 2011.
91. K. Pramanik, *Renewable Energy,* Vol. 28, p. 239–248, 2003.
92. M. Canakci, A.N. Ozsezen, E. Arcaklioglu, and A. Erdil, International Journal of Vehicle Design, Vol. 50, p. 213–228, 2009.
93. B. Sialve, N. Bernet, and O. Bernard, *Biotechnology Advances,* Vol. 27, p. 409–416, 2009.
94. D. Feng, Z. Chen, S. Xue, and W. Zhang, *Bioresource Technology,* Vol. 102, p. 6710–6716, 2011.
95. L. Batan, J. Quinn, B. Willson, and T. Bradley, *Environmental Science & Technology,* Vol. 44, p. 7975–7980, 2010.
96. B.J. Krohn, C.V. McNeff, B. Yan, and D. Nowlan, *Bioresource Technology,* Vol. 102, p. 94–100, 2011.
97. H. Xu, X. Miao, and Q. Wu, *Biotechnology Advances,* Vol. 126, p. 499–507, 2006.
98. S. Pinzi, F.J. Lopez-Gimenez, J.J. Ruiz, and M.P. Dorado, *Bioresource Technology,* Vol. 101, p. 9587–9593, 2010.
99. I.M. Atadashi, M.K. Aroua, and A.A. Aziz, *Renewable and Sustainable Energy Reviews,* Vol. 14, p. 1999–2008, 2010.
100. Y. Zhang, M.A. Dube, D.D. McLean, and M. Kates, *Bioresource Technology,* Vol. 90, p. 229–240, 2003.
101. K. Janda-Ulfi, K. Ulfi, J. Cano, and J. Guarro, *Annuals of Agricultural and Environmental Medicine,* Vol. 15, p. 45–49, 2008.
102. J.W. Goodrum, and D.P. Geller, *Bioresource Technology,* Vol. 96, p. 851–855, 2005.
103. C. Aparicio, B.R. Guignon, I.M.R. Antoan, and P.D. Sanz, *Journal of Agricultural and Food chemistry,* Vol. 55, p. 7398–7394, 2007.
104. P. Campbell, T. Beer, and D. Batten, *Bioresource Technology,* Vol. 102, p. 50–56, 2011.
105. M.B. Leite, M.M. de Araújo, I.A. Nascimento, A.C. da Cruz, S.A. Pereira, et al., *Environmental Toxicology and Chemistry,* Vol. 30, p. 893–897, 2011.
106. M.K. Lam, and K.T. *Biotechnology Advances,* Vol. 29, p. 124–141, 2011.

11

Bio-Hydrogen Production: Current Scenarios and Future Prospects

Sumita Srivastav[1], Prashant Anthwal[2], Tribhuwan Chandra[2]
and Ashish Thapliyal[2,*]

*[1]Department of Physics, Government Post Graduate College,
Uttarkashi, Uttarakhand
[2]Department of Biotechnology, Graphic Era University, Dehradun, Uttarakhand*

Abstract

Hydrogen is an important energy carrier and when used as a fuel, can be considered an alternate to the major fossil fuels and their derivatives. Its major advantage is that the product of its combustion with oxygen is water, rather than CO and CO_2, which are greenhouse gases. Currently, 95% of hydrogen is produced through conventional fossil sources, like coal, oil, natural gas etc. There are many pathways to produce hydrogen from renewable sources. All these methods can be broadly classified into two categories; one belongs to hydrogen production through biomass and all other methods in which hydrogen is produced from renewable sources fall into another category. Hydrogen production by biomass basically involves biomass-derived fuels through reforming, gasification processes and biological-biomimetic hydrogen production. In this chapter, various methods of hydrogen production through biomass are discussed in detail. Recent advancements in the production of bio-hydrogen from bio-oil, syngas, photo fermentation, dark fermentation and through integrated technologies are also presented. An overall comparison among various methods and their status has been summarized.

Keywords: Hydrogen, reforming, gasification, photo fermentation, dark fermentation

Corresponding author: ashish.thapliyal@geu.ac.in

Dr.Vikash Babu, Dr. Ashish Thapliyal & Dr. Girijesh Kumar Patel (eds.) Biofuels Production,
(309–332) 2014 © Scrivener Publishing LLC

11.1 Introduction

Hydrogen is a clean, reliable, renewable and affordable non-conventional energy source. It is an important energy carrier and has versatile applications in industry, automobile and as liquid fuel in rockets. Hydrogen may be produced from water and after combustion the product is also water. Hence, it is totally renewable in nature.

It is known that hydrogen can be used directly in internal reciprocating combustion engines with minor modifications. It can also be used in hydrogen/oxygen fuel cells to directly produce electricity, with the only product 'water'. The energy efficiency of fuel cells can be as high as 60%. Fossil fuel systems on the other hand, are typically about 34% efficient. When high temperature fuel cells are used, it is possible to obtain electricity and also to use the heat generated in the fuel cell for heating purposes. In this manner, total energy efficiency reaches a value of 80%. In electrically powered vehicles, electric motors can have energy efficiency of about 90%, whereas typical internal combustion engines are about 25% efficient. Presently fuel cells cost 100 times more than the equivalent internal combustion engines of comparable power. However, one can expect a cost reduction with the mass production and the further development of fuel cells [1].

Because of its attractive features and versatile use, hydrogen is expected to play a major role in future energy systems and the world will adopt a 'Hydrogen Economy'. It reacts with oxygen, either by combustion or in fuel cells, to give energy, and the only product is water. It can be regenerated directly from water by electrolysis. Hence, it is a closed chemical cycle, in which no chemical compounds are created or destroyed, but there is a net flow of energy.

11.2 Conventional Methods of Hydrogen Production

The major and least expensive way to obtain hydrogen is to extract it from natural gas, which is primarily methane. The most common method for the conversion of methane to hydrogen involves the use of steam reforming, followed by a water-gas shift reaction.

It provides about 95% of all the hydrogen products in the world [1]. Some of the important methods of hydrogen production are as follows.

11.2.1 The Steam Reforming Process

The first step in this procedure is the elimination of impurities, such as sulfur, from the methane-rich natural gas. The methane is then reacted with steam at a relatively high temperature, using nickel oxide as a catalyst. This process is called steam reforming. It can be written as:

$$CH_4 + H_2O \longrightarrow CO + 3H_2 \qquad (11.1)$$

This can be followed by a second step in which air is added to convert any residual methane that did not react during the steam reforming:

$$2CH_4 + O_2 \longrightarrow 2CO + 4H_2 \qquad (11.2)$$

This is then followed by the water-gas shift reaction at a somewhat lower temperature that produces more hydrogen from the CO and steam:

$$CO + H_2O \longrightarrow CO_2 + H_2 \qquad (11.3)$$

11.2.2 The Reaction of Steam with Carbon

Countries having large amounts of coal can generate hydrogen from the coal. It is possible to react steam with solid carbon instead of methane. In this case, the product is syngas (synthetic gas), a mixture of CO and hydrogen. The resultant CO then reacts with steam in the water-gas shift reaction, similar to the case of steam reforming methane.

11.2.3 Electrolytic Production of Hydrogen

The second major method for the production of hydrogen involves the electrolysis of water by imposing a voltage between two electrodes. This results in the evolution of hydrogen gas at the negative electrode, and oxygen gas at the positive electrode. Relatively pure hydrogen can be produced by this method. Since there is an enormous amount of water on earth, it can provide a limitless supply of hydrogen. About 4% of the hydrogen currently used in the world is produced by electrolysis. However, the cost of production is comparatively expensive.

11.2.4   Thermal Decomposition of Water to Produce Hydrogen

In this method, water is thermally decomposed by heating it to a very high temperature. It requires a temperature of about 4,300K. Therefore this method is not practically viable.

11.2.5   Chemical Interaction of Hydrogen from Water

It is also possible to produce hydrogen by chemically decomposing water. Many species form an oxide when in contact with water. In general, the oxide produces a protective surface layer that prevents further reaction with the water. In a few cases, like lithium, an aluminium hydroxide is formed and hydrogen is produced.

$$Al + 3H_2O \longrightarrow 3/2\ H_2 + Al(OH)_3 + Heat \qquad (11.4)$$

11.3   Hydrogen from Renewables Sources

At present, 95% of the hydrogen needed is produced through conventional fossil fuel sources like coal, oil, natural gas, etc. [2]. Two major problems with fossil fuels are CO_2, which creates pollution, and the limited availability of fossil fuel sources. Therefore, attention in being given to alternate and renewable sources such as solar, wind, thermal, hydroelectric, biomass etc.

There are many pathways to produce hydrogen from renewable sources. All these methods are categorized into two streams. The first belongs to hydrogen production through biomass and the second category involves all other methods of hydrogen production from renewable sources other than biomass

11.3.1   Hydrogen Production by Biomass

Biomass encompasses all the living matter present on the earth. It is derived from growing plants including algae, crops and animal manure. Among all the renewable source of energy, biomass is unique as it effectively stores solar energy. All the processes of biological hydrogen production depend on the presence of hydrogen producing enzymes. The quantity of inherent activity in these enzymes can limit the overall process. All the enzymes contain complex metallo-clusters as active sites. These active sites are synthesized in a complex process involving auxiliary enzymes and protein maturation steps. These enzymes, which are involved in this type of reaction, are nitrogenase, Fe-Hydrogenase and Ni-Fe-Hydrogenase. There are mainly three processes by which hydrogen can be produced through biomass, which are described below.

11.3.1.1   Hydrogen from Biomass-Derived Fuels Through Reforming

In this method, biomass is first converted into a liquid fuel like ethanol, butanol, etc. In the next step, this bio-fuel is processed with the same mechanisms as conventional fossil fuels like natural gas to produce hydrogen through steam reforming.

11.3.1.2   Hydrogen from Biomass through Gasification

In this process, biomass is converted into synthetic gas (syngas) through combustion. This gas contains CO, CO_2 and H_2. Through a water-gas shift reaction, more hydrogen can be produced by synthetic gas.

11.3.1.3   Biological and Biomimetic Hydrogen Production (Biophotolysis and Fermentation)

These processes are usually split into two varieties. One variety requires incident sunlight (bio-photolysis and photo-fermentation) and the other can be carried out in the absence of light (dark fermentation).

The starting process of bio-photolysis is photosynthesis, in which water is split into oxygen, protons and electrons (water splitting) with the aid of sunlight. In photo-fermentation, hydrogen is produced by anaerobic photosynthetic bacteria employing nitrogenase instead of hydrogenase to enhance reducing power of organic substances, preferably organic acids.

In dark fermentation hydrogen, along with organic acids, CO_2 is also produced in the absence of sunlight from organic substrates. In particular, hydrogenase bacteria produce hydrogen from wet biomass under anaerobic conditions.

11.3.2   Hydrogen Production from Renewable Sources Other than Biomass

In this category methods are incorporated which use renewable energy sources other than biomass.

11.3.2.1   Hydrogen from Renewable Electricity Sources (Wind, Solar, Geothermal, Hydro, Wave) Through Electrolysis

This method is similar to the method of hydrogen produced by water electrolysis. The only difference is that the electricity being used in this method is generated through renewable sources like wind, solar, geothermal, hydropower, etc.

11.3.2.2   High Temperature Solar Thermochemical Production – Thermolysis

Thermolytic water splitting is the generic term for a multi-step thermochemical process that uses high-temperature heat to split water into hydrogen and oxygen. The interest in this route comes from the theoretical potential that such a process could convert high-temperature-heat into hydrogen with 50% efficiency, thereby increasing the efficiency of the hydrogen production in comparison to electricity/electrolysis pathway and offering an alternative to electrolysis for renewable hydrogen generation. Thermolysis is mostly proposed in the context of advanced nuclear reactors.

11.3.2.3   Photo-Electrochemical Production- Photoelectrolysis

Photo-electrochemical processes can produce hydrogen in one step, splitting water by illuminating a water-immersed semiconductor with sunlight. This is basically the combination of photovoltaic cells (PV) with electrolysis of water. There are two types of photochemical processes. The first uses soluble metal complexes as catalysts and mimics photosynthesis. The second method uses semiconducting electrodes in a photochemical cell to convert light energy into chemical energy. The semiconductor surface serves two functions, to absorb solar energy and to act as an electrode.

11.4 Methods of Hydrogen Production through Bio-Routes involving Biochemical Processes

In this section, different methods of hydrogen production from biomass, various biochemical processes and the major steps involved during the hydrogen production from biomass are discussed in detail.

11.4.1 Hydrogen Production from Biomass-Derived Fuels

Renewable biomass is an attractive alternative to fossil feedstocks because of the potential for essentially zero CO_2 impact. Unfortunately, hydrogen content in biomass is only 6–6.5% compared to almost 25% in natural gas. For this reason, on a cost basis, producing hydrogen by the biomass gasification/water-gas shift process cannot compete with the well-developed technology for steam reforming of natural gas. However, an integrated process, in which part of the biomass is used to produce more valuable materials or chemicals and only residual fractions are used to generate hydrogen, can be an economically viable option.

This method combines two steps: fast pyrolysis of biomass to generate bio-oil and the catalytic steam reforming of the bio-oil to hydrogen and carbon dioxide. Fast pyrolysis is a thermal decomposition process that requires a high heat transfer rate to the biomass particles and a short vapor residence time in the reaction zone. Several reactor configurations have been shown to assure this condition and to achieve yields of liquid product as high as 70–80% based on the starting biomass weight. They include bubbling fluid beds, transport reactors and cyclonic reactors [3].

In the 1990s, several fast pyrolysis technologies had reached near-commercial status. The concept of pyrolysis/steam reforming has several advantages over the traditional gasification/water-gas shift technology. First, bio-oil is much easier to transport than solid biomass and therefore pyrolysis and reforming can be carried out at different locations to improve the economic cost. A second advantage that could significantly impact the economics of the entire process is the potential production and recovery of higher value co-products from bio-oil. In this concept, the lignin-derived fraction would be separated from bio-oil and used either directly or, after specific treatments, as a phenol substitute that is a phenol-formaldehyde adhesive. The carbohydrate-derived fraction would be catalytically steam reformed to produce hydrogen.

11.4.2 Hydrogen from Biomass through Gasification

Synthesis gas (syngas) produced from biomass gasification contains hydrogen (H_2), carbon monoxide (CO), carbon dioxide (CO_2), water (H_2O), nitrogen (N_2), methane (CH_4) and trace amounts of other hydrocarbons. The relative proportion of each component in syngas depends on the gasification operating conditions, i.e. temperature, pressure, biomass type, etc. Among those conditions, the gasification agent is the most influential. Different biomass gasification technologies include the use of air, steam, or a steam O_2 mixture as a gasification agent.

Further increases in hydrogen content in the gas product require the adjustment of the H_2/CO ratio. The most widely used process is the Water Gas Shift (WGS) reaction, which allows conversion of CO into CO_2 and H_2 in the presence of steam.

$$CO + H_2O \longrightarrow H_2 + CO_2 \qquad (11.5)$$

This reaction is reversible and mildly exothermic ($\Delta H^0_{298} = -41$ kJ mol^{-1}). At high temperatures the reaction is equilibrium limited and at low temperature it is kinetically limited, so the reaction requires the use of a catalyst. Industrial WGS processes usually include two catalytic reaction steps: one at high temperature, in the range of $350 - 450°C$, using Fe-Cr based catalysts and another at low temperature, e.g. $250°C$, with Cu-Zn based catalysts.

11.4.3 Hydrogen Production via Bio-photolysis

In direct photolysis, cyanobacteria decompose water to generate hydrogen and oxygen in the presence of light according to the equation:

$$2H_2O \longrightarrow 2H_2(g) + O_2(g) \qquad \Delta G^0 = +1498 \text{ kJ} \quad (11.6)$$

Since the end product of hydrogen utilization is water this process is considered sustainable. However, the rate of this reaction is low because a large amount of free energy has to be overcome. As such, its large-scale application has been hindered by the high cost of photo-bioreactors and the low solar conversion efficiencies [4]. The separation of hydrogen from oxygen can further add to the costs.

In indirect biophotolysis, the photosynthesis reaction during hydrogen production in light by algae, purple sulfur and non-sulfur bacteria is described as:

$$H_2O + CO_2 + \text{Sunlight} + \text{Hydrogenase / nitrogenase} \longrightarrow$$
$$\text{Carbohydrates} + H_2 + O_2 \qquad (11.7)$$

Many phototropic organisms such as purple bacteria, green bacteria, Cyanobacteria and several algae can produce hydrogen with the aid of solar energy. Microalgae, such as green algae and Cyanobacteria, absorb light energy and generate electrons. The electrons are then transferred to ferredoxin (FD) using the solar energy absorbed by the photo-system. However, the mechanism varies from organism to organism but the main steps are similar to those of photosynthesis. Some strains like *Rhodopseudomonas capsulata* produce hydrogen during irradiation in the presence of organic compounds such as maleate, succinate, etc. Cyanobacteria (B-G algae) represent the most numerous group of phototropic prokaryotes that can produce hydrogen. However, direct bio-photolysis is sensitive to oxygen and thus difficult to sustain hydrogen production. The indirect bio-photolysis can overcome the problem by producing hydrogen and oxygen at different stages to resolve the issue of oxygen sensitivity.

11.4.4   Hydrogen Production through Photo-Fermentation

In photofermentation, anoxygenic photoheterophic bacteria utilize organic feedstock to produce hydrogen in the presence of light according to the equation:

$$C_6H_{12}O_6 \text{ (glucose)} + 6H_2O \longrightarrow 12H_2\,(g) + 6CO_2\,(g)$$
$$\Delta G^0 = +75.2 \text{ kJ} \qquad (11.8)$$

Merits of this process include moderate energy needs, high theoretical yields of hydrogen (often reported as moles H_2/mole feedstock) and the ability to stabilize organic waste streams. Drawbacks include low solar conversion efficiency (<10%) resulting from the inefficient nitrogenase enzyme, and low light intensities at which photosynthesis saturates [4].

11.4.5   Hydrogen Production from Dark Fermentation

Many scientists have focused on the studies of bio-hydrogen production through anaerobic dark-fermentation with mixed microorganisms. The carbohydrates are converted into H_2, CO_2, volatile fatty acid (VFA) and other products by hydrogen producing bacteria. Based on the constitution of fermentation products, three kinds of fermentation types are defined, which are the propionic-type, the butyric-type and the ethanol-type fermentation. The predominant products of propionic-type fermentation are propionic and acetic acids, with no hydrogen produced. For the purpose of hydrogen production, propionic-type fermentation must be avoided. The butyric-type fermentation results in the formation of H_2, CO_2, butyric and acetic acids as the dominant products. The ethanol-type fermentation mainly produces H_2, CO_2, ethanol and acetic acid. Different fermentations result in different products, thus the constitution of fermentation products could reflect fermentation types.

In dark fermentation, anaerobic bacteria and some microalgae (such as green algae on carbohydrate-rich substrates), can produce hydrogen in a dark environment. Hydrogen can be produced by dark, anaerobic bacterial growth on carbohydrate rich substrates giving organic fermentation end products, hydrogen and carbon dioxide. Pure cultures found to produce hydrogen from carbohydrates include species of Enterobactor, Bacillus and Clostridium. The pure substrates used include glucose, starch and cellulose. Process conditions including inoculums have a significant effect on hydrogen yield as they influence the fermentation end products. Carbohydrates are the preferred organic carbon source for hydrogen production. Glucose in biomass gives a yield of 4 mol of hydrogen per glucose when acetic acid is the byproduct:

$$C_6H_{12}O_6 + 2H_2O \longrightarrow 2CH_3COOH + 2CO_2 + 4H_2 \qquad (11.9)$$

$$C_6H_{12}O_6 \longrightarrow CH_3CH_2CH_2COOH + 2CO_2 + 2H_2, \qquad (11.10)$$

The following process can explain anaerobic fermentation in the dark by heterotrophic bacteria:

Biomass $\longrightarrow$ Carbohydrate $\longrightarrow$ Dark fermentation $\longrightarrow$
Organic acids $\longrightarrow$ Hydrogenease $\longrightarrow$ $H_2 + CO_2$

11.4.6 Hydrogen Production by Biological Water Gas Shift Reaction

It is a new route to hydrogen production via the biological water-gas shift reaction (BWGS). Certain photo-heterotrophic bacteria, such as *Rubrivivax gelatinosus* are capable of performing water gas shift reactions at ambient temperature and atmospheric pressure. These bacteria can survive in the dark by using CO as the sole carbon source to generate adenosine triphosphate (ATP) coupling the oxidation of CO with the reduction of H+ to H_2:

$$\text{Biomass} \longrightarrow CO(g) + H_2O(l) \longrightarrow CO_2 + H_2(g);$$
$$\Delta G^0 = -20 \text{ kJ mol}^{-1} \qquad (11.11)$$

The purple non-sulfur bacteria perform CO–water gas shift reaction in the darkness by converting 100% CO into a near stoichiometric amount of hydrogen. It is also able to utilize CO in the presence of other organic substrates. The biomass available in nature can easily be converted to water gas (CO and H_2O) by thermo-chemical conversion. The studies have indicated that 1 kg of cells can produce 1 kg of hydrogen per day in a bubble column or trickling bed bioreactor. The hydrogen synthesis rate by bio water gas shift reaction was reported as 96 mmol H_2 l^{-1} h^{-1} as compared to 20–50 mmol H_2 l^{-1} h^{-1} by dark anaerobic fermentation process. It has been estimated that the processing cost of hydrogen production by bio water-gas shift processes would be \$3.4 kg^{-1} as compared to other biological processing cost at around \$ 12–20 kg^{-1} [5].

11.5 Recent Advancement in Production of Bio-Hydrogen

Currently, world hydrogen production is around 5×10^6 m^3, 95% of which is derived from fossil fuels with a net negative energy gain. Production of hydrogen, for example, by methane-steam reforming yields 2.95 mol H_2 per mole of methane at best, with a negative net energy gain of −16 MJ/kg of H_2. Production by electrolysis of water using electricity generated by a natural gas-fired combined cycle power plant yields 1.37 mol H_2 per mole of methane at best, with a negative net energy gain of −172 MJ/kg of H_2 [4]. If hydrogen is to be widely accepted as a sustainable substitute for fossil fuels, it has to be produced from a renewable feedstock other than fossil fuels. Hence the process of hydrogen production through fossil feedstock should be replaced with bioprocesses with a net positive energy gain. Biological processes have been proposed as promising approaches for the cleaner and sustainable production of hydrogen known as bio-hydrogen. In contrast to current hydrogen producing technologies, biological technologies can be engineered to produce hydrogen utilizing renewable feedstocks under mild operating conditions without generating any harmful byproducts.

11.5.1 Hydrogen Production from Bio-Oil

Czernik *et al.* have demonstrated that hydrogen could be efficiently produced by catalytic steam reforming of carbohydrate-derived bio-oil fractions in a fluidized bed reactor using a commercial nickel-based catalyst [3]. Greater steam excess than that used for natural gas reforming is necessary to minimize the formation of char and coke (or to gasify these carbonaceous solids) resulting from the thermal decomposition of complex carbohydrate-derived compounds. At 850°C and a molar ratio of steam to carbon of 9, the hydrogen yield was 90% of that possible for stoichiometric conversion during eight hours of the catalyst on-stream time. This yield could be 5–7% greater if a secondary water- gas shift reactor followed the reformer. Finally, coke deposits were efficiently removed from the catalyst by steam and carbon dioxide gasification, which restored the initial catalytic activity.

Techno-economic analysis for the process assumed a hydrogen production capacity of 35.5 t/day, which could require 930 t/day of dry biomass. Biomass was considered available at a cost of $ 25 /dry tone. A 15% internal rate of return was assumed for both the pyrolysis and reforming installations. If the phenolic fraction of bio-oil could be sold for $ 0.44 /kg (approximately half of the price of phenol), the estimated cost of hydrogen from this conceptual process would be $ 7.7 /GJ, which is at the low end of the current hydrogen selling price [3].

Valle *et al.* (2013) reported on steam reforming, in a continuous regime, of the aqueous fraction of bio-oil obtained by flash pyrolysis of lignocellulosic biomass (sawdust) [6]. The reaction system was provided with two steps in series: i) thermal step at 200°C, for the pyrolytic lignin retention, and ii) reforming in-line of the treated bio-oil in a fluidized bed reactor, in the range 600–800°C, with space-time between 0.10 and 0.45 $g_{catalyst}$ h $(g_{bio-oil})^{-1}$. La_2O_3 was incorporated to the Ni/α-Al_2O_3 catalyst to increase kinetic behavior (bio-oil conversion, yield and selectivity of hydrogen) and deactivation was determined. The significant role of temperature in gasifying coke precursors was also analyzed. Complete conversion of bio-oil is achieved with the Ni/La_2O_3-αAl_2O_3 catalyst, at 700°C and space-time of 0.22 $g_{catalyst}$ h ($g_{bio-oil}$)$^{-1}$. The catalyst deactivation was low and the hydrogen yield and selectivity achieved were 96% and 70%, respectively.

11.5.2 Hydrogen Production from Syngas

Further work has been reported in last few years on hydrogen production through syngas in comparison to the method of steam reforming bio-oil. In the gasification process the biomass is completely converted into cleaner gaseous products. Gasification with steam since the gasifying agent favors the production of hydrogen enriched gas. The steam reformation and water gas shift reaction that occurs in the presence of steam results in a higher yield of hydrogen. Biomass can be considered a negative carbon dioxide emitter, if carbon dioxide produced during gasification is captured during the process. Thus, steam gasification of biomass within the process of carbon dioxide capture could be a promising option for sustainable hydrogen production. Solid based sorbents are found to have various operational advantages in absorbing carbon dioxide. They have higher capture efficiency at high temperature where other liquid based sorbents cannot work. Such types of sorbents are rhodium, aluminium oxide, nickel based solid absorbent, dolomites and CaO. The metal based sorbents are expensive and are not economically feasible, whereas calcium oxide is cheap and abundant. It can be an effective sorbent to capture carbon dioxide at a very high temperature. Acharya *et al.* have carried out an experiment to find out the potential of hydrogen production from steam gasification of biomass in the presence of sorbents CaO and its effect on different operating parameters (steam to biomass ratio, temperature, and CaO/biomass ratio) [7]. Product gas with hydrogen concentration up to 54.43% was obtained at steam/biomass = 0.83, CaO/biomass = 2 and T = 670°C. A drop of 93.33% in carbon dioxide concentration was found at CaO/biomass = 2 as compared to the gasification without CaO.

One of the most advanced facilities demonstrates the feasibility of biomass gasification technology in the Växjö Värnamo Biomass

Gasification Center (VVBGC) in Sweden, which has a biomass–fueled pressure IGCC (Integrated Gasification Combined Cycle) CHP (Combined heat and Power) pilot plant of 18 MWh. This plant demonstrates the production of clean hydrogen rich synthetic gas (Syngas) based on pressured steam/oxygen blown gasification of biomass, followed by cleaning and upgrading. Under these conditions, hydrogen content in the syngas can reach values ranging from 35% to 45% vol. Maroño et al. (2010) have studied the catalytic activity of a Fe-Cr WGC catalyst in terms of CO conversion and H_2 production [8]. The influence of the main operating parameters including feed composition, space velocity and steam to CO ratio, on the activity of the catalyst has been studied. In this study, despite the high CO content in the feed gas (44 – 60% dry basis), the catalyst has shown very good performance at intermediate temperatures 350–450°C, providing hydrogen content increases in the exit gas in the range of 10–17% (dry basis) and CO concentration at the reactor outlet lower than 3% v/v.

A series of nickel-based catalysts supported on silica (Ni/SiO_2) with different loadings of Ce/Ni (molar ratio ranging from 0.17 to 0.84) were prepared using conventional co-impregnation method and were applied to synthesis gas production in the combination of CO_2 reforming with partial oxidation of methane [9]. Among the cerium-containing catalysts, the cerium-rich ones exhibited the higher activity and stability than the cerium-low ones. The temperature-programmed reduction (TPR) and UV–vis diffuse reflectance spectroscopy (UV–vis DRS) analysis revealed that the addition of CeO_2 reduced the chemical interaction between Ni and support, resulting in an increase in reducibility and dispersion of Ni. Over NiCe-x/SiO_2 (x = 0.17, 0.50, 0.67, 0.84) catalysts, the reduction peak in TPR profiles shifted to higher temperatures with an increasing Ce/Ni molar ratio, which was attributed to the smaller metallic nickel size of the reduced catalysts. The transmission electron microscopy (TEM) and X-ray diffraction (XRD) for the post-reaction catalysts confirmed that the promoter retained the metallic nickel species and prevented the metal particle growth at high reaction temperature. The NiCe-0.84/SiO_2 catalyst with a small Ni particle size exhibited stable activity with the constant H_2/CO molar ratio of 1.2 during 6-h reaction in the combination of CO_2 reforming with partial oxidation of methane at 850°C and atmospheric pressure.

Replacing batteries with fuel cells is a promising approach for powering portable devices; however, hydrogen fuel generation and storage are challenges to the acceptance of this technology. A potential solution to this problem is on-site fuel reforming, in which a rich fuel/air mixture is converted to a hydrogen-rich syngas. Smith *et al.* (2012) have studied the conversion of jet fuel (Jet-A) and butanol to syngas by non-catalytic filtration combustion in a porous media reactor operating over a wide range of equivalence ratios and inlet velocities [10]. This study was focused on the production of syngas, the hydrogen yield, the carbon monoxide yield, and the energy conversion efficiency. This study was intended to increase the understanding of filtration combustion for syngas production and to illuminate the potential of these fuels for conversion to syngas by non-catalytic methods.

Gasification is a promising alternative process for sewage sludge energy utilization. CaO has been identified as an effective additive that can increase the H2 content of syngas produced by coal, biomass, and sludge gasification. Considering that lime (CaO) is a widely applied conditioner for sewage sludge dewatering in the filter press, Liu *et al.* investigated the enhanced efficiency of syngas, especially regarding H_2 yield, in the catalytic steam gasification of dry dewatered sludge with physically mixed CaO and dry sludge dewatered with CaO as a conditioner [11]. The experiments were conducted in an electrically heated reactor at 873 K, 973 K and 1073 K, respectively. According to the results, the conditioner CaO improved the H_2 and syngas production more remarkably than additive CaO. It was identified by XRD and SEM-EDX that conditioner CaO was completely converted into $Ca(OH)_2$ while additive CaO was still presented mainly as CaO. Furthermore, the Ca species of conditioner CaO was evenly distributed over the sludge matrix while the Ca species of additive CaO maintained the original state with uneven distribution, both of which could increase the formation of H_2 through interacting with produced gas and catalyzing thermal cracking of tar to some extent. In addition, the pore structure tests and XPS analyses revealed that, compared to additive CaO, conditioner CaO was more favorable for the formation of pores, and it had a greater potential to encourage partial cleavages of C–C bonds and C–H bonds, resulting in the decomposition of organic macromolecules into relatively small molecules, which might be more easily converted to the gaseous products. These indicated that it is valuable to reuse the Ca in lime-conditioned sludge during the gasification process.

11.5.3   Hydrogen Production from Photofermentation

The decomposition of organic compounds in the presence of heterotrophic bacteria under illumination of visible light with a simultaneous evolution of hydrogen and carbon dioxide is known as the photofermentation process. The ideal substrates for this process can be organic wastes with low concentrations of total nitrogen (both organic and inorganic). It is well known that purple non-sulfur bacteria of *Rhodobacter sphaeroides* O.U.001 are very efficient biocatalysts in the hydrogen generation process from wastes originating from food, dairy, sugar or the alcohol-distilling industries [12]. A very high concentration of organic substances (average Chemical Oxygen Demand-COD: 0.8–2.5 kg/hl) in wastes from breweries suggests an application of these wastes in hydrogen production. The amount of waste during beer production is enormous. A chemical composition of waste strongly depends on the kind of beer produced and the degree of fermentation. Such waste can contain amino acids, proteins, organic acids, sugars, alcohols, as well as vitamins of the B group. All these substrates can be efficiently used in photobiological hydrogen production. Seifert *et al.* (2010) applied *Rhodobacter sphaeroides* O.U.001 (concentration of inoculums-0.36 g dry wt/l) and brewery wastewaters to the photobiogeneration of hydrogen under illumination of 116 W/m² [12]. The best results were obtained with filtered wastewater sterilized at 120°C for 20 min and a maximal concentration of waste in medium equal 10% v/v. The main product in generated biogas was hydrogen (90%). After sterilization, the amount of generated hydrogen was tripled (from 0.76 to 2.2 l H_2/l medium and light conversion efficiency reached a value of 1.7%.

11.5.4   Hydrogen Production from Dark Fermentation (DF)

While dark fermentation is spontaneous and has the highest conversion rate among the three bioprocesses, the yield is low and the conversion is not complete, resulting in volatile fatty acids and alcohols as end products. Thus, the feedstock-to-energy capture efficiency of DF (based on the energy content of glucose and hydrogen of 16 MJ/kg and 120 MJ/kg, respectively) is about 33%. The fundamental factor that has to be evaluated when comparing biohydrogen production processes is the number of electrons transferred from the feedstock to the end product that serves as the energy carrier and the net energy gain of the process. While nearly all the electron equivalence of the substrate is routed to hydrogen in biophotolysis and photofermentation, only a fraction of the electrons in the feedstock are routed to hydrogen in DF as most of the electrons are routed to organic end products such as volatile fatty acids and alcohols. In spite of these disadvantages, DF has been suggested more practical than the other two phototrophic processes because it does not require external energy to drive the processes or large surface area to capture the necessary light. It can take advantage of existing reactor technologies to utilize organic wastes as feedstock, serving dual functions of energy production and waste stabilization. Additional advantages of the DF process over the phototrophic processes include its ability to utilize particulate organic feedstock and to run throughout the day.

In DF various wastes such as palm oil mill, effluent household solid waste, food waste, and cheese whey wastewater can be used as feedstock. Anaerobic sludge, sludge compost, sewage sludge and soil have been widely used as inoculums for the fermentative hydrogen production process since they contain various bacteria from which hydrogen producing bacteria can be obtained after appropriate enrichment.

The fermentation products are affected by the operation parameters, such as temperature and pH. Four parameters, including pH, temperature, hydraulic retention time (HRT) and organic loading concentration or rate have frequently been investigated in the process of hydrogen production. Wu et al. (2010) used cow dung compost as a source of microorganisms [13]. The reported optimal pH value is 4.0 to 9.0 due to the different strains used. HRT is considered a major factor influencing the performance of continuous operation. Shorter HRT is preferred when considering economic benefits. Temperature affects hydrogen evolution because the hydrogenase is active in a narrow range of temperature. In most studies, the temperature for hydrogen production is set between 30 and 37°C.

Guo et al. (2010) have produced cost efficient cellulose hydrogen by anaerobic fermentation [14]. First, cellulase used for hydrolyzing cellulose was prepared by solid-state fermentation (SSF) on a cheap biomass from Trichoderma viride. Several cultural conditions for cellulose production on cheap biomass such as moisture content, inoculums size and culture time were studied. The maximal hydrogen yield of 122 ml/g-TVS was obtained at the substrate concentration of 20 g/l and at a culture time of 53 h. The value was about 45-fold that of raw corn stalk wastes. The hydrogen contained in the biogas was 44–57% v/v and there was no significant methane gas observed.

Antonopoulou *et al.* (2010) have focused their study on the influence of pH in fermentative hydrogen production from the sugars of sweet sorghum extract, in a continuous stirred tank bioreactor [15]. The maximum hydrogen production rate and yield was obtained at pH 5.3. In a different study, Cui *et al.* (2010) have carried out the experiments to convert poplar leaves pretreated by different methods into hydrogen using anaeorobic mixed bacteria at 35°C [16]. The effects of acid, alkaline and enzymatic pretreatments on the saccharification of poplar leaves were studied. A maximal cumulative hydrogen yield of 44.92 mL/g-dry poplar leaves was achieved.

In another work, Castello *et al.* (2013) applied an upflow anaerobic packed bed reactor configuration to produce hydrogen using cheese whey as the substrate [17]. The microbiological composition was linked to the reactor operation. Three different organic loading rates of 22, 33, and 37 g COD/l-d were used at a fixed hydraulic retention time of 24 h. The increase of the organic loading rate from the initial value of 22 g COD/l-d to a value of 37 g COD/l-d and the adjustment of the pH to values greater than 5 had a positive effect on the hydrogen production and values of up to 1 L H_2/l-d were achieved. The production of hydrogen was stable under all of the investigated conditions, and the problems frequently reported with this kind of reactor (bed clogging, methanogenesis and solvent production) were not observed during operation. The highest hydrogen yield was 1.1 mol H_2/mol lactose, far from the maximum theoretical value (8 mol H_2/mol lactose), but similar to previous work using cheese whey as substrate. The microbiological analysis showed a mixed population with a low proportion of hydrogen-producing fermenters (Clostridium and Klebsiella) and other non-hydrogen producing organisms.

The conversion of glycerol to biodiesel waste streams to valuable products (e.g. hydrogen and 1,3-propanediol (1,3-PD)) was studied by Liu *et al.*(2013) through batch-mode anaerobic fermentation with organic soil as inoculums [18]. The production of hydrogen in headspace and 1,3-PD in the liquid phase was examined at different hydrogen retention times (HyRTs), which were controlled by gas-collection intervals (GCIs) and initial gas-collection time points (IGCTs). Two purification stages of biodiesel glycerol (P2 and P3) were tested at three concentrations (3, 5 and 7 g/l). Longer HyRT (longer GCI and longer IGCT) led to lower hydrogen yield but higher 1,3-PD yield. The P3 glycerol at the concentration of 7 g/l had the highest 1,3-PD yield (0.65 mol/mol glycerol$_{consumed}$) at the GCI/IGCT of 20 h/65 h and the highest hydrogen yield (0.75 mol/mol glycerol$_{consumed}$) at the GCI/IGCT of 2.5 h/20 h), respectively. A mixed-order kinetic model was developed to simulate the effects of GCI/IGCT on the production of hydrogen and 1,3-PD. The results showed that the production of hydrogen and 1,3-PD can be optimized by adjusting HyRT in the anaerobic fermentation of glycerol.

11.5.5 Hydrogen Production from Integrated Technologies

Recently integrated technologies have emerged as better options for biohydrogen production since the yield of a pure method is not sufficient for commercial production. When more than two methods are integrated, byproducts may also be utilized, increasing the overall efficiency of the integrated process.

Combined dark and photo fermentation was a rather new approach in biological hydrogen gas production. It has certain advantages over single stage dark-fermentation or photo-fermentation processes. A three step process scheme consisting of pre-treatment-hydrolysis, dark fermentation and photo-fermentation could be used for this purpose. The first step of pre-treatment includes grinding, acid hydrolysis, neutralization and nutrient balancing to produce carbohydrate solution from the biomass. Fermentable sugars were converted to organic acids, CO_2 and hydrogen in the dark fermentation phase. Light-fermentation was used for the production of hydrogen from organic acids under anaerobic conditions in the presence of light. The effluent of dark fermentation in hydrogen production provides sufficient amount of organic acids for the photo-fermentation. Therefore, the limitation by the organic acid availability would be eliminated. Using dark and photo fermentative bioreactors, hybrid fermentation technology might be one of the more promising routes for the enhancement of H_2 production yields. The synergy of the process lies in the maximum conversion of the substrate which otherwise fails to achieve a complete conversion due to thermodynamic and other limitations [19]. Thus, in this system the light independent bacteria and light dependent bacteria provide an integrated system for maximizing the H_2 yield. The further utilization of organic acids by photofermentative bacteria could provide better effluent quality in terms of COD. However, the system should be well controlled to provide optimum media composition and environmental conditions for the two microbial components of the process. In such systems, the anaerobic fermentation of carbohydrates (or organic wastes) produces intermediates, such as low molecular weight organic acids, which are then converted into H_2 by the photosynthetic bacteria in the second step in a photobioreactor. Higher hydrogen production yields can be obtained when two systems are combined.

One such method is subjected to dark fermentation effluent of wheat powder solution to light fermentation for biohydrogen production using different light source and intensities to increase the hydrogen production [20].

Another integrated process is hydrogen from biofuels in Brazil, and it has becoming more important each year [21]. Within this context, biohydrogen production could exploit residual streams from first generation ethanol (ethanol 1G), second generation ethanol (ethanol 2G) and biodiesel production. Therefore hexoses, pentoses and glycerin were tested as substrates for hydrogen production. Firstly, the effects of different inoculum pretreatments (acid, alkaline and heat) on bacterial communities' performance were evaluated through the levels of *Clostridium* hydrogenase expression. The heat-pretreated inoculum provided the highest yield of H_2 (4.62 mol H_2/mol sucrose) and also the highest level of hydrogenase expression, 64 times higher when compared with untreated inoculum after 72 h. C5 and C6 sugars and glycerin were tested for H_2 production (35°C and pH 5.5), which resulted in promising yields of H_2: sucrose (4.24 mol H_2/mol sucrose), glucose (2.19 mol H_2/mol glucose), fructose (2.09 mol H_2/mol fructose), xylose (1.88 mol H_2/mol xylose) and glycerin (0.80 mol H_2/mol glycerin).

One more integrated technology is the simultaneous production of hydrogen and ethanol from waste glycerol by a newly isolated bacterium *Enterobacter aerogenes* KKU-S1 employing response surface methodology (RSM) with central composite design (CCD) [22]. The Plackett-Burman design was first used to screen the factors influencing simultaneous hydrogen and ethanol production, i.e., initial pH, temperature, amount of vitamin solution, yeast extract (YE) concentration and glycerol concentration. Results indicated that initial pH, temperature, YE concentration, and glycerol concentration had a statistically significant effect ($p \leq 0.05$) on the hydrogen production rate (HPR) and ethanol production. The significant factors were further optimized using CCD. Optimum conditions for simultaneously maximizing HPR and ethanol production were YE concentration of 1.00 g/l, glycerol concentration of 37 g/l, initial pH of 8.14, and temperature of 37°C in which a maximum HPR and ethanol production of 0.24 mmol H_2/l h and 120 mmol/l were achieved.

11.6   Status of Biohydrogen Production

In 2005, 48% of the global demand for hydrogen was produced from the steam reforming of natural gas, about 30 % from oil /naphtha reforming refinery/chemical industrial off- gases, 18% from coal gasification, 3.9% from water electrolysis and 0.1% from other sources [19]. Due to the increasing trend of hydrogen demand, the development of cost effective and efficient hydrogen production technologies has gained significant attention. Biohydrogen production technology can utilize renewable energy sources like biomass for the generation of hydrogen, the cleanest energy carrier for human use. Hydrogen is considered a dream fuel by virtue of the fact that it has the highest energy content per unit mass of any known fuel (142 MJ/kg), is easily converted to electricity by fuel cells and during combustion, it gives water as the only by-product. Presently, 40 % of H_2 is produced from natural gas, 30 % from heavy oils and naphtha, 18 % from coal, and 4 % from electrolysis and about 1 % is produced from biomass [19]. In Table 11.1, various processes are compared by topic, such as raw material used, unit cost to fuel and total output cost.

11.7   Conclusions

Over the last seven years, the production rate of biohydrogen has increased 10 times. To save the environment from fossil sources of energy, the only option will be renewable hydrogen production from bioroutes.

Among various options of biohydrogen production, dark fermentation is the most cost effective because the reactor is operated

Table 11.1 Comparison of the energy conversion efficiency and unit cost of various biological hydrogen production processes with that of conventional processes.

Name of the process	Raw materials Energy conversion Efficiency	Unit cost of energy content of the fuel	US $/MBTU
Photo biological hydrogen [23]	H_2O and organic acids	10	10
Fermentative hydrogen [24]	Molasses	28.34	30
Fast pyrolysis for hydrogen production [25]	Coal biomass	–	4
H_2 from advanced electrolysis [25]	Water	–	11
H_2 from thermal decomposition of stream [25]	Water	–	13
H_2 from photo-chemical [25]	Organic acid	–	21
Fermentative ethanol [26]	Molasses	–	31.5
Gasoline [25]	Crude petroleum	–	6

for whole day, it accomplishes the dual goals of waste reduction and energy production. It also requires less electricity or energy input than other hydrogen-producing processes. This process can utilize a wide range of carbon sources and when it is coupled with photo fermentation, it may become an even more cost effective process.

An interdisciplinary approach towards biohydrogen production through integrated technologies will be useful in future. Hybrid processes that involve scientific advancements such as the genetic modification of organisms, the engineering of hydrogenase enzymes and the development of bioreactors may increase the yield of biohydrogen production and reduce the costs.

References

1. R. A. Huggins, *Energy Storage*, Springer Press, New York, p. 95–107, 2010.
2. N. Lymberopoulos: *Assessment of Hydrogen Energy for Sustainable Development*, Springer Publishers, 51–57.2006.
3. R. Czernik, R. French, C. Feik, and E. Chornet, "Advances in Hydrogen Energy, in E. Catherine, P. Gregoire and L. francis eds., Kluwer Academic/ Plenum Publishers, New York, pp87–91, 2000.
4. K. R. J. Parera, B. Ketheesan, V. Gadhamshetty, and N. Nirmalakhandan. *International Journal of Hydrogen Energy*, Vol. 35, p. 12224, 2010.
5. R. C. Saxena, D. K. Adhikari, and H. B. Goyal. *Renewable and Sustainable Energy Reviews*. Vol. 13 p. 167, 2009.
6. B. Valle, A. Remiro, A. T. Aguayo, J Bilbao, and A. G. Gayubo. *International Journal of Energy Hydrogen*, Vol. 38, p.1307, 2013.
7. B. Acharya, A. Dutta, and P. Basu. *International Journal of Hydrogen Energy*, Vol. 35 p.1582, 2010.
8. M Marono, JM Sanchez, and E Ruiz. *International Journal of Hydrogen Energy*, Vol. 35, p. 37, 2010.
9. B. Li, X. Xu, and S. Zhang. *International Journal of Hydrogen Energy*, Vol. 38, p. 890, 2013.
10. C. H. Smith, D. I. Pineda, C. D. Zak, and J. L. Ellzey. *International Journal of Hydrogen Energy*, Vol.38, p.879, 2012.
11. H. Liu, H. Hu, G. Luo, A. Li, M. Xu, and H. Yao. *International Journal of Hydrogen Energy*, Vol. 38, p.1332, 2013.
12. K. Seifert, M. Waligorska, and M. Laniecki. *International Journal of Hydrogen Energy*, Vol. 35 p.4085, 2010.
13. X. Wu, H. Yang, and L. Guo. *International Journal of Hydrogen Energy*, Vol. 35, 46, 2010.
14. Y. P. Guo, S. Q. Fan, Y. T. Fan, C. M. Pan, and H. W. Hou. *International Journal of Hydrogen Energy*, Vol. 35, p.459, 2010.
15. G. Antonopoulou, H. N. Gavala, and I. V. Skiadas. *International Journal of Hydrogen Energy*, Vol. 35, p.1921, 2010.
16. M. Cui, Z. Yuan, X. Zhi, L. Wei, and J. Shen. *International Journal of Hydrogen Energy*, Vol. 35 p.4041, 2010.
17. V. Perna, E. Castello´, J. Wenzel, C. Zampol, D. M. Lima, L. Borzacconi, M. B. Varesche, and M. Zaiat, C. Etchebehere. *International Journal of Hydrogen Energy*, Vol. 38 p.54, 2013.
18. B. Liu, K. Christiansen, R. Parnas, Z. Xu, and B. Li. *International Journal of Hydrogen Energy*, Vol. 38 p.3196, 2013.
19. K. Pandu, and S. Joseph. *International Journal of Advances in Engineering & Technology*, Vol. 2 (1) p.342–356, 2012.
20. H. Argun, and F. Kargi. *International Journal of Hydrogen Energy*, Vol. 35, p.1595, 2010.
21. L. R. V. de Sá,, M. C. Cammarota, T. C. Oliveira, E. M. M. Oliveira, A. Matos, and V. S. Ferreira-Leitão. *International Journal of Hydrogen Energy*, Vol. 38, p.2986, 2013.
22. A. Reungsang, S. Sittijunda, and I. Angelidaki, *International Journal of Hydrogen Energy*, Vol. 38 p.1813, 2013.

23. B. E. Logan, *Journal of Environment Science and Technology*, Vol.38 (9), p.160A, 2004.
24. K. Nath, and D. Das. *Bio Energy News*, Vol. 7, p. 15, 2003.
25. J. O' M. Bockris, *International Journal of Hydrogen Energy* Vol. 6, p.223, 1981.
26. S. Tanisho, "Feasibility study of biological hydrogen production from sugarcane by fermentation", In Veziroglu T. N. Winter C. J. Basselt J. P. Kreysa G eds., *Hydrogen Energy Progress*. XI Proceedings of 11th WHEC, Stuttgart, Vol. 3 pp.2601, 1996.

12

Biomethane Production

Ruchika Goyal, Vikash Babu and Girijesh Kumar Patel*

Department of Biotechnology, Graphic Era University,
Dehradun-248002, India

Abstract

Currently, there is extensive utilization and consequently, widespread depletion of non-renewable energy stocks. The whole world demands safe, renewable and integrated energy sources to meet future energy needs. Keeping this notion in mind, biomethane seems to be one of the best energy sources that can be utilized in different fields in place of natural gas. The anaerobic decay of the organic waste, sewage sludge and manure by methanogenic bacteria produces biogas. A major part of biogas is made of biomethane, which is cleaned and upgraded. Biomethane is the cleanest and greenest of all renewable fuels with the highest BTU value. Being a finite but widely available resource, biomass can play an important role in the future realization of a sustainable energy system. In fact, with standard anaerobic digestion approximately 20–30% of the organic matter is mineralized. Therefore, to increase hydrolysis and digestibility, the efficient use of biomass is a prerequisite to resolve energy problems. The generation of biogas from biomass is one of the most auspicious applications of biomass and hence, biomethane is referred to as a renewable or green gas. This chapter describes different aspects of biomethane production, the global scenario and its utilization.

Keywords: Biomethane, BTU, biogas, methanogen, organic waste, CHP unit, renewable fuel, wobbe-index.

Corresponding author: patelgirijesh@gmail.com

Dr. Vikash Babu, Dr. Ashish Thapliyal & Dr. Girijesh Kumar Patel (eds.) Biofuels Production, (333–356) 2014 © Scrivener Publishing LLC

12.1   Introduction

Biomethane is the renewable natural gas derived from biogas. Chemically the same as conventional natural gas, biomethane is produced from the anaerobic digestion of organic matter such as animal manure, sewage and municipal solid (Table 12.1) waste through a process called methanogenesis i.e. the production of CH_4 and CO_2 by biological processes that are carried out by methanogens [1]. The three main reasons for this new tendency in the treatment of solid wastes are: i) the need to apply a process to dispose of organic solid wastes in a more eco-friendly way than in landfills ii) the opportunity to obtain a renewable fuel called biogas which may be an alternative to fossil fuels; and iii) the advantages of relatively low costs in the start up and management process [2]. Biomethane extracted from biomass can replace fossil-based natural gas. In this way, it can abate the emissions from greenhouse gases and offer an important contribution to a sustainable and eco-friendly energy economy. Moreover, the utilization of agricultural waste for biogas generation can make a further contribution to climate protection. It is for this reason that biogas and biomethane can be seen as having a more positive influence on the global climate balance than other forms of biomass currently in use. The production of biogas and biomethane make an important

Table 12.1 Typical sources and characteristics of biogas generation.

Sources	General Characteristics
Landfills	Methane (CH_4), carbon dioxide (CO_2), Hydrogen sulfide (H_2S), water vapor, other sulfide and mercaptans, siloxane, oxygen, nitrogen, ammonia and other trace gases
Wastewater treatment plants	CH_4, CO_2, H_2S, water vapor, nitrogen, ammonia, siloxane and other trace gases
Dairy manure	CH_4, CO_2, H_2S, water vapor
Food processing byproducts	CH_4, CO_2, H_2S, water vapor, nitrogen, ammonia, siloxane and other trace gases
Municipal Organic wastes	CH_4, CO_2, H_2S, water vapor, nitrogen, ammonia, siloxane and other trace gases

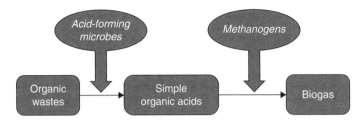

Figure 12.1 Basic flow diagram of biogas production.

contribution to a stable and reliable energy supply by utilizing slurry, manure and organic waste [3].

Anaerobic digestion of biomass generates biogas, which can be used as a potent energy source. It is a multi-step biological process. In biogas production, initially the organic waste matter containing carbohydrates, proteins and fats are converted into organic acids (e.g. acetic, butyric, formic, propionic acids etc.) by the acid-forming bacteria [3–5]. These organic acids are further converted into components like methane, carbon dioxide, ammonia, hydrogen sulfide (H_2S) and water vapor, which constitute the biogas (Figure 12.1) leaving behind a semi-solid material, called digestate, rich in nutrients and suitable for utilization in farming as a bio-fertilizer [6]. Many parameters, besides the anaerobic conditions, like temperature between 37 and 42°C (mesophilic conditions) or 50 to 60°C (thermophilic conditions), an adequate biomass mix or the provision of the microorganisms with nutrient matter influence the efficiency and the stability of the biogas production process.

Biogas produced from different sources will have different concentrations of methane. The general components of biogas are methane (50–80%), carbon dioxide (20–50%), nitrogen (1–4%), hydrogen (0–1%), other hydrocarbons (0–1%), oxygen (0–1%), hydrogen sulfide (50–5000 ppm) and ammonia (0–300 ppm) [7–11]. Natural gas delivered to customers is effectively 100% methane and has an energy value of about 1,000 British Thermal Units (BTU) per cubic foot while the biogas energy values are in a range of 500 to 650 BTU per cubic foot. Biogas is processed and cleaned to the required standards of purity in a process called as "Biogas to Biomethane" which removes the impurities in biogas such as carbon dioxide, ammonia, oxygen, nitrogen, siloxanes and H_2S (Figure 12.2) and the BTU content can be improved to that of natural gas [11]. In this process, along with the removal of impurities or trace elements, the raw biogas is

Figure 12.2 General flow diagram of Biomethane separation.

dried and compressed to a pressure required for the utilization of the resulting gas and tasks like odorisation or adjustment of the heating value by propane dosing might have to be performed. The removal of carbon dioxide results into the enhancement of the heating value and Wobbe-Index.

During processing, the raw biogas is divided into two streams, first the methane rich biomethane stream and second, the CO_2 rich off gas stream. Since no technology is fully efficient, this off gas stream may also contain small amounts of methane. Depending on the concentration of methane in the off gas stream, it is further treated. There are certain compositional parameters that are to be maintained or checked during processing. After processing, biomethane becomes a renewable substitute for natural gas and, once compressed or liquefied, can be used to fuel natural gas vehicles. This biogas upgrading is a gas separation task, finally yielding a methane-rich product with required specifications. It is the cleanest and greenest of all renewable fuels [12]. Therefore, it is of pipeline-quality and is a readily available low-carbon alternative fuel that can be produced locally from organic waste. Presently, biomethane is in the category of low carbon fuels but it is soon to be re-classified from this category to the "super low carbon fuel" category because it has been recognized as the greenest of all biofuels and soon it will completely replace the "natural gas" that is typically transported to markets via vast underground pipeline systems.

As the waste decomposes a greenhouse gas twenty times more destructive than carbon dioxide i.e., methane goes into the stratosphere, putting our future in the dark. Biomethane production avoids the release of this destructive greenhouse gas, as biomethane in an internal combustion engine vehicle emits fewer emissions than electricity in a very efficient electric vehicle. Currently, biofuels have gathered significant opposition in much of the world but biomethane has avoided the food for fuels controversy associated with ethanol from corn, biodiesel from soy and palm oil, while biomethane is normally processed from waste. It has been reported that biomethane has four times the energy production than corn ethanol from an acre of land [13].

Waste management landfills contain significant organic waste that is suited for anaerobic digestion. This gives various opportunities for the construction of waste-to-energy plants. The need to dispose of ever-increasing quantities of waste is diminished by its conversion to biogas. This gives vision to a zero-waste society where anything no longer used is converted to something valuable, be it recycled paper, electricity, heat or fuel. With these ideas and a portfolio of solutions, we can achieve energy independence and avoid a climate crisis, leading us to a near zero-emissions future [14]. Energy efficient buildings, transportation and sustainable living make bigger differences. Considering the importance of biomethane we can put the emphasis on the following words "reduce, reuse, and recycle" [15–17].

12.2   Features of Biomethane

Methane is an odorless gas in its pure form. It easily forms a mixture with air and is explosive in nature at a concentration of 5–15 %. It is a non-toxic gas but in confined environments or spaces, it can cause death due to asphyxiation by displacing oxygen. The heating value of pure methane gas is 1,000 BTU per cubic foot. Additionally, methane is considered a powerful greenhouse gas that can remain in the atmosphere for up to 15 years and is about 20 times more effective in trapping heat in the Earth's atmosphere than CO_2 [1].

12.2.1   Advantages

Some of the advantages of biomethane are the following:

- Biomethane from renewable sources like dairy digesters and landfills can be a reliable source of renewable fuel that can power the cleanest and most efficient electricity generation facilities
- Reduce dependence on fossil fuels and foreign sources of petroleum
- Eco-friendly and can reduce CO_2 generation

- Creates local jobs by keeping local power plants operational and keeps electricity costs lower, thus, helping local businesses to prosper
- Cut down on greenhouse gas emissions by displacing natural gas and reducing the release of methane into the atmosphere
- Biomethane, in many instances, is the low-cost option among renewable products
- Highly flexible in use because of transportation and storage capabilities
- Increases productive development and the use of renewable resources from organic wastes
- Pipeline biomethane utilizes existing natural gas pipelines and the most efficient generation resources, thus preserving valuable transmission capacity for the delivery of wind and solar energy and the optimization of public investments
- Requires no new expensive transmission or other grid infrastructure
- The opportunity to use a low cost renewable fuel (biomethane) when firing combined cycle generators, producing renewable electrons at higher efficiencies
- Unlike wind and solar, generation from biomethane is not intermittent and can be dispatched to fill in gaps and promote system reliability

The choice of purification technology to be used extensively depends on the quality and quantity of the raw biogas to be upgraded along with the extent to which purified biomethane is to be produced according to its final usage, whether in a vehicle utility or in a gas grid system. Additionally, factors like the operation of the anaerobic digestion plant, types of used substrates in the plant and their continuity, local circumstances at the plant site also contribute to the choice of purification technology to be used. Generally, this choice is made by the planner and future operator.

12.2.2   Advantages over Methane

Biomethane has a few advantages over methane and natural gas, which make it more in demand. Firstly, there is an unlimited supply of biomethane, whereas the methane sold by gas companies is limited in supply and secondly, biomethane is renewable, whereas the methane sold by gas utility companies is not [12, 13, 17, 18]. Biomethane and biogas recovery, use and production generate "green tags" or a "renewable energy credits" for the owners and are therefore, beneficial for our environment.

12.2.3   Cost Estimation

Biomethane, after installation of the biomethane equipment is essentially free, as opposed to buying natural gas, which currently costs around $10.00/mm BTU. Unlike the price of natural gas, which wildly fluctuates and is very unstable, biomethane prices will tend to be more stable over the years if biomethane can be produced in excessive amounts and through reliable and sustainable methods can fuel our energy needs now and in the future [12, 19].

12.2.4   Utility

The use of biomethane as a vehicle fuel can reduce greenhouse gas emissions up to 88% compared to gasoline and is an important domestic resource that can reduce any national dependence on foreign oil [20, 21].

12.3   Global Scenario of Biomethane

On November 21, 2012, the Prince of Wales inaugurated the UK's 'first' commercial-scale anaerobic digestion (AD) and biomethane-to-grid plant at Rainbarrow Farm in Poundbury, Dorset. This plant is owned and operated by J. V. Energen, a joint venture between local farmers and the Duchy of Cornwall. This plant is reportedly the first fully operational commercial-scale plant in the UK to inject biomethane (produced from the breakdown of organic waste) directly into the gas grid on-site. First, they clean up the biogas produced by the AD process and then inject the resulting biomethane directly into the gas network. For this plant, it is estimated that it will generate approximately 400 cubic meters per hour (cu m/h) of 96 per cent pure methane (mixed with propane and an odoriser) into the local gas grid which is enough to supply heat for up to 4,000 houses in the local community in midwinter and 56,000 houses in midsummer. This plant will produce around 850 cu m/h of biogas by using around 41,000 tonnes of maize, grass and potato waste grown by local farmers and organic waste from nearby factories. The resultant biogas will be used in a combined heat and power plant to produce approximately 10 mega watts of electricity per day [22]. According to the report of the IEA Bioenergy Task 37 [17], currently, there are more that 220 biomethane upgrading plants present in the different countries depicted in Figure 12.3.

Moreover, to promote these practices there are celebrations happening under the banner of 'UK Biomethane Day', which was started in 2012. This year, in 2013, this day was celebrated on May 21, 2013 at the National Motorcycle Museum in Birmingham by the Renewable Energy Association (REA), London.

In addition, to meet the standards for biomethane usage without waste management control, the Environment Agency has recently revised the draft quality protocol for biomethane through the European Pathway to Zero Waste (EPOW). This is also called the "New Waste Protocol" for biomethane. Presently, the revised draft quality protocol is being developed and is available for public comment under the following names:

- Revised draft Quality Protocol for biomethane
- Environmental Risk Assessment
- Explanation of the revisions made to the draft Quality Protocol

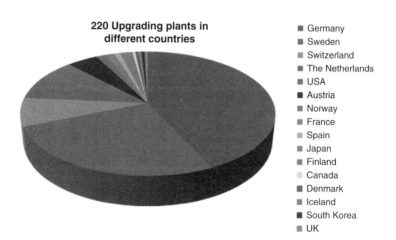

Figure 12.3 Data was adopted from IEA, Task 37 [17].

The final version of the revised protocol will come up later this year (2013). Today, biogas produced through the treatment of waste by anaerobic digestion in landfills and its subsequent upgradation to biomethane, may be considered waste. This new quality protocol will bring regulatory clarity about the point at which biomethane is no longer regarded as waste. It will also help to establish a consistent set of standards required in order to facilitate the direct injection of waste-derived biomethane into the gas grid and its use in other appliances. By reducing the regulatory burden on businesses using biomethane from waste sources, we hope to encourage its increased production and use.

12.4 Biomethane Production – Waste to Fuel Technology

As mentioned earlier, biomethane is 'naturally' produced from organic materials by anaerobic decay. Sources of biomethane include: landfills, POTW's/wastewater treatment systems and every tree or agricultural product that is no longer living. Biomethane is also generated from animal operations where manure can be collected and the biomethane is generated from anaerobic digesters where the manure decomposes. Making money from meadow muffins also gives an opportunity to dairy farmers to stay in business [23, 24].

12.4.1 Components of the Biogas Plant

A typical biogas system consists of a manure collection tank, slurry tank, anaerobic digestion tank and effluent storage tank (Figure 12.4) along with gas handling and gas use equipment [25–30]. The basic components of the digesters are as follows:

Manure collection tank- A key consideration in the system design is the amount of water and inorganic solids that mix with the manure during collection and handling.

Slurry tank- This is also called as mixing tank. In this tank, the waste material to be digested is first mixed with water to form slurry and then passed to the digester tank.

Digester tank- In this tank, the fed slurry is left for anaerobic digestion (AD) by the anaerobic micro- flora. The anaerobic degradation decomposes the fed waste material into some simpler compounds

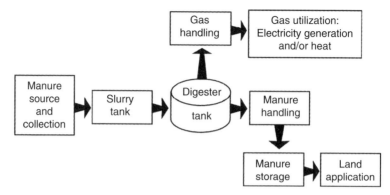

Figure 12.4 Basic flow chart of material flow in biogas production.

or into various gases, which are collected into separate vessels and processed for further use.

Collection tank- The biomass left after anaerobic digestion is pumped into this tank and used as manure in fields as a biofertilizer.

12.4.2   Pretreatment

The substrate composition is the major factor in methane production and yield determination from anaerobic digestion. On the basis of substrate source and its composition, different pretreatment approaches have been adopted to increase the digestibility by altering the physical and chemical properties. The different methods of pretreatments are physical (mechanical, thermal, ultrasonic) chemical, biological or combinations of it [26]. For the details about the pretreatment methods, please refer section 4.2.1.

12.4.3   Biogas Production

The anaerobic digestion depends on the structural components of the different organic materials as well as microbial community. After pretreatment, the formation of methane occurs in four steps: hydrolysis, acidogenesis, acetogenesis and methanogenesis. A number of microorganisms are involved in these processes. The microbial composition and viability depends on different factors such as temperature, pH, ammonia, concentration of volatile fatty acids and also on the presence of macro- and micronutrients etc. The extracellular hydrolytic enzymes produced by the microorganisms, responsible for the degradation of the biomass into macromolecules e.g. carbohydrates, proteins and lipids and subsequently to simpler subunits such as glucose, fructose, amino acids and lipid derived acids. The microorganisms involved in the hydrolytic process belong to different taxonomic phyla like *Clostridia, Firmicutes, Bacilli*.

In acidogenesis the simple organic compounds are converted to the volatile fatty acids acids e. g. propionic acids which are responsible for acidosis raveling acetic, butyric, formic, acids along with ethanol, hydrogen, and CO_2. The microbes related to the acidosis belong mainly to the *Clostridia*.

The capacity of methanogenesis is limited to the *Archaea*. On the basis of main metabolic precursors utilized in methane production, methanogens are divided into two groups, first acetoclastic methanogen, which strictly metabolize acetate and contribute approximately 70% of the total methane production and second hydrogenotrphic methanogen which use H_2 or formate as an electron donor and CO_2 as a carbon source. Some methanogen are able to convert H_2/CO_2 and acetate e.g. *Methanosarcina* spp. beside these, certain methanogens are also able to convert methyl group, methylamine or methanol to methane [26].

12.5   Biogas Cleaning and Upgrading

The utility of the end product i.e., biomethane serves as the basis for biogas upgrading. Biogas can be upgraded into various higher-grade fuels like compressed biomethane (CBM), liquefied biomethane (LBM), hydrogen only, methanol etc. CBM and LBM are equivalent to CNG and LPG respectively. The process of biogas upgrading is also called sweetening, in which most of the CO_2, H_2O, H_2S and other impurities are removed. There is a slight difference between the two terms, biogas cleaning and biogas upgrading. Biogas cleaning refers to the removal of H_2S, water vapor, NH_3, particulate matter, etc., whereas biogas upgrading generally refers to the removal of CO_2 [31–33].

The biogas must be pressurized for natural gas injection or compressed for fuel applications and thus high-pressure processes are particularly suitable for biogas upgrading. For instance, among the various methods to upgrade biogas to biomethane, High-Pressure Reactive Solvent Scrubbing (HPRSS) technology uses the separation process based on elevated pressure which has significant effect on CO_2 separation rates as it increases CO_2 removal efficiency and decreases dimensions of scrubbing reactors. This technology uses reactive solvents like NH_3 and KOH which have promising CO_2 separation efficiencies and also allow production of valuable fertilizer additives, i.e. ammonium bicarbonate and potassium carbonate, respectively [34, 35]. This technology reveals the following ecological benefits:

   i. the permanent lock-up of carbon in its highest possible oxidation state (as CO_2) in soil
   ii. soil fertilization by N, K, and P elements which are enriched in digestate and
   iii. soil fertilization by N or K elements contained in the CO_2 saturated solvents from the biogas scrubbing units

The undesired components like H_2S, CO_2, N_2, NH_3 and water vapor from biogas are removed sequentially for biogas up gradation through various processes. Some of them are described as follows:

12.5.1   Removal of Hydrogen Sulfide (H_2S)

The removal of H_2S is preferred due to its unpleasant odor and corrosive nature as it corrodes the digester, gas pipes, gas storage vessels and gas burners or engines. It is also removed concurrently with CO_2. Some of the technologies used for removal of H_2S [34–36] and other sulfides like mercaptans from biogas are as follows:

   • Air injected into the digester biogas holder
   • Application of activated-carbon sieve
   • Precipitation
   • Reaction with iron oxide or hydroxide (iron sponge)
   • Sodium hydroxide scrubbing
   • Water scrubbing

12.5.1.1   Air/Oxygen Injection

The injection of air into the collected biogas results in the oxidation of H_2S into the hydrogen or water and yellow clusters of elemental sulphur. This oxidation occurs with the help of sulphur oxidizing (chemoautotrophic) bacteria of genus *Thiobacilli* or *Sulfolobus*, which are naturally occurring bacteria on the surface of digestate. This method is an example of biological fixation reaction and has low maintenance or operational costs because it is less expensive and this makes the upgrading process more economical. However, the proportion of air injected is highly critical because in excess it leads to the synthesis of sulphuric acid or other explosive gas mixtures and residual oxygen would also be a concern for a pressurized gas.

12.5.1.2 Application of Activated Carbon

H_2S is removed by using an activated carbon filter impregnated or doped with potassium iodide (KI) or potassium permanganate or potassium carbonate (K_2CO_3) or zinc oxide (ZnO) or sulphuric acid (H_2SO_4), as catalyzers. This impregnation or doping is done to increase the rate of reaction. The removal is based on the adsorption of H_2S molecules in the carbon sieve. For grid injection or vehicle utilization, low levels of O_2 are allowed in the gas. Hence, oxidation of sulphur is not preferred. So KI-doped carbon or permanganate impregnated carbon is used because the addition of oxygen is not required in the case of KI under reduced loading whereas, ZnO impregnated carbon is expensive, so it is also not welcomed.

Along with the H_2S, other molecules like CO_2, water vapors also are adsorbed. This method is also used in combination with air injection methods to increase the efficiency since the elemental sulphur formed is adsorbed rapidly by the activated carbon. However, this selective adsorption is achieved by applying pressure to the carbon sieve in the pressure-swing adsoption systems and the saturated carbon sieves are regenerated by desorption of the contaminants for reuse.

12.5.1.3 Sulfide Precipitation

The application of liquid iron chloride solution or addition of Fe^{2+} ions or Fe^{3+} ions in the form of salts like $FeCl_2$, $FeCl_3$ or $FeSO_4$ into the influent mixing tank or into the digester results into the formation of insoluble iron sulfide salt particles [$Fe_2(SO_4)_3$] by reacting with H_2S within the biogas fermenter. Since they are heavier, these formed iron sulfide particles get precipitated and removed together with the digestate. They have a positive effect on reducing odors. This process is highly efficient in reducing high levels of H_2S, but less effective in maintaining the low and stable H_2S levels needed for vehicular fuel applications. By this approach, ammonia is also removed along with H_2S. Sulfide precipitation is a relatively cheap method of desulphurization.

12.5.1.4   Iron Oxide or Hydroxide Bed (Chemical Adsorption)

In this method, biogas is passed through a reaction bed media which is composed of rust-covered steel wool or pelleted 'red mud' (a byproduct of aluminum production) or woodchips impregnated with iron oxide or hydroxide on which the H_2S reacts endothermically to form iron sulfide. Impregnated woodchips are most preferred due to easy availability, low cost, larger surface-to-volume ratio than steel wool and a lower surface-to-weight ratio due to the low density of wood. In anaerobic digesters of municipal waste treatment plants, scrubbing of biogas with iron-impregnated wood chips is generally a common practice to remove H_2S in the gas pipeline.

This reaction requires water; therefore, the biogas is not dried prior to this stage. Moreover, the condensation in the iron sponge bed is avoided since water can reduce the reactive surface area by binding with iron oxide material.

In this technology, the regeneration of iron oxide is quite easy and can be done by flowing oxygen (air) over the bed material. Generally for this, two reaction beds are installed, with one bed undergoing regeneration while the other bed is operating to remove H_2S from the biogas. The regenerative reaction in this technology is highly exothermic and can result in the self-ignition of the wood chips, if temperature and airflow are not precisely controlled. This is the main disadvantage of this technology. Thus, some small-scale operations that have low levels of H_2S do not prefer regeneration.

Iron oxide is also the desulphurizing agent in SOXSIA® (Sulphur Oxidation and Siloxanes Adsorption), a catalyst developed by Gastreatment Services B.V. SOXSIA® that adsorbs siloxanes and removes H_2S from the biogas. Sulfa Treat® is another commercially available product for adsorption of H_2S from biogas.

12.5.1.5   Sodium Hydroxide Scrubbing (Chemical Adsorption)

When biogas is bubbled in a sodium hydroxide (NaOH) solution, it forms sodium sulfide or sodium hydrogen sulfide, which is insoluble and non-regenerative. A solution of NaOH and water increases the potential of scrubbing due to the enhancement in physical adsorption of water by chemical reaction of NaOH and H_2S. This is also called lime scrubbing.

12.5.1.6   Water Scrubbing

This scrubbing process is based on the solubility of gases in water. For example, H_2S and CO_2 are more soluble in water than in methane. Taking the advantage of this property, these gases are selectively removed from biogas. As H_2S more soluble in water than CO_2, it can be selectively removed first. In this method a solution of H_2S in water is made by feeding the biogas through a counter flow of water. Due to its simplicity, this method is routinely used for biogas cleaning.

12.5.2   Removal of Oxygen

Being an anaerobic process, biogas generally does not house oxygen in high amounts but it is a common contaminant in biogas. It is found in high concentrations only if introduced through biological fixation for H_2S removal. Most of the oxygen present in biogas is consumed by the facultative aerobic microorganisms in the digester and is therefore present in traces. But it has the dilution effect on the final output of biogas energy content. There are no specific methods aimed at removing oxygen. However, it can be removed by adsorption with activated carbon, membrane separation and low pressure PSA. To some extent, it is also removed in desulphurization or other upgrading methods.

12.5.3   Removal of Nitrogen

Nitrogen is an inert gas and present in landfill gas in high amounts but absent in farm-derived biogas unless introduced as air for H_2S removal. Its removal is quite difficult and expensive as it utilizes the PSA and cryogenic systems for its removal. Hence its presence is avoided by limiting its introduction before or during biogas cleaning processes, unless biogas is used for CHPs or boilers. The common methods of membrane separation and adsorption with activated carbon are also used for its removal to some extent. Like oxygen, it also has the dilution effect on biogas energy content and its presence in biogas may lead to increased

NOx combustion emissions. Therefore, it is preferred that nitrogen is removed from biogas.

12.5.4   Removal of Ammonia

Biogas is formed by the anaerobic digestion of organic waste matter, the degradation of proteins into amino acids and their subsequent degradation leads to ammonia production. The amount of ammonia that is present in the biogas depends on the substrate composition, temperature and pH in the digester. However, its amount is in minute terms. Due to its highly corrosive nature, pungent smell and its combustible products i.e. nitrogen oxides (NOx), its removal is done by its separation from biogas while drying or upgrading the biogas, by refrigerated water vapor removal methods and by water scrubbing technology where the biogas is passed through a counter flow of water. These last two processes take the advantage of solubility of ammonia in water.

12.5.5   Removal of Siloxanes

As the name suggests, siloxanes are the compounds having a silicon-oxygen bond. They are widely used in cosmetics, food additives, deodorants, soaps and shampoos. They are found in biogas derived from sewage sludge or wastewater treatment plants and in landfill gas but not in farm biogas from organic feedstocks. The siloxanes in biogas can lead to abrasive siloxane deposits on hardware like plant pipelines, pistons and cylinder heads of engines, etc which reduces their life. In addition, when siloxanes are burned, silicon oxide is formed which is a whitish powder that can create problems in gas engines. Hence their removal is done by cooling the gas, by adsorption on activated carbon, activated aluminum or silica gel, or on liquid mixtures of hydrocarbons [36, 37]. The commercial products like SOXSIA® (Sulphur Oxidation and Siloxanes Adsorption), a catalyst developed by Gastreatment Services B.V. SOXSIA® are also available for removal of siloxanes through their adsorption.

12.5.6   Removal of Halogenated Hydrocarbons

Halogenated hydrocarbons are present in landfill gas but rarely in biogas from sewage sludge and not in farm biogas derived from organic feedstock. Their presence in biogas leads to the formation of dioxins and furans during combustion [31]. In addition, the halogens present lead to the corrosion of the mechanical parts and reduce their life. The most common method used for removal of the halogenated hydrocarbons is their adsorption on activated carbon.

12.5.7 Removal of Particulates

Particulates consist of some dust and oil particles, which can be present in biogas and in landfill gas. They cause mechanical wear in gas engines and turbines. They are separated by using mechanical filters which have 2–5µm pore size and are made of paper or fabric. After saturation, they need replacement at regular intervals [36].

12.5.8 Removal of Water Vapor

When biogas leaves the digester, it is saturated with water vapor and this water causes corrosion by reacting with CO_2 or H_2S forming carbonic or sulphuric acid respectively and also leads to pressure regulated clogging in gas conveyance systems or in other control devices by condensing. The water content in non-saturated biogas can also enter during some cleaning and upgrading processes. Most biogas utilization devices work on dry gas and for grid injection, dry biogas is preferred. So removal of this water content from the biogas is a pre-requisite and is done by its condensation through altering the temperature and pressure i.e., by cooling, compression, absorption or adsorption [31]. Various mechanical techniques like the insertion of tees, U-pipes or siphons are also employed for removal of condensation from the pipelines. This process of removing water content from biogas is called the drying of gas. For cooling, two practices are followed. The first is refrigeration in which heat exchangers are used which cool the biogas to desired dew point at which water vapor condenses and more biogas is pressurized for the desired drying. The other is passive gas cooling in which gas is lead underground for a short time to lower the temperature, which condenses the present water. The condensed water is further collected and removed. For adsorption activated carbon, silica, aluminum oxide or molecular sieves are used. Absorption can be done by using glycol solutions, calcium chloride, hygroscopic salts or other organic absorbents in the lines before the point of biogas utilization.

12.5.9 Removal of Carbon Dioxide

The upgrading of biogas or landfill gas is defined as the removal of carbon dioxide from the gas. This will result in the enhancement of energy density due to increment in the concentration of methane. There are many processes or technologies for CO_2 removal from raw biogas [17, 31–36]. Some of them are the following:

  i. Water scrubbing (absorption)
  ii. Pressure swing adsorption (PSA)
  iii. Chemical scrubbing with amines (absorption)
  iv. Chemical scrubbing with glycols (absorption)
  v. Membrane separation
  vi. Cryogenic separation
  vii. Other processes

12.5.9.1 Water Scrubbing (Absorption)

This is the most common technique for biogas upgradation. It requires a very low maintenance cost. This method is based on the fact that CO_2 has a higher solubility in water than methane and therefore gets dissolved to a higher extent, particularly at lower temperatures. In this method, the raw biogas meets a counter flow of water in a plastic packed column. The water that leaves the column will have more CO_2 while the gas leaving the scrubber has an increased methane concentration. It has about 95% pure methane. The water can be regenerated by contact with air at atmospheric pressures, either in a packed bed column or in a passive system such as a flash tank or pond, where pressure is abruptly reduced. The regenerated water is piped back to the inlet and used as a fresh scrubbing media. Companies like Biorega AB, Metener have also developed some advanced versions of this technology. This type of system was apparently first used in the USA for stripping CO_2 from biogas at a wastewater treatment plant in Modesto, California.

It is the most common biogas clean-up process used worldwide as it is simple, well established and inexpensive and through this, there is relatively little CH_4 loss because of the large difference in solubility of CO_2 and CH_4, although methane losses can be larger if the process is not optimized. When compared to other chemicals, water is easy to dispose off when used, because it doesn't need special handling and disposal. This is another advantage of this method.

The main drawback of this process is that the biomethane produced requires gas drying at the final step since it is saturated with water. Moreover, during regeneration air gases like O_2 and N_2 get dissolved in water and are transported to the upgraded biomethane gas stream. Therefore, the biomethane produced will always have O_2 and N_2.

12.5.9.2   Adsorption: Pressure Swing Adsorption (PSA)

In this technique, the adsorption of gases on the surface of adsorbent under elevated pressure, serve as the basis for gas separation. The adsorption behavior of different gases under elevated pressure is different and therefore eases their separation. This technique is widely employed for the separation of CO_2 from biogas. For adsorption, different types of activated carbon or molecular sieves (zeolites) are used as the adsorbing material [17, 31–33]. By selectively removing CO_2, the enrichment of biogas with methane is done. At the pilot scale, there are multiple adsorbent loaded columns, generally four, six or nine, working in parallel at different positions ensuring a continuous operation. These columns are regenerated by a sequential decrease in pressure in several steps, before the column is reloaded again. When one column becomes saturated, the raw biogas flow is switched to another column in which the adsorbing material has been regenerated. During regeneration, there is desorption of the adsorbed gases, which constitute the offgas. To reduce the methane slip, the offgas from the first decompression is piped back to raw biogas inlet and then in second decompression stage it is released in atmosphere or sent to the offgas treatment unit depending on the contents in it. If the methane content is high in the offgas, then recycling in the second decompression stage is the additional step. Moreover, the components like water and hydrogen sulfide irreversibly harm or foul the adsorbent material; therefore, they are removed prior to the operation. There are companies like Quest Air and Acrion Systems that have developed various other upgradation systems based on this technology. Various advanced setups for automated cycling of multiple columns are also used today. For instance, Air Products, Inc. at the Olinda Landfill in California uses one such setup.

12.5.9.3   Chemical Scrubbing with Amines (Absorption)

Amine scrubbing is a chemical process for the removal of CO_2 from biogas through adsorption. The compounds, organic amines, for example, mono ethanol amine (MEA), di-methyl ethanol amine (DMEA) and diglycolamines (DGA) are used in this scrubbing process as absorbers for CO_2 at slightly elevated pressures (<150 psi). The methane slip in this process is less than 0.1% as amines selectively react with CO_2 and due to this selectivity, there is a highly reduced volume of the process i.e. high amounts of CO_2 can be dissolved per unit volume. Nonetheless, due to evaporation, a part of the liquid is lost and has to be replaced. The high capacity and high selectivity are the main advantages of this process. At high temperatures (around 160°C), the solvent is regenerated via heating and CO_2 is recovered as an essentially pure by-product at the end of the process [17, 31, 32]. This method is preferred for large-scale systems that recover CO_2 from natural gas wells because at the small-scale, this process is difficult to apply due to the problems of amine breakdown, contaminant buildup and corrosion. However, biogas from farm source has fewer contaminants of concern than biogas from landfills and the corrosion is minimized using steel pipelines. A Dutch company Cirmac has developed a variant of it, a proprietary amine (COOAB™) scrubbing process based on a complex technology and is in operation at the Gasslosa biogas plant in Boras, Sweden.

12.5.9.4　*Organic Physical Scrubbing (Absorption)*

Organic physical scrubbing is another method for CO_2 removal by physical absorption [17]. It uses an organic solvent such as polyethylene glycol (PEG) for absorption of CO_2. CO_2 and H_2S are more soluble relative to CH_4 in polyethylene glycol than in water. This leads to lower solvent demand and reduced pumping. Selexol® and Genosorb® are the easily available commercialized solvents for this method and are used extensively in the natural gas industry and in other applications. By scrubbing with selexol, contaminants other than CO_2 i.e. H_2S, water and halogenated hydrocarbons also get removed, if present. However, prior to upgrading, it is preferred to remove them. By heating and/or depressurizing the solution of polyethylene, glycol is regenerated.

12.5.9.5　*Membrane Separation: Gas Permeation*

Membranes are also used for upgrading the biogas. They are highly selective due to their specific pore sizes. Membrane separation is one of the classical methods for landfill gas upgradation. Generally, membrane separation processes uses pressure and a selective membrane, which allows preferential passage of a particular gas [32]. Due to imperfect separation, the process is often performed in several stages. The numbers and interconnection of the applied membrane stages are determined by the desired methane recovery and specific compression energy demand. Filters are also used before membranes, to enhance their performance. In some cases for achieving high efficiency, membranes are used in a series. They are often made up of polymers like polysulfone, polyimide or polydimethylsiloxane. Usually, membranes are in the form of hollow fibers bundled together to provide a sufficient membrane surface area within compact plant dimensions. In hollow fiber bundles, these are combined to assemble a number of parallel membrane modules for increasing the efficiency of the separation process. They are permeable for ammonia, carbon dioxide and water, whereas hydrogen sulfide and oxygen pass through the membrane to a certain extent and methane and nitrogen passes only to a very low extent. By cooling, the nitrogen is separated from methane due to the difference in their boiling points. Low-pressure membranes seem to be more effective for removal of CO_2. The first plants were built in US during late 1970's and later in the Netherlands. These early designs suffered methane loses up to 25% due to elevated pressures but later, newer designs have showed far lower methane losses or high methane recoveries and relatively low energy demands due to reduced pressures. The operation pressure and compressor speed are highly controlled for the desired quality and quantity of the produced biomethane stream. Companies like Axiom have paved a great role in this technology.

12.5.9.6   Cryogenic Separation

This is another method for the separation of CO_2 and methane to upgrade biogas. This process is carried out in two stages. First, the raw biogas is cooled down to the temperature of sublimation or condensation of CO_2 where 30–40% of CO_2 is removed as a liquid and secondly, the remaining CO_2 is separated as a solid. The sublimation temperature of pure CO_2 is 194.65 K. This two-stage process leads to the accumulation of methane in the gas phase and eases the separation of CO_2 as a solid or a liquid fraction. During the cooling process, undesirable contents like water and siloxanes are also removed. The optimization of energy recovery and the removal of different undesired gases in the biogas are done by cooling the gas in several steps [31–33].

12.6   Conclusions

Anaerobic digestion is one of the most effective biological processes to treat a wide variety of the substrates leading to low cast production of biogas. However, different factors such as substrate and co-substrate composition and quality, environmental factors (temperature, pH, organic loading rate) and microbial dynamics must be optimized to achieve maximum benefit from this technology in terms of both energy production and organic waste treatment. Biomass recalcitrance poses another problem in enzymatic as well as microbial digestion, which leads to a low yield of biomethane. It is therefore important to understand different factors in order to have effective retreatment and digestion, increase the available surface and reduce the lignin content. In general, the thermal method (physical or thermophilic biological) increases biogas production. Ultrasonication is an emerging method to enhance biodegradability by disrupting the physical, chemical and biological properties of sludge leading to enhanced biogas production. The production of biomethane necessitates a comprehensive understanding of the preservation of biodiversity contextualized within a multifaceted landscape. The cultivation of fuel crops for the production of biogas is capable of being integrated into existing agro-ecosystems and provides opportunities for the responsible use of natural resources.

References

1. A.O Jactone., W. Zhiyou, I. John, B. Eric, E.R. Collins, *Virginia Cooperative Extension Publication* p 442–881, 2009.
2. E.Giovanni, F.Luigi, L. Flavia, P. Antonio, P. Francesco, *The Open Environmental Engineering Journal*, 5, 1–8, 2012.
3. "Biomethane: The energy system's all-rounder". German Energy Agency, Article No. 5012, Dena, Germany.
4. J. C. Barker, *EBAE* pp 71–80, 2001.
5. M.H. Gerardi, *The Microbiology of Anaerobic Digesters*, Wiley-Interscience. N.J. 2003
6. F.Tambone, P. Genevini, G.D. Imporzano and F. Adani, *Bioresource Technology*, 100, 3140–3142, 2009.

7. URL: www.biogas.psu.edu Accessed on *25–04-2013*.
8. J. Friehe, P.Weiland, A. Schattauer, Biogas guide, Gülzow, 2010.
9. www.cleanenergyfuels.com/why/biomethane.html
10. E.R.Collins, H.A. Hughes, *Energy Management Series Publication* 718, Virginia Cooperative Extension, 1976.
11. J. Jensen, "Biomethane for Transportation: Opportunites for Washington State", January 2013.
12. http://www.biomethane.com (Accessed on Jan 25, 2013).
13. "Biogas to biomethane technology review". Deliverable Reference: Task 3.1.1, 2012.
14. R.D. Treloar, Gas Installation Technology. Blackwell. p. 24. ISBN 978-1-4051-1880-4, 2005.
15. EPA AgStar (URL: www.epa.gov/agstar/) Accessed on 25-04-2013.
16. ASTM International. *"Annual book of ASTM standards: Waste Management"*, vol. 04.11, 2011.
17. A. Petersson and A. Wellinger, *IEA Bioenergy* pp 1–20. 2009.
18. E. Ryckebosch, M. Drouillon, H. Vervaeren, *Biomass and Bioenergy* 35, pp 1633–1645, 2011.
19. "Financial Analysis of Biomethane Production in Biomethane from Dairy Waste": *A Sourcebook for the Production and Use of Renewable Natural Gas in California*, pp 147–163.
20. T. Persson and M. Svensso, *IEA Bioenergy Conference*, 2012.
21. M. Kaltschmitt, H. Hofbauer, S. Stucki, M. Seiffert, *Presentation at the Energy Convent*, Stockholm, 2007.
22. http://www.biomethane.org.uk/ Accessed on 25-04-2013.
23. A. Karagiannidis and G. Perkoulidis, *Bioresource Technology*, 100, 2355–2360, 2009.
24. F. Raposo, V. Fernandez-Cegrı, M.A. De la Rubia, *et al.*, *Journal of Chemical Technology and Biotechnology*, Vol. 86, p1088–1098, 2011.
25. R.A. Parsons *NRAES-20*, Cooperative Extension, Northern Regional Agricultural Engineering Service. 1984.
26. T.S. Quinones M. Plochl *et al.*, "Hydrolytic enzyme enhancing anaerobic digestion in: Biogas Production" A. mudhoo (ed), Scrivener-Wiley, Beverly, MA pp. 157–198, 2012.
27. F. Graf, S. Bajohr *et al.*, "Biogas – Production, preparation, supply", Munich, 2011.
28. "Demonstration of the production and utilization of synthetic natural gas (SNG) from Solid Biofuels", *Specific Targeted Research Project*, 2006.
29. G. Plaza, P. Robredo, O. Pacheco, T. A. Saravia, *Water Science and Technology*, 33,169–175, 1996.
30. T. Forster-Carneiro, M. Pérez and L. I. Romero, *Bioresource Technology*, Vol. 99, p 6994–7002, 2008.
31. K. Krich, D. Augenstein, J. P. Batmale, J. Benemann, B. Rutledge, D. Salour, *A Sourcebook for the Production and Use of Renewable Natural Gas in California*, pp 47–69.
32. A. Makaruk, M. Miltner, M. Harasek, *Separation and Purification Technology*, 74 pp. 83–92, 2010.

33. M. Miltner, A. Makaruk, J. Krischan, M. Harasek, *Water Science and Technology*, vol. 66 pp.1354–1360, 2012.
34. W. F. Owen, D. C. Stuckey, J. B. Healy Jr., L. Y. Young and P. L McCarty., *Water Resources*, Vol. 13, pp 485–492, 1979.
35. http://www.environment-agency.gov.uk/epow accessed on 25-04-2013.
36. "Biomethane Regions: Introduction to Production of Biomethane from Biogas". *IEE* pp 1–47.
37. P. Hook, "Siloxanes in Biomethane". (URL: www.sgn.co.uk/greening-the-gas) accessed on 25-04-2013.

Index

Also of Interest

Check out these other related titles from Scrivener Publishing

Biogas Production, Edited by Ackmez Mudhoo, ISBN 9781118062852. This volume covers the most cutting-edge pretreatment processes being used and studied today for the production of biogas during anaerobic digestion processes using different feedstocks, in the most efficient and economical methods possible. *NOW AVAILABLE!*

Bioremediation and Sustainability: Research and Applications, Edited by Romeela Mohee and Ackmez Mudhoo, ISBN 9781118062845. Bioremediation and Sustainability is an up-to-date and comprehensive treatment of research and applications for some of the most important low-cost, "green," emerging technologies in chemical and environmental engineering. *NOW AVAILABLE!*

Sustainable Energy Pricing, by Gary Zatzman, ISBN 9780470901632. In this controversial new volume, the author explores a new science of energy pricing and how it can be done in a way that is sustainable for the world's economy and environment. *NOW AVAILABLE!*

Green Chemistry and Environmental Remediation, Edited by Rashmi Sanghi and Vandana Singh, ISBN 9780470943083. Presents high quality research papers as well as in depth review articles on the new emerging green face of multidimensional environmental chemistry. *NOW AVAILABLE!*

Ethics in Engineering, by James Speight and Russell Foote, ISBN 9780470626023. Covers the most thought-provoking ethical questions in engineering. *NOW AVAILABLE!*

Energy Storage: A New Approach, by Ralph Zito, ISBN 9780470625910. Exploring the potential of reversible concentrations cells, the author of this groundbreaking volume reveals new technologies to solve the global crisis of energy storage. *NOW AVAILABLE!*

Bioremediation of Petroleum and Petroleum Products, by James Speight and Karuna Arjoon, ISBN 9780470938492. With petroleum-related spills, explosions, and health issues in the headlines almost every day, the issue of remediation of petroleum and petroleum products is taking on increasing importance, for the survival of our environment, our planet, and our future. This book is the first of its kind to explore this difficult issue from an engineering and scientific point of view and offer solutions and reasonable courses of action. *NOW AVAILABLE!*

Also of Interest

Check out these other related titles from Scrivener Publishing

Biogas Production, Edited by Ackmez Mudhoo, ISBN 9781118062852. This volume covers the most cutting-edge pretreatment processes being used and studied today for the production of biogas during anaerobic digestion processes using different feedstocks, in the most efficient and economical methods possible. *NOW AVAILABLE!*

Bioremediation and Sustainability: Research and Applications, Edited by Romeela Mohee and Ackmez Mudhoo, ISBN 9781118062845. Bioremediation and Sustainability is an up-to-date and comprehensive treatment of research and applications for some of the most important low-cost, "green," emerging technologies in chemical and environmental engineering. *NOW AVAILABLE!*

Sustainable Energy Pricing, by Gary Zatzman, ISBN 9780470901632. In this controversial new volume, the author explores a new science of energy pricing and how it can be done in a way that is sustainable for the world's economy and environment. *NOW AVAILABLE!*

Green Chemistry and Environmental Remediation, Edited by Rashmi Sanghi and Vandana Singh, ISBN 9780470943083. Presents high quality research papers as well as in depth review articles on the new emerging green face of multidimensional environmental chemistry. *NOW AVAILABLE!*

Ethics in Engineering, by James Speight and Russell Foote, ISBN 9780470626023. Covers the most thought-provoking ethical questions in engineering. *NOW AVAILABLE!*

Energy Storage: A New Approach, by Ralph Zito, ISBN 9780470625910. Exploring the potential of reversible concentrations cells, the author of this groundbreaking volume reveals new technologies to solve the global crisis of energy storage. *NOW AVAILABLE!*

Bioremediation of Petroleum and Petroleum Products, by James Speight and Karuna Arjoon, ISBN 9780470938492. With petroleum-related spills, explosions, and health issues in the headlines almost every day, the issue of remediation of petroleum and petroleum products is taking on increasing importance, for the survival of our environment, our planet, and our future. This book is the first of its kind to explore this difficult issue from an engineering and scientific point of view and offer solutions and reasonable courses of action. *NOW AVAILABLE!*